Springer Proceedings in Mathematics & Statistics

Volume 417

This book series features volumes composed of selected contributions from workshops and conferences in all areas of current research in mathematics and statistics, including data science, operations research and optimization. In addition to an overall evaluation of the interest, scientific quality, and timeliness of each proposal at the hands of the publisher, individual contributions are all refereed to the high quality standards of leading journals in the field. Thus, this series provides the research community with well-edited, authoritative reports on developments in the most exciting areas of mathematical and statistical research today.

Madhusudhan Mishra • Nishtha Kesswani •
Imene Brigui
Editors

Applications of Computational Intelligence in Management & Mathematics

8th ICCM, Nirjuli, AP, India,
July 29–30, 2022

Springer

Editors
Madhusudhan Mishra
Electronics & Communication Engineering
North Eastern Regional Institute of Science and Technology (NERIST)
Itanagar, Nirjuli
Arunachal Pradesh, India

Nishtha Kesswani
Department of Computer Science
Central University of Rajasthan
Ajmer
Rajasthan, India

Imene Brigui
EMLYON Business School
Écully, France

ISSN 2194-1009 ISSN 2194-1017 (electronic)
Springer Proceedings in Mathematics & Statistics
ISBN 978-3-031-25193-1 ISBN 978-3-031-25194-8 (eBook)
https://doi.org/10.1007/978-3-031-25194-8

Mathematics Subject Classification: 68Txx

This Springer imprint is published by the registered company Springer Nature Switzerland AG
The registered company address is: Gewerbestrasse 11, 6330 Cham, Switzerland

Preface

We are very pleased to introduce the Proceedings of the 8th International Conference on Computers, Management & Mathematical Sciences (ICCM) 2022 which brings together researchers, scientists, academics, and engineers of different digital technologies based on Computational Intelligence, Management, and Mathematical Science. Through sharing and networking, ICCM 2022 provides an opportunity for researchers, practitioners, and educators to exchange research evidence, practical experiences, and innovative ideas on issues related to the areas of information and management systems.

The success of the conference is due to the collective efforts of everyone involved. We would like to express and record our gratitude and appreciation to the authors for their contributions. Many thanks go as well to all of the reviewers who helped us maintain the high quality of manuscripts included in the Proceedings published by Springer PROMS. We also express our sincere thanks to the members of the conference committees and IAASSE team for their hard work.

We wish that all the authors and delegates find ICCM 2022 interesting, exciting, and inspiring.

Organization

Program Committee Chairs

Bora, Joyatri	
Brigui, Imène	Emlyon Business School, Ecully, France
Das, Piyali	North Eastern Regional Institute of Science and Technology, Electrical Engineering, Nirjuli, India
Kesswani, Nishtha	Central University of Rajasthan, Computer Science, Kishangarh (Ajmer), India
Mishra, Madhusudhan	NERIST, Electronics and Communication Engineering, Nirjuli, India

Program Committee Members

Abdul Kareem, Shaymaa Amer	
Acharjee, Raktim	Indian Institute of Technology Guwahati, Electronics and Electrical Engineering, Guwahati, India
Agarwal, Pallavi	
Agarwal, Pooja	IIT Delhi, Design, Delhi, India
Balasundaram, Saveetha	
Bhavnani, Mansi	
Bora, Joyatri	
Brigui, Imène	Emlyon Business School, Ecully, France
Chapi, Sharanappa	

Chauhan, Prathvi Raj	Indian Institute of Technology Delhi, 110016, India, Department of Energy Science and Engineering, New Delhi, India
Ch Rajendra Prasad, Dr.	SR University, ECE, Warangal, India
D, Renuka Devi	
Diaz, Rowell	Nueva Ecija University of Science and Technology, College of Management and Business Technology, San Isidro, Philippines
Das, Piyali	North Eastern Regional Institute of Science and Technology, Electrical Engineering, Nirjuli, India
G, Dr Lavanya	
G A, Sivasankar	KIT Kalaignar Karunanidhi Institute of Technology, Aeronautical Engineering, Coimbatore, India
J, Ramarajan	
Johri, Era	
Jain, Shelendra	
Kesswani, Nishtha	Central University of Rajasthan, Computer Science, Kishangarh (Ajmer), India
Kumar, Vikash	
Kumhar, Malaram	
Lo, Man Fung	The University of Hong Kong, Hong Kong, Hong Kong
Mishra, Shivanchal	NIT Kurukshetra, Civil Engineering, Kurukshetra, India
Majiwala, Hardik	
Majumder, Swanirbhar	Tripura University, Information Technology, Agartala, India
Manhas, Dr Pratima	
Mishra, Madhusudhan	NERIST, Electronics and Communication Engineering, Nirjuli, India
Mishra, Shubhranshu	Dr B R Ambedkar National Institute of Technology, Jalandhar, Department of Mechanical Engineering, Jalandhar, India
Mishra, Abhaya Raj	
Nayyar, Anand	
Neeraj, Bheesetti	
R, Sreeraj	
Ray, Ashok	
Sahu, Sanat	
Saxena, Dipanshu	
Sharma, Madhu	
Thakar, Pooja	Vivekananda Institute of Professional Studies - Technical Campus, Affiliated to GGSIPU, Delhi, Information Technology, Delhi, India
Yadav, Anil Singh	Lakshmi Narain College of Technology, Mechanical Engineering, Bhopal, India
Rana, Aneri	

Reviewers

Acharjee, Raktim	Indian Institute of Technology Guwahati, Electronics and Electrical Engineering, Guwahati, India
Agarwal, Pallavi	
Agarwal, Pooja	IIT Delhi, Design, Delhi, India
Baruah, Smriti	
Bhavnani, Mansi	
Bora, Joyatri	
Borah, Janmoni	
Brigui, Imène	Emlyon Business School, Ecully, France
Chauhan, Prathvi Raj	Indian Institute of Technology Delhi, 110016, India, Department of Energy Science and Engineering, New Delhi, India
Choudhury, Shibabrata	NERIST, Nirjuli, India
Das, Piyali	North Eastern Regional Institute of Science and Technology, Electrical Engineering, Nirjuli, India
Divvala, Chiranjevulu	
J, Ramarajan	
Jain, Shelendra	
Kamboj, Pradeep	
Kar, Mithun	
Kesswani, Nishtha	Central University of Rajasthan, Computer Science, Kishangarh (Ajmer), India
Kumar, Manish	
Lo, Man Fung	The University of Hong Kong, Hong Kong, Hong Kong
Mishra, Shivanchal	NIT Kurukshetra, Civil Engineering, Kurukshetra, India
Majiwala, Hardik	
Majumder, Swanirbhar	Tripura University, Information Technology, Agartala, India
Mall, Manmohan	
Mishra, Madhusudhan	NERIST, Electronics and Communication Engineering, Nirjuli, India
Mishra, Shubhranshu	Dr B R Ambedkar National Institute of Technology, Jalandhar, Department of Mechanical Engineering, Jalandhar, India
Mishra, Abhaya Raj	
Mohan, Yogendra	
Nath, Malaya Kumar	
Neeraj, Bheesetti	
Patra, Aswini Kumar	
R, Sreeraj	
Ray, Ashok	
Saxena, Dipanshu	

Sharma, Madhu	
Singh, Hemarjit	
Singh, M Edison	
Tamang, Santosh	
Vakamullu, Venkatesh	NERIST, Electronics and Communication Engineering, NIRJULI, India
Yadav, Ajit Kumar Singh	North Eastern Regional Institute of Science and Technology, CSE, Papumpare, India
Rana, Aneri	

Contents

Development of Energy Efficient Routing Protocol for Lifespan Optimization in Wireless Sensor Networks (EERP)

Ayam Heniber Meitei and Ajit Kr. Singh Yadav

1 Introduction

Wireless sensor networks consist of several small sensors that have limited battery capacity. Sensor nodes can perceive, store, send, and receive data, and they can communicate with each other via a wireless channel such as radio waves. WSN research and implementation are complicated by high power consumption, poor energy efficiency, memory restrictions, and the difficulty of recharging or changing batteries. The network's lifespan is determined by the energy consumption of sensor nodes. Sensor nodes are usually powered by a small amount of energy. When the battery dies, the node is no longer functional. Consequently, researchers have started to create many energy-efficient strategies for routing nodes and extending sensor nodes' lifespans in order to conserve energy. Even though many procedures have already been paid for in this way, there are still ways that we can try to improve them. In the design of a WSN, energy consumption is a critical factor. When sensor nodes are dispersed over a complicated environment, replacing or charging batteries is very challenging. To extend the network's lifespan, energy conservation via various data transmission protocols is necessary to redirect information from nodes to the base station.

A. H. Meitei · A. K. S. Yadav (✉)
Department of CSE, NERIST, Itanagar, Arunachal Pradesh, India

M. Mishra et al. (eds.), *Applications of Computational Intelligence in Management & Mathematics*, Springer Proceedings in Mathematics & Statistics 417,
https://doi.org/10.1007/978-3-031-25194-8_1

2 Literature Review

Low Energy Adaptive Clustering Hierarchy (LEACH) was the first widely used wireless sensor network clustering protocol introduced by W. B. Heinzelman et al. [1] Clusters are formed by groups of sensor nodes, and cluster heads (CHs) are picked at random, with the possibility of becoming cluster heads each round. The cluster head's role is alternated among nodes. Balances between the nodes' energy consumption are achieved by the network's rotation duty. The total number of CHs in any particular round has no fixed limit, despite the fact that LEACH is a distributed method. The distributed technique allows each node to pick a random number as the CH. Due to the random number generator's randomness property, it is conceivable that each node chooses the same CH number. Thus, the number of CHs changes from round to round.

W. B. Heinzelman et al. [2] proposed LEACH-C, and this protocol used a centralised method to provide information about node position and level of energy to BS. BS ensures that such nodes have insufficient energy to become cluster heads. Due to the difficulty of conveying the position of a node that is far away from the base station, this protocol is unsuitable for large-scale networks. Information cannot be sent on a regular basis due to the fact that the function of the cluster head changes every time.

M.J. Handy et al. [3] developed an extension of LEACH that incorporates deterministic CH selection, and an equation that accounts for residual energy has been suggested; nonetheless, the equation is ineffective. S. Georgios et al. [4] proposed the Stable Election Protocol (SEP) and introduced heterogeneity, which elongates the duration of stability before the first node dies. The remaining energy in each node affects its weighted election probability to become the CH. In this scenario, there are two types of nodes: normal and advanced. In this technique, no global energy information is required to choose a CH. Except for heterogeneity awareness, the authors expanded the LEACH procedure. This algorithm's cluster count is changeable, and the algorithm's unstable period isn't ideal. A. Femi et al. [5] proposed an enhanced SEP and also introduced it by assuming that the sensor nodes contain three distinct forms of initial energy.

In ZSEP [6], there are three zones in the network field; advanced nodes are put in two zones, and normal nodes are put in one zone. Only extended advanced nodes create clustering and cluster head elections since it takes a lot of energy to transmit directly to the BS. CHs gather information from cluster members and transmit the information to the BS. Unstable networks develop as a consequence of the high energy consumption since these nodes soon exhaust themselves. As a consequence, the WSN's overall service life is shortened. As the deployed field area expands, the sensor network under the Z-SEP protocol becomes unstable and the cluster's head nodes die faster leading to increasing power consumption. Since distance and energy use are linked, normal nodes and advanced nodes with expanded cluster heads are also directly connected to BS.

Kham et al. [7] proposed the Advanced Zonal-Stable Election Protocol (Az-SEP), which is an upgraded version of Z-SEP that additionally employs simultaneous cluster head selection. As with AZ-SEP, data is sent from the first CH to the second CH and so on until all of the data has been received at the base station. O. Younis et al. [8] introduced Hybrid Energy-Efficient Distributed Clustering (HEED), which modifies the LEACH protocol to accomplish power balancing by employing node degree, residual energy, or density as key cluster formation factors. Four key parameters were provided for this protocol: (1) increasing network lifespan by dispersing energy consumption, (2) stopping clustering after a specified number of iterations, (3) distributing CH properly, and (4) clustering ends after a specified number of iterations. This protocol uses algorithms to choose cluster heads on a regular basis, depending on two primary features. The main parameter creates a probability-based starting set of cluster heads, whereas the secondary parameter breaks ties. HEED can't fix the number of clusters in each cycle because it doesn't know about heterogeneity. Duan and Fan [9] proposed "Distributed Energy Balance Clustering" (DEBC), and the probabilities based on radio transmissions between the remaining energy of the node and the average remaining energy of the whole network are used to choose the cluster heads. When nodes have a considerable amount of initial and remaining energy, CHs are more likely to form. By addressing two levels of heterogeneity and extending to multihop heterogeneity, LEACH and SEP protocols have been improved by this protocol. Distributed Cluster Head Election (DCHE) has been developed by Dilip Kumar et al. [10] CHs are selected using a weighted probability distribution. The selected CH communicates with the cluster's member nodes, and the collected data is then sent to the BS via the cluster heads. The authors examined three alternative kinds of nodes, each with a different threshold. The cluster head chosen for each kind is determined by the weight allocated to each node. The DCHE system outperforms the LEACH, DEEC, and Direct Transmission schemes in terms of longevity and stability.

An improved LEACH was suggested by Said et al. [11] that employs randomization to distribute the energy consumption throughout the network's sensors more equitably. Node Cluster Gateways (NCG) are the system's CHs, gathering information from cluster members and sending it to gateways that use the least amount of energy for communication, minimising cluster head energy consumption and decreasing the risk of node failure.

Md. G. Rashed et al. [12] created a wireless energy-efficient protocol (WEP) to improve sensor network reliability. The author uses clustering to improve the energy and stability of the period restrictions. The best probability for each node is given a weight. Using the LEACH algorithm once the weighted probability has been allocated, CHs and cluster numbers are determined. The programme created a chain between chosen cluster heads. After that, a chain leader is chosen at random from the CHs. Data is sent to the CHs via other nodes. Each cluster's head node then aggregated information and relayed it to the BS.

Kyounghwa Lee et al. [13] introduced Distributed Cluster Head Election (DDCHS) by dividing the cluster area into four quadrants; each quadrant selects the next CH depending on the node density of its group and its proximity to the CH.

The energy usage for communication between complete nodes and cluster heads at various cluster locations was analysed by the author to assess the LEACH and HEED protocols. Compared to LEACH and HEED, this technique performs better. It is a centralised method that necessitates the position of each node.

Sanjeev Kumar Gupta et al. [14] introduced Node Degree-based Clustering (NDBC), and the energy level of advanced nodes is higher than that of normal nodes. Because of their energy and network node degree, CHs are selected from the most advanced nodes. NDBC was used by the authors to minimise transmission costs between sensor nodes that broadcast and receive cluster head selection signals.

Ant Colony Optimization (ACO)-based routing has been presented [15] to increase network longevity. Using fuzzy logic control, cluster heads may be effectively selected and unequal clusters can be partitioned, depending on remaining energy, distance to the BS, node degree, and other parameters. Multiple hop communication between CHs and the sink may be made more efficient by using the ACO-based routing approach.

Using a corona-based structure, an energy-balanced data gathering (EBDG) has been suggested in [16]. Using "corona-based network division and hybrid routing algorithms," which simultaneously analyze nodes in the same corona layer with an equal probability using both hybrid multi-hop and single-hop transmission, the characteristics of optimal data accumulation are designed. Yu and Ku [17] developed a WSN routing technique which is a hybrid of "single-hop and multi-hop routing" in order to decrease considerable routing traffic in the Sink Connectivity Area (SCA). By optimising the network lifetime, "the min optimization and min-max optimization approaches" may be utilised to virtually accomplish both efficiency and utilisation in terms of energy design objectives. Hasan and Al-Turjman [18] provides a heuristic method for detecting several unrelated paths at nodes, and the suggested method takes advantage of the BS to build multipaths between any source node and any destination node in a centralised manner.

The approach described in [19] employs a flooding strategy to generate a large number of routes with the required cost values between the source nodes and destination nodes, with routes with lower costs being retained in the routing table. In addition, the inverse probability of a defined cost is used to determine the final selected path for packet transmission. Severin et al. [20] proposed the extension of Z-SEP, and the authors have divided the network field into two zones. The results show that their protocol performed better than the others.

3 Methodology and Proposed System

Two behaviours are considered in our proposal for energy management: the first is how a network field is configured, and the second is to determine the closest node to the BS; it can be either a cluster head or a cluster member, depending on their lesser distance from the BS. There are three zones in the Z-SEP network field: zone 1, zone 2, and zone 3. BS is put in the centre of the network field. Zone 1 is for normal nodes, whereas zones 2 and 3 are for advanced nodes. Advanced nodes are positioned far from the BS. Zones 2 and 3 have sensor nodes that are distant from the BS, but zone 1 nodes are nearer to the BS. This results in a large amount of energy consumption since sensor nodes located far from the BS spend a lot of energy sending data packets. Consequently, our proposed work will help to decrease the amount of energy used and the number of zones. We created two zones in our network's field, and only a few nodes will be deployed far from the BS. Normal nodes will be put in zone 1 and away from the BS. All advanced nodes will be clustered together in zone 2, with BS at the centre. In a network of 100 coordinates, we split X coordinates into two zones: zone 1 and zone 2 for X and Y coordinates, respectively. Zone 1 has a $0 < X <= 50$ coordinate, in which all normal nodes will be randomly arranged. The $50 < X <= 100$ is the coordinate of zone 2, and in this zone, advanced nodes are randomly placed with a base station in the centre (75, 25). The network architecture of our proposed network is as indicated in Fig. 1.

3.1 *Cluster Head Selection and Data Transmission*

Advanced nodes are used in this system to send data to the CH first, and either CH or cluster member transmits the data to the BS based on their lesser distance from the CH to the BS and the lowest distance from a cluster member of this current cluster to the BS. Normal node data is sent straight to the BS. Sensor nodes sensed data in the same cluster that the CH sent and received data from. After that, we calculated the distance of all sensor nodes, including CH, in the same cluster from BS. If the CH is a short distance from the BS, then the CH transmits the collected data to the BS directly without the assistant cluster head; otherwise, select a sensor

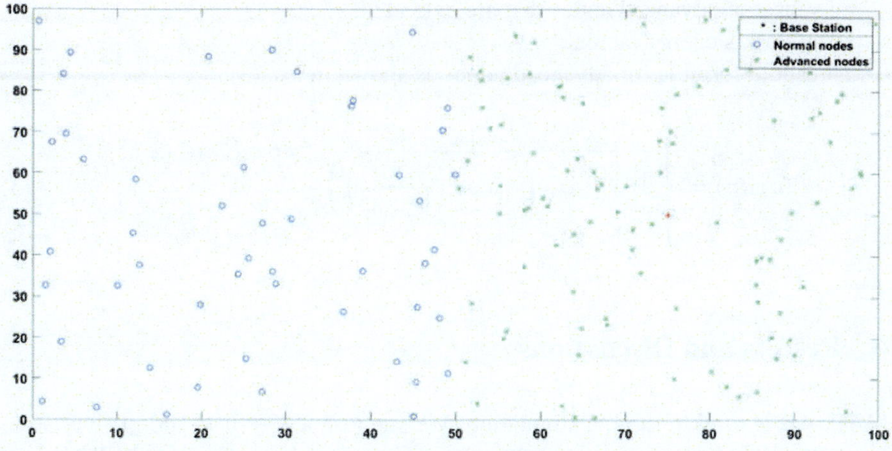

Fig. 1 Deployment of 100 sensor nodes in a network field of $100 \times 100\ m^2$

node which has the lowest distance from the BS, which is known as the assistant CH, and the CH transmits data to the assistant CH, then data is sent by the assistant cluster head to the BS. Clusters are formed only at advanced nodes that are located in zone 2. After a given number of rounds, normal nodes in zone 1 will run out of energy. Advanced nodes have higher energy than normal nodes, and BS is in zone 2, so their energy will slowly deplete in zone 2. Zone 2 will thus re-elect cluster heads and relay data to the BS. As compared to ZSEP, the advanced node in zone 2 of our proposed work has a bigger throughput than the advanced node in zones 1 and 2 of ZSEP and improves the lifespan of the network.

Using the Z-SEP method, Eq. (1) shows the threshold formula that is used to cluster head election.

$$T(n) = \begin{cases} \dfrac{P_{optp}}{1 - P_{optp}\left(r \bmod \left(\dfrac{1}{P_{optp}}\right)\right)} & if\ n \in G \\ 0 & Otherwise \end{cases} \tag{1}$$

where P_{optp}: cluster head probability,

G: set of sensor nodes which have not been chosen as cluster head in the last $1/P_{optp}$ rounds,

r: number of rounds.

n: number of advance nodes.

The advanced nodes elect the cluster head using the following probability formula, as shown in Eq. (2):

$$\boldsymbol{P_{advn}} = \frac{P_{optp}}{1 + a \times m} \times (1 + \mathrm{a}) \tag{2}$$

where P_{optp}: cluster head's optimal probability,

m: subset of advanced nodes that are extended,

a: network's energy factor.

Threshold formula for advanced nodes as shown in Eq. (3):

$$T(avdn) = \begin{cases} \dfrac{P_{avdn}}{1 - P_{avdn}\left(r \bmod \left(\dfrac{1}{P_{avdn}}\right)\right)} & if\ advn \in G' \\ 0 & Otherwise \end{cases} \tag{3}$$

4 Results and Discussions

MATLAB is used to run the simulation. We suppose a 100 m × 100 m simulation area with 100 sensor nodes is randomly distributed in this network field. There are two zones of our proposed method in the network field. The base station is put at the centre of zone 2 (75,50). In comparison to the three WSN protocols, the aforementioned variables are compared to the package delivery rate, death nodes,

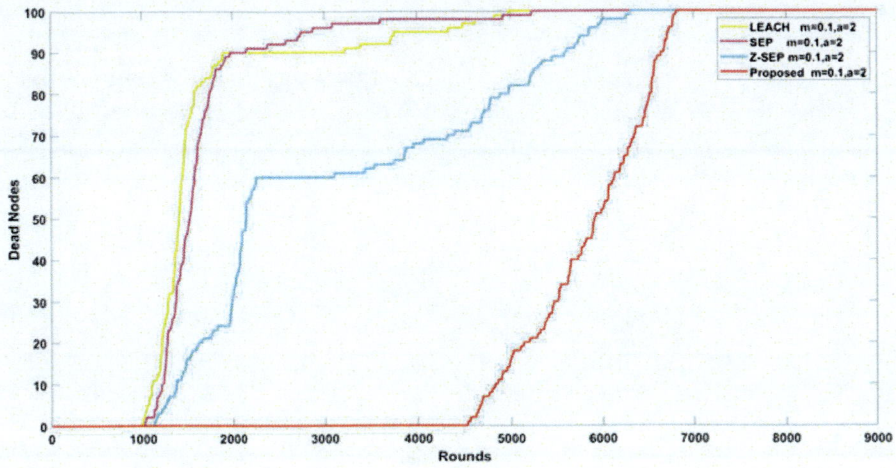

Fig. 2 Dead nodes when m = 0.1, a = 2

and active nodes. The LEACH protocol is represented by a yellow line, the SEP protocol is represented by a magenta line, the Z-SEP algorithm by a cyan line, and the proposed protocol by a red line. The graphs in Fig. 2 show that the LEACH method has the lowest performance at the beginning of the network when compared to others. The first death node and instability in the network in the LEACH occur in the 992nd round, but in the SEP protocol, it occurs in the 1021st round, the first node death in Z-SEP occurred at the 1132nd round, and the first death node in the proposed protocol occurred at the 4530th round, which demonstrates a longer longevity than the other three protocols. Figure 2 shows that our suggested method outperforms other protocols in every round.

Figure 3 shows that our suggested protocol has a longer network lifespan than the three existing protocols. The fact that the BS is placed at the centre of an advanced node zone is an advantage of our proposed protocol. When the data is sent to BS by a cluster member or cluster head, it will take less energy because of the shorter distance. So, the energy of advanced nodes will reduce slowly, and they will die later.

Figure 4 depicts the ratio of packet delivery to BS. A WSN's main function is to relay data from nodes to base stations. More data is detected and more environmental data is collected and transmitted the data to BS when the network has been functioning for a long period and its efficiency is better suited to long-term use. Figure 4 depicts data packet delivery to BS via all three protocols. Our proposed method outperforms the other protocols by sending over 5.8×10^5 packets to BS, as seen in the graph Fig. 4. The proposed protocol outperforms LEACH, SEP, and Z-SEP in terms of throughput (Fig. 5).

When m = 0.2 and a = 1, our proposed method outperforms the others.

In Fig. 6, the first node death and network instability occur at the 1078th round in the LEACH protocol, but in the SEP protocol, they occur at the 1094th round.

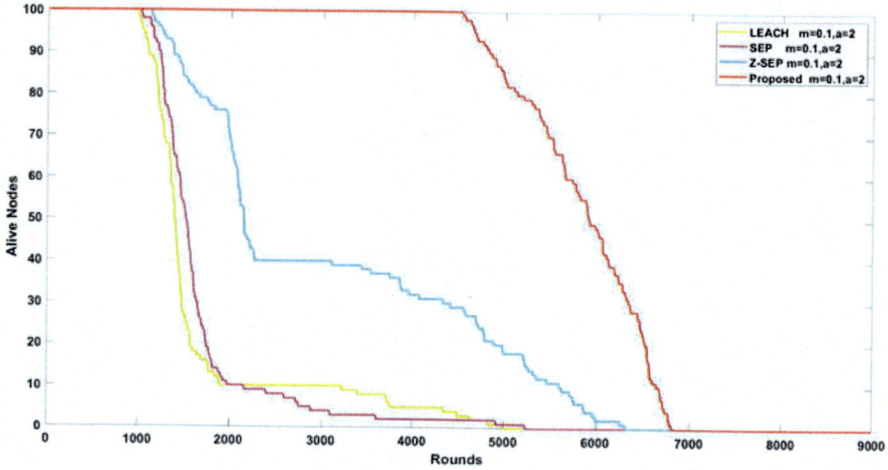

Fig. 3 Alive nodes when m = 0.1, a = 2

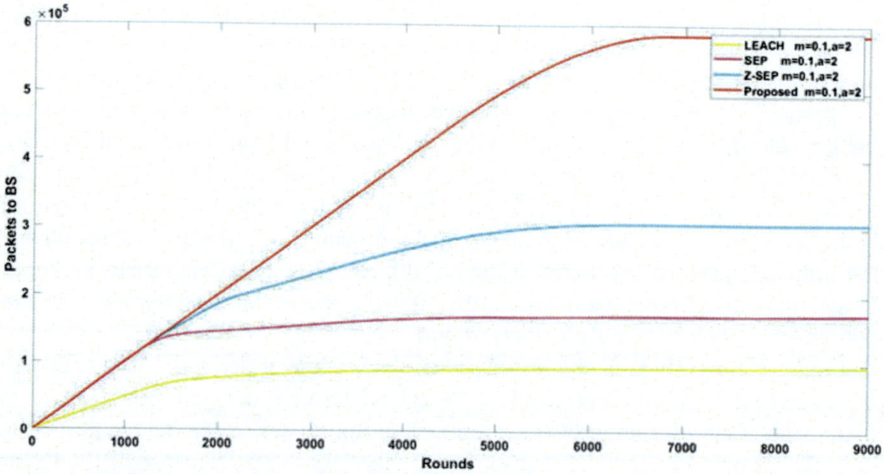

Fig. 4 Ratio of packet delivery to BS (where m = 0.1, a = 2)

The first node death in Z-SEP occurred at round 1131, and the first death node in the proposed method occurred at the 3007th round. In comparison to the other three protocols, our proposed method has a longer lifespan for the first node death (Fig. 7).

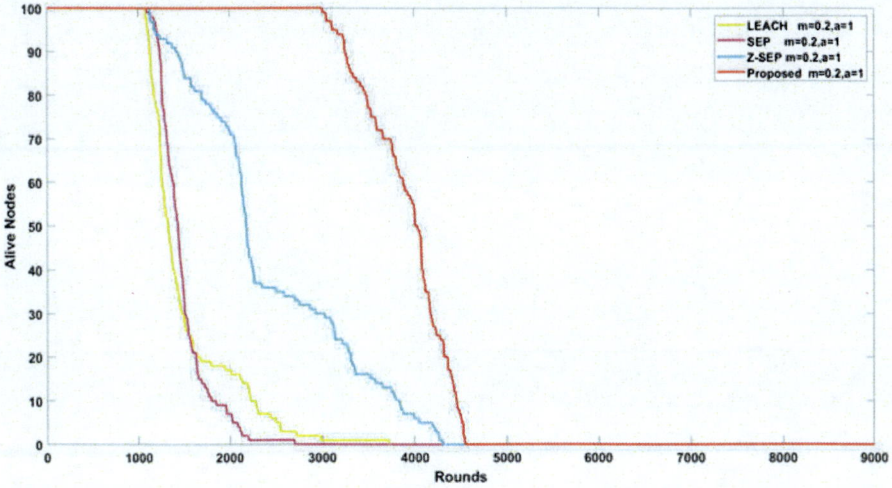

Fig. 5 Alive nodes vs rounds when m = 0.2, a = 1

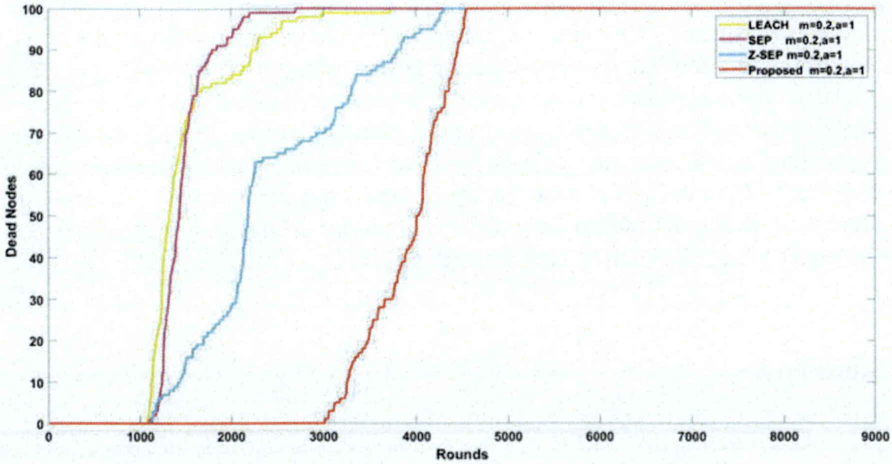

Fig. 6 Dead nodes vs rounds (when m = 0.2, a = 1)

5 Conclusion

Two of the most essential issues in the domain of a network routing algorithm are energy efficiency and network longevity. It is challenging to create a protocol that is low in energy consumption and also distributes the load throughout the network. However, Z-SEP is a good method for this; it has several drawbacks. The goal of this is to enhance the lifespan of the network by accounting for the network's energy use. When compared to the parent Z-SEP approach, the cluster head might be chosen from the nodes using the remaining residual energy, resulting in an enhanced

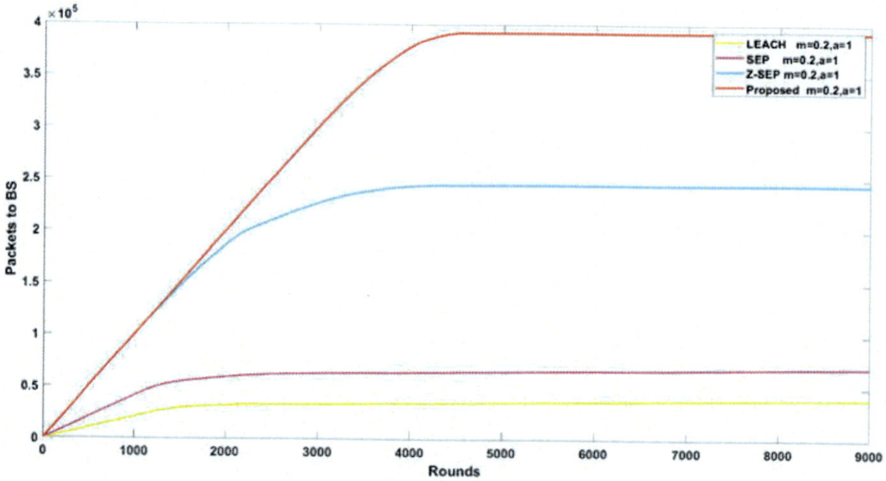

Fig. 7 Packet delivery ratio to BS vs rounds where m = 0.2, a = 1

routing mechanism. The increased number of live nodes and prolonged duration of stability before the first node dies are the key advantages of this technique. There are more environmental events observed, sensed, and supplied to BS as a result of the improved packet transmission ratios. Network factors like lifespan, packets transmitted to BS, and energy usage all show improvements in simulation results. SEP, LEACH, and Z-SEP have been evaluated and compared to the suggested algorithm, with the graphics demonstrating that our proposed system outperforms the others while maintaining high network stability.

References

1. W. R. Heinzelman, A. C. and. H. B., "Energy-Efficient Communication Protocol for Wireless Microsensor Networks," in *Proceeding of the 33rd Hawali International Conference on System Science*, 2000.
2. W. B. Heinzelman, A. P. Chandrakasan and H. B., "An application specific protocol architecture for wireless microsensor networks," *IEEE Transactions on Wireless,* vol. VOL. 1, pp. 660–670, 2002.
3. M. J. Handy, M. H. and D. T., "Low Energy Adaptive Clustering Hierarchy with Deterministic Cluster-Head Selection," *4th International Workshop on Mobile and Wireless Communications Network,* 2002.
4. G. S. I. M. and A. B., "SEP: A Stable Election Protocol for Clustered Heterogeneous Wireless Sensor Networks," 2004.
5. F. A. Aderohunmu and J. D. Deng, "An Enhanced Stable Election Protocol (SEP) for Clustered Heterogeneous WSN," 2017.
6. S. F. N. J.A. J. M. A. Khan, S. H. Bouk and Z. A. Khan, "Z-SEP: Zonal-Stable Election Protocol for Wireless Sensor Networks," *Journal of Basic and Applied Scientific Research (JBASR),* 2013.

7. F. A. Kham, M. K. M. A. A. K. and I. U. Haq, "Hybrid and Multi-hop Advenced Zonal-Stable Election Protocol for Wireless Sensor Networks," *Center of Excellence in Information Technology, Institute of Management Sciences,* 2019.
8. O. Y. and S. F. , "HEED: A Hybrid, Energy Efficient, Distributed clustering approach for Ad Hoc sensor networks," *IEEE Transactions on Mobile Computing,* Vols. 3, NO. 4, 2004.
9. C. D. and H. F., "A Distributed Energy Balance Clustering Protocol for Heterogeneous Wireless Sensor Networks," *IEEE WiCon.,* 2007.
10. D. K. T. C. Aseri and R.B. Patel, "Distributed Cluster Head Election (DCHE) Scheme for Improving Lifetime of Heterogeneous Sensor Networks," *Tamkang Journal of Science and Engineering,* 2010.
11. B. A. S. et al., "Improved and Balanced LEACH for heterogeneous wireless sensor networks," *(IJCSE) International Journal on Computer Science and Engineering,* Vol. 02, No. 08, 2010.
12. M. G. Rashed, M. H. Kabir and S. E. Ullah, "WEP: an Energy Efficient Protocol for Cluster Based Heterogeneous Wireless Sensor Network," *International Journal of Distributed and Parallel Systems (IJDPS),*Vol.2, No.2, 2011.
13. K. L.J. . L. . H. L. and Y. S. , "A Density and Distance based Cluster Head Selection Algorithm in Sensor Networks," *ICACT,* 2010.
14. S. K. Gupta, N. J. and P. S., "Node Degree Based Clustering for WSN," *International Journal of Computer Applications (IJCA),* vol. 40– No.16, 2012.
15. S. A. and P. S., "Lifetime Maximization of Wireless Sensor Network Using Fuzzy based Unequal Clustering and ACO based Routing Hybrid Protocol," *Appl Intell,* vol. 48, pp. 2229–2246, 2018.
16. H.Z. and H. S., "Balancing Energy Consumption to Maximize Network Lifetime in Data-Gathering Sensor Networks," *IEEE Trans. Parallel Distrib. Syst.,* vol. 20, no.10, pp. 1526–1539, 2009.
17. C.-M. Y. and M.-L. K., "Joint Hybrid Transmission and Adaptive Routing for Lifetime Extension of WSNs," *IEEE Access,* vol.6, pp. 21658–21667, 2018.
18. M. Z. Hasan and F. A.-T., "Optimizing Multipath Routing with Guaranteed Fault Tolerance in Internet of Things," *IEEE Sensors J.,* vol. 17, pp. 6463–6473, 2017.
19. P. G. et al., "Path Finding for Maximum Value of Information in Multi-modal Underwater Wireless Sensor Networks," *IEEE Trans. Mobile Comput.,* vol. 17, no. 2, pp. 404–418, 2018.
20. N.S. and M. I. zuzu Iragena, "Enhanced Clustering Protocol in Zonal-Stable Election Protocol for WSN," *International Journal of Scientific Research and Modern Technology,* vol. 2, no. 2, 2022.

Sentiment Analysis Using Social Media and Product Reviews

Divyansh Bose, Divyanshi Chitravanshi, and Santosh Kumar

1 Introduction

In today's era, the Internet is becoming very popular for every activity, thus globalization has helped to link people of various cultures across the world, and this communication through the Internet has enabled people to understand each other's traditions and culture. In today's world, a large number of people are already on social networking sites such as Facebook, Twitter, and WhatsApp, and these sites are responsible for spreading not only informative knowledge but also hate speech [1]; for example, Twitter, a very rapidly growing social networking platform, is used by people to tweet (a short message) their views and thoughts about a particular issue. Nowadays, there is also increase in the number of people buying and selling products. So, for companies that sell products, the reviews or feedback of customers matter a lot for the development of the product's quality. Hence, to analyze people's reviews of a particular product and the opinions they provide on social networking sites, a technique called sentimental analysis is used [2, 3].

It is a field of study which helps to evaluate attitudes, emotions, and opinions of the public and sentiments on various social networking sites using various techniques like NLP [4].

It is an effective and efficient way which can help find people's views and opinions about a concerned product [5]. Sentiment analysis retrieves the insight information by various text classification processes. The real-life applications of

D. Bose · D. Chitravanshi
Department of Computer Science and Engineering, BES Engineering College, Ghaziabad, Uttar Pradesh, India
e-mail: divyansh.20b0101023@abes.ac.in; divyanshi.20b0101052@abes.ac.in

S. Kumar (✉)
Galgotias University, Greater Noida, Uttar Pradesh, India

M. Mishra et al. (eds.), *Applications of Computational Intelligence in Management & Mathematics*, Springer Proceedings in Mathematics & Statistics 417,
https://doi.org/10.1007/978-3-031-25194-8_2

sentimental analysis are evaluating customer support, performing market research, analyzing product reviews, checking texts on social media, etc. [4]. Sentimental analysis has a wide scope in different fields, and thus it can help in analyzing the views and opinions of people on various issues. One such issue was covid. Right from the onset of the virus, various news stories about the virus circulated through different social media platforms. There was a challenge in understanding people's views and sentiments regarding the disease, but sentimental analysis can help in overcoming this problem as the author A.H. Alamoodi and his team presented in their paper [6]. It can also be used to identify and analyze people's reactions and beliefs about vaccinations as many rumors and misinformation are spreading. This analysis will help the government and the administration in coordinating the situation and conducting the drives. It has applications in politics as well. In a democratic country like India, where public opinion is very important for the elections and the parties can prepare their propaganda on the basis of the feedback they received from the public, sentimental analysis plays a very important role in shaping the future of the country, and platforms like Twitter, Facebook, and WhatsApp can be very useful, but at the same time, they can be a medium to propagate false information. For this sentimental analysis, it can be very useful to analyze people's views regarding the parties, their opinions on the problem, and their trust in the government. Sentimental analysis can be very helpful for companies as this allows them to analyze their customers' feedback on their products and services. It can be beneficial for the reputational risk management of the companies.

In this paper, sentimental analysis of product-based reviews and social media reviews and comments has been implemented to categorize Amazon reviews into negative, positive, and neutral categories. Amazon product reviews and feedback given by the public have been collected and analyzed in order to reach a meaningful conclusion. Section 2 discusses the previous work done on sentimental analysis based on reviews of e-commerce and social media websites. In Sect. 3, a proposed approach and framework have been discussed. Results have been discussed in Sect. 4.

2 Related Work

In [1], the author aims to know the strength of hatred and extremism in polyglot written data through sentimental analysis of various networking sites. In this paper, the views have been categorized into moderate extreme, high extreme, neutral extreme, and low extreme, and approaches like lexicon and machine learning methods are used for performing sentimental analysis. It also uses the NB and LSVC algorithms for categorization purposes. This paper concludes that on the fundamental polyglot dataset, LSVC has achieved high accuracy. In [2], the author has proposed a process for determining a tweet's polarity and then differentiating them as negative or positive tweets. In this paper, the author has also proposed an ensemble classifier to form a single classifier to advance the functioning of the sentiment classification technique. The author also depicted the techniques for feature representation and data preprocessing in sentiment classification. In [3], sentiment analysis has been performed on a large data set of customer reviews using the MLlib library. Three classifiers—NB, SVM, and LR—have been used where SVM provides better classification. In [4], various types of sentimental analysis techniques are discussed—emotion detection, intent analysis, fine-grained analysis, and aspect-based analysis—of which intent analysis stands out as it provides users with a deep understanding of the intention of the user.

Haider et al. [5] in their paper concentrated on the effect of various types of adverbs that are not considered for sentimental analysis such as degree comparative adverbs (RGR), general adverbs (RR), locative adverbs (RL), adverbs of time (RT), adverbs (RA), general comparative adverbs (RRR), degree adverbs (RG), and prep. Adverbs (RP). The result of this paper shows that RL and RP are important polarities for positive and negative opinions, respectively, while RRR and RR are for neutral opinions.

Alamoodi and his team [6] analyzed the sentiment of people regarding the pandemic, and through this paper, the author reviewed articles about the origin of a variety of infectious diseases.

In [7], the author described a sentimental analysis study for more than 1000 Facebook posts which aims to compare the sentiments for RAI, a public broadcasting service (Italy). The study has been done using various analyses and search engines: lexical and semantic analyses, natural language search, semantic role search, etc. Its results have accurately defined the reality of using Facebook as an online marketing platform. In [8], a thorough study of techniques used in the sentimental analysis is done, and the author has proposed his own technique in this paper too. Key considerations of this newly proposed technique are that it is part of speech and that it has been tested on a standard Stanford dataset, with the help of six known supervised classifiers. It is noticed that the adverb, verb, and adjective combination turned out to be the best combination among various combinations of the parts of speech. In [9], the author has used two methods to compare reviews on sentimental analysis, i.e., lexicon-based sentiment classification and a model based on an NB multinomial event. The author has found the accuracy of both models to be interestingly similar. So, the author has proposed to combine both methods, which aims to improve the result; hence, a combined approach is valuable to identify the phenomenon. In [10], the author has discussed different techniques for sentiment analysis of Twitter reviews, such as lexicon-based approaches, ensemble learning, and machine learning. Additionally, hybrid and ensemble Twitter reviews of sentimental techniques have been also discussed. Results of the researches have shown that the SVM and MNB provide the greatest precision when including multiple features. SVM classifiers can be considered a standard learning strategy, while lexicon-based techniques can be considered very feasible at the time and need very few human efforts. ML algorithms, such as the maximum entropy, SVM, and NB, achieved approximately 80% accuracy, but ensemble and hybrid-based algorithms for sentiment analysis of Twitter reviews perform better than supervised ML techniques by achieving accuracy of 85%.

3 Methodology

Machine learning algorithms such as NB, SVM, multinomial text, and logistic regression have been used for sentiment analysis on reviews given on Amazon products. Both unsupervised and supervised machine learning approaches have been used and are depicted in Fig. 1. In supervised learning, the data with labels is provided as positive and negative. The NB classifier is used as it is one of the most naive and efficient algorithms for classification which aims to build fast machine learning models and, thus, results in quick predictions about the data. It is also known as the probabilistic classifier as it predicts the data based on the probability that an object possesses certain properties. An SVM classifier is used with a set of labeled data, in which the data is classified into different classes with high accuracy. The reason behind using logistic regression is that in the feature space, no assumptions are made about the distribution of classes; it gives provision to extend to multiple classes and is very efficient while training the data.

3.1 Naïve Bayes Classifier

It is a supervised machine learning model which uses Naïve Bayes theorem to calculate probability of the class and is used to solve problems that requires classification like classification of texts. NB is one of the most naïve and proficient algorithms for classification, which aims to build fast machine learning models and, thus, results in quick predictions about the data. It is also known as probabilistic classifier as it predicts the data based on the probability that an object possesses certain properties.

3.2 Support Vector Machine Classifier

SVM, which stands for support vector machine, is an algorithm based on supervised machine learning that is mainly used for classification-based problems but can also be used for problems that require regression methodology.

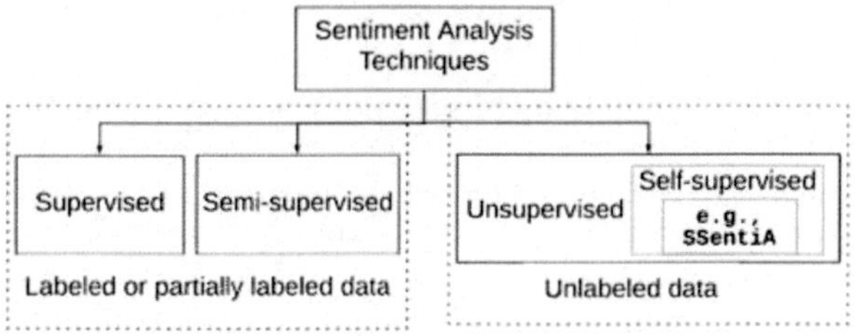

Fig. 1 Sentimental analysis techniques

The main objective behind using the SVM is that for an N-dimensional space (where N denotes the count of features present), it can efficiently find a decision boundary or a hyperplane which classifies the data.

SVM is generally used in face detection, image detection, text categorization, etc.

Furthermore, SVM can be segregated into two parts:

- Linear SVM.
- Nonlinear SVM.

Although it is one of the most efficient and versatile algorithms, it does not yield the probability estimations directly; rather, a cross validation method with five folds is used for calculation, which is an expensive method.

3.3 Logistic Regression

LR is an algorithm that uses a supervised learning approach for classification in which for a given set of input, the output procured can only be distinct values. The probability is predicted using the model of regression that the input given either belongs to group "I" or group "0" if it has only two classes. LR models use the sigmoid function that acts as an activation function. LR can be categorized into different categories based on the number of classes present, i.e.:

- Binomial LR: Where the number of target classes is, for example, 0 or 1.
- Multinomial LR: Where the number of target classes is more than two.
- Ordinal: Where the target classes are defined in a particular ordered structure.

The dataset is divided into two parts in the ratio of 7:3: 70% of the total dataset has been used as a training input and the rest of the 30% as a testing input. The cross validation method with the value of folds set to 10 is used for validating the data set.

In the proposed work, the unprocessed raw data is considered. First, the data is preprocessed to convert the strings into a nominal dataset as classifiers work only on nominal datasets. For this purpose, an unsupervised algorithmic string-to-word vector filter is used. It takes a string as input and converts it into a set of numeric attributes that represent the occurrence information of the words present in the text. The working of the filter has been demonstrated in Fig. 2.

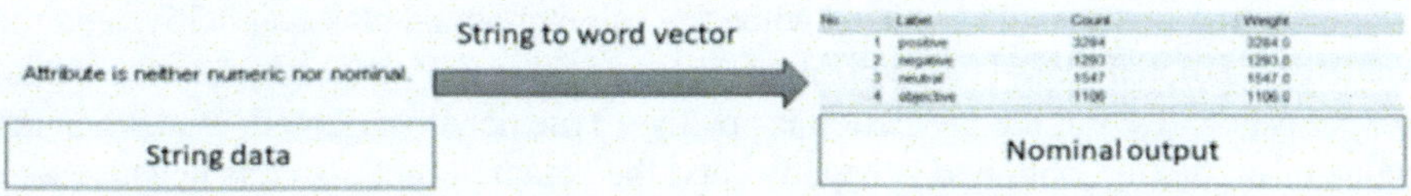

Fig. 2 Preprocessing of word to nominal

Once the preprocessed data is obtained, it is then passed to various classification algorithms. The first classifier is SVM; in this algorithm, a suitable hyperplane is discovered for performing the classification and is used to segregate the classes as positive and negative.

SVM uses the "kernel trick" to draw the hyperplane. Our work is based on linear data, i.e., positive and negative; thus, a linear kernel is used for this purpose. The formula for SVM is shown in $y = w.x + b$, where w denotes the vector that is normal to the hyperplane and b denotes the threshold.

If y > 0, then the class is positive, and if y < 0 then it belongs to the negative class.

Then, the second algorithm used is Naïve Bayes. This algorithm is based on Bayes' theorem of probability to find the unknown class. It calculates the probability that a given data belongs to a specific class. The higher the probability of the class, the more likely the data belonged to that particular class. The formula for Naïve Bayes is given in Eq. 1.

$$P\left(\frac{x}{y}\right) = \frac{P\,(y/x)\,P(x)}{P(y)} \tag{1}$$

The third algorithm used is logistic regression. In this algorithm, the data is represented as a multidimensional vector of the same size as the size of text. In the vector, 1 is indicated against the words that are present in the data or text and 0 otherwise. The sigmoid function is used as an activation function to obtain the probability of the output in the range of 0 and 1. It provides nonlinearity to the model. If the value of the output is greater than 0, then the value predicted is considered to be 1, and if the value is less than 0, then the predicted value is considered to be 0. The formula of the sigmoid function is shown in Eq. 2.

$$\varphi(x) = \frac{1}{1 + e^{-x}} \tag{2}$$

where x is given by:

$$x = \sum W_i y_i + b \tag{3}$$

where w is weight, y is input, and b is bias.

Once the data is passed into the classifiers and classified with various classifiers, the accuracy, false positive, true positive, precision, and recall have been obtained. The term true positive (TP) is used when the value of actual class is positive and the value of predicted class is also positive. False positive (FP) is used when the actual output is negative but the prediction is positive. True negative (TN) is used when the value of prediction obtained is negative and the actual output also has negative as a value or class. False negative (FN) is used when the prediction obtained is negative but the actual output is positive in value. The formulas for accuracy, precision, and recall are given in Eqs. 4, 5, and 6.

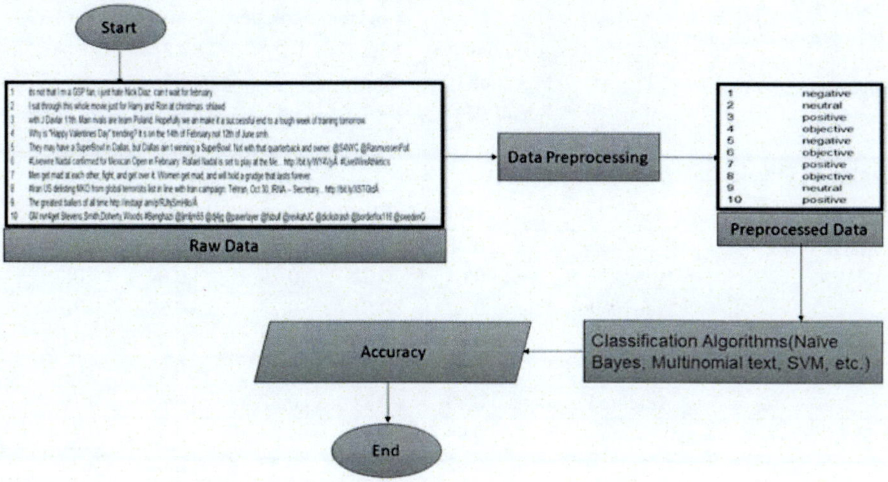

Fig. 3 Framework of the proposed approach

$$accuracy = \frac{TP + TN}{(TP + TN + FP + FN)} \tag{4}$$

$$precision = \frac{TP}{(TP + FP)} \tag{5}$$

$$recall = \frac{TP}{(TP + FN)} \tag{6}$$

The complete flow of our approach has been shown in Fig. 3.

4 Result and Discussion

The proposed work has been carried out on the Amazon reviews dataset, which is available as open source. The framework has been carried out on a system with the following configuration: x64-based, AMD Ryzen 55600H @ 3301 MHz, 6 Core system with 8GB RAM. The dataset consists of 119,845 instances. It is categorized into two classes, namely, 0 and 1, where 0 signifies a negative class and 1 signifies a positive class. The dataset has been trained using 70% of the actual data, and the remaining 30% of the data has been used as a testing data. Table 1 shows the comparative accuracy, precision, and recall of the different classifiers used. Table 2 provides the detailed analysis of different classification algorithms used in the proposed system. It shows highest value for all the parameters such as TP Rate, FP Rate, Precision, Recall, and F1-score for Logistic Regression as compared to Naïve Bayes and SVM.

Table 1 Comparison between accuracy, precision, and recall of various algorithms

	Logistic regression	Naïve Bayes	SVM
Accuracy (%)	83.37	82.2	82.78
Precision (%)	82.2	74.4	78.3
Recall (%)	83.4	82.2	82.8

Table 2 Detailed analysis of classifiers

	Class	TP rate	FP rate	Precision	Recall	F-1 score
Naive Bayes	0	0.98	0.96	0.84	0.98	0.9
	1	0.04	0.02	0.29	0.04	0.08
	Weighted average	0.82	0.8	0.74	0.82	0.76
Logistic regression	0	0.99	0.98	0.83	0.99	0.91
	1	0.01	0	0.76	0.01	0.02
	Weighted average	0.83	0.82	0.82	0.83	0.76
SVM	0	0.98	0.97	0.84	0.98	0.9
	1	0.02	0.01	0.53	0.02	0.05
	Weighted average	0.82	0.81	0.78	0.82	0.76

5 Conclusion

This paper presents a framework for classifying different reviews of products and social networking sites using Naïve Bayes, logistic regression, and SVM. In this, the reviews are classified as negative or positive based on the sentiments that are being predicted using the classifiers. As it has been shown in the result, the accuracy procured is 82.20% for Naïve Bayes, 83.37% for logistic regression, and 82.8% for SVM. Logistic regression achieved better accuracy as compared to SVM and Naïve Bayes because it maximizes the posterior class probability and thus increases the accuracy. The future scope of this is that, as the government of India has strict laws regarding the posts and comments on social networking and blogging sites, this approach can be used to enhance the filtration of the comments so that fewer negative comments are posted and the regulation over the messaging can be sustained. Also, the companies can use this method to review the feedback of their customers so that they can provide more precise suggestions to their customers related to their product.

References

1. Asif, Muhammad & Ishtiaq, Atiab & Ahmad, Haseeb & Aljuaid, Hanan & Shah, Jalal. (2020). Sentiment Analysis of Extremism in Social Media from Textual Information. Telematics and Informatics. 48. 101345. https://doi.org/10.1016/j.tele.2020.101345.
2. Wan, Yun & Gao, Qigang. (2015). An Ensemble Sentiment Classification System of Twitter Data for Airline Services Analysis. 1318–1325. https://doi.org/10.1109/ICDMW.2015.7.

3. Alsaqqa, Samar & Al-Naymat, Ghazi & Awajan, Arafat. (2018). A Large-Scale Sentiment Data Classification for Online Reviews Under Apache Spark. Procedia Computer Science. 141. 183–189. https://doi.org/10.1016/j.procs.2018.10.166.
4. Singh S., Kaur H. (2021). Comparative Sentiment Analysis Through Traditional and Machine Learning-Based Approach.
5. Haider, Sajjad & Afzal, Muhammad & Asif, Muhammad & Maurer, Hermann & Ahmad, Awais & Abuarqoub, Abdelrahman. (2018). Impact analysis of adverbs for sentiment classification on Twitter product reviews. Concurrency and Computation: Practice and Experience. 33. e4956. https://doi.org/10.1002/cpe.4956.
6. A.H. Alamoodi, B.B. Zaidan, A.A. Zaidan, O.S. Albahri, K.I. Mohammed, R.Q. Malik, E.M. Almahdi, M.A. Chyad, Z. Tareq, A.S. Albahri, Hamsa Hameed, Musaab Alaa, Sentiment analysis and its applications in fighting COVID-19 and infectious diseases: A systematic review, Expert Systems with Applications, Volume 167, 2021, 114155, ISSN 0957-4174, https://doi.org/10.1016/j.eswa.2020.114155.
7. Glesias, Carlos A.; Moreno, Antonio (2019). Sentiment Analysis for Social Media. Applied Sciences, 9(23), 5037. doi:https://doi.org/10.3390/app9235037
8. Najma Sultana, Pintukumar, Monika Rani Patra, Sourabh Chandra and S.K. Safikul Khan. Sentimental analysis on product review.
9. Zulfadzli Drus, Haliyana Khalid* (2019). Sentiment Analysis on Social Media and Its Application: Systematic Literature Review.
10. Abdullah Alsaeedi, Mohammad Zubair Khan (IJACSA) International Journal of Advanced Computer Science and Applications, Vol. 10, No. 2, 2019. A Study on Sentiment Analysis Techniques of Twitter Data.

A Stable Model for Learning and Memorization in Over-Parameterized Networks

Eshan Pandey and Santosh Kumar

1 Introduction

If the total number of trainable parameters in a deep neural network is much higher than the total number of training data samples, the network is said to be over-parameterized. Over-parameterized deep neural networks have huge model capacity and can completely shatter the entire training dataset. The question of what a deep model can generalize and what it cannot was highlighted by [1, 2]. They demonstrated that over-parameterized neural networks have a huge capacity and can memorize an entire dataset. Most modern deep neural networks can fit random labels and/or complete random noise and reduce the training error to almost zero. The work in [1–4] provides an experimental study on deep neural networks and suggests that optimization remains an easy process even when generalization is not possible, e.g., in cases where the labels are randomized. The work in this paper is hugely inspired by [1, 2].

2 Related Work

Numerous studies try to provide a framework for controlling generalization errors: VC dimensions [5], Rademacher complexity [6], and uniform stability [7–9].

E. Pandey
Department of Computer Science & Engineering, ABES Engineering College, Ghaziabad, Uttar Pradesh, India
e-mail: eshan.19m101004@abes.ac.in

S. Kumar (✉)
Galgotias University, Greater Noida, Uttar Pradesh, India

M. Mishra et al. (eds.), *Applications of Computational Intelligence in Management & Mathematics*, Springer Proceedings in Mathematics & Statistics 417,
https://doi.org/10.1007/978-3-031-25194-8_3

Numerous studies try to formulate some bounds on the network design and learnability of deep neural networks. It is known that a two-layer network can optimize the regularized loss to a global minimum in polynomial iterations using noisy gradient descent if it has infinite width [10]. The expressive power of a deep neural network is greater than the expressive power of a shallow neural network [11]. In [22], authors argued with experiments that the size of the network may affect the capacity of neural networks, but the size is not the primary form of capacity control. The universal approximation theorem defines the upper bound on the approximation capability of a two-layer network. Any continuous and bounded function can be approximated by a two-layer network having nonlinear activation [12–14]. A 2-layer network can optimize the regularized loss to a global minimum in polynomial iterations using noisy gradient descent if it has infinite width. The problem with such theoretical bounds is that such upper bounds fail to capture any meaningful real-life application. Under such presumptions, the width of the network is infinite, and hypothesis space is infinite.

SGD can achieve the global minimum with near-zero training error in polynomial time [16]. If the labels are true, SGD learns and generalizes, but when the labels are random, SGD finds a network that memorizes the data. Over-parameterized deep networks can boost generalization by making use of huge network sizes. The trained network spread the information evenly among the neurons [17].

The global minimum will most likely have error values close to the local minima [18], and looking for the global minima might lead to overfitting [19]. A small batch size with an SGD optimizer produces flat minimizers which generalize well, while a large batch size with an SGD optimizer produces sharp minima which fail to generalize well [20]. However, [21] suggests large batches cause optimization difficulty, but when addressed, trained networks can achieve good generalization. For fully connected networks and large convolutional networks, the critical layers lie towards the end of the networks (towards the output layer), and the initial layers are robust to reinitialization.

Spatial transformer networks [23] transform the training images spatially in the networks themselves and achieve better generalization. By increasing the number of hidden units, the generalization capability of the network is improved even when the training error does not decrease [15]. This might indicate how implicit and explicit regularization can be helpful for better generalization.

This paper extends the work of Zhang et al. [1]. The aim of the experiments carried out was to study the generalization and memorization capacity of over-parameterized deep neural networks. The networks that were trained on true images and true labels showed high generalization ability, demonstrating that the model learns the classification. When the bytes of the images were shuffled, the models showed some initial learning with true labels. In the initial epochs, the training and test accuracy improved in sync before the models started to overfit. The validation accuracy achieved is far greater than mere random guessing. This implies that even in complete distortion of localized information, over-parameterized deep networks show signs of learning. The question remains: how can an over-parameterized model, having the capacity to memorize the entire data set, generalize? Even in cases where overfitting is easier than generalization, the model demonstrates initial learning before overfitting.

Shuffling the bytes distorts the semantic and localized relations in the data (image). Each pixel in the image consists of three bytes. Only image bytes were shuffled while keeping the labels true.

3 Experimental Setup

CIFAR-10 data is used for ten-class classification tasks. The dataset contains 5000 training samples per class. For the binary class, two classes were randomly selected from the ten-class dataset; the results remain similar for any two classes randomly picked. Four deep convolutional networks were trained on true images and true labels, shuffled bytes and true labels, and randomly generated noise. Shuffling of bytes distorts/modifies the semantic and localized relations in the data. Figure 1 shows how an image appears before and after byte shuffling. Each pixel in the image is made of three bytes. Shuffling of bytes compared to the shuffling of pixels provides more distortion as the color code is also modified.

Figure 2 shows four models that have been used for the study, namely, modified VGG, smallNet, tNet, and convNet. The dense layers of the VGG model are modified to have 512 parameters each. VGG (modified) has convolutional blocks followed by max pooling. SmallNet is the smallest architecture, in terms of total trainable parameters and makes use of dropout for regularization. tNet and convNet additionally make use of batch normalization along with dropout for regularization. tNet does not use any max pooling. tNet has the highest number of trainable parameters, 15, 245, 130; followed by VGG (modified), at 67, 716, 666. ConvNet and smallNet are relatively smaller nets, with convNet having 2,395,242 and smallNet having 1,250,858 trainable parameters.

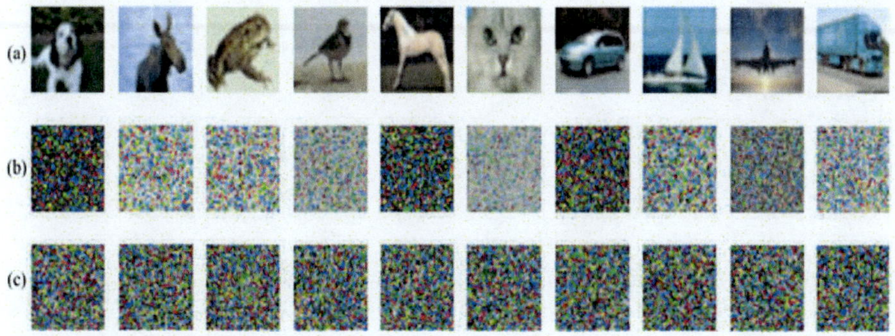

Fig. 1 Sample images from CIFAR-10 dataset. (**a**) True images. (**b**) Corresponding images from (a) with shuffled bytes. (**c**) Randomly generated noise

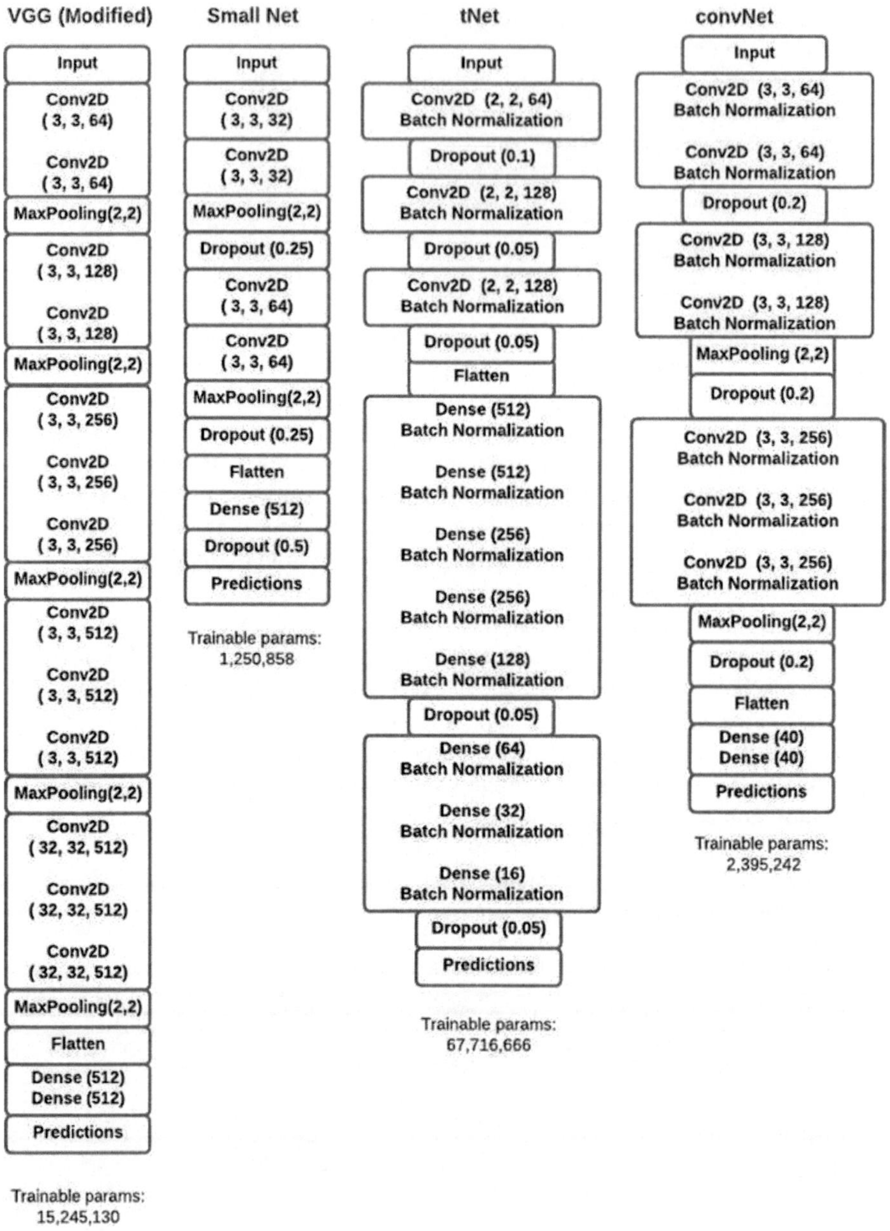

Fig. 2 Architecture design of VGG (modified), smallNet, tNet, and convNet

Little to no extra effort was made to train the models better. Each model was trained for 250 epochs. The categorical cross entropy loss function was used with stochastic gradient descent and ReLU activations for VGG (modified). A learning rate of 0.01 with no decay was used. For smallNet, RMSprop and ReLU activations with a learning rate of 0.001 and a decay of 1e-06 were used. Each model was trained for 250 epochs. The training settings remained the same for true images, shuffled bytes, and random noise.

If a model achieves high training accuracy and high validation accuracy, the model is said to have learned the classification task. Low training error and low generalization error are indications of good learning by the model, and the model is said to have good generalization ability. If the training accuracy is high with low validation accuracy, the model is said to have memorized the training data. 0 (or approximately 0) training error with high generalization error is the indication of memorization by the model. Memorization is the overfitting of the data. Most modern deep neural networks are over-parameterized and can memorize entire training data, and randomly generated noise, yet are able to generalize to true data and true labels [1].

4 Results

Low generalization error with high accuracy is observed in all models for true data images. VGG (modified), smallNet, tNet, and convNet all achieve near 100% training accuracy on binary classification tasks trained on true image data with true labels. These models also achieve high test accuracy of approximately 95% or higher, indicating high generalization ability. For ten-class classification on true images, smallNet achieved the lowest training accuracy but achieved the lowest effective generalization error. VGG (modified), tNet, and convNet achieved approx. 100% training accuracy and approx. 75% test accuracy or higher. All modes show leaning (high training accuracy and good generalization ability) on true image data.

As observed by [1], we observed that the models could easily memorize the randomly generated noise. This shows that the over-parameterized networks have a huge model capacity. However, convNet failed to memorize the randomly generated binary class data. ConvNet and smallNet failed to memorize randomly generated ten-class noise. As there is no relation in the data in case of randomly generated noise, the test accuracy remains equivalent to mere random guessing. There is no feature to be extracted or learned from randomly generated noise. In case of shuffled bytes of the images, for a binary classification task, VGG (modified), tNet, and convNet achieve 100% training accuracy, while smallNet achieves 75% training accuracy. All models achieve approximately 74% test accuracy or higher. A similar trend is observed on the ten-class classification task of shuffled pixels.

It is to be noted that the test accuracy achieved by all the models for shuffled bytes is higher than random guessing. For a binary classification and a ten-class classification, the accuracy of a random guess is 50% and 10%, respectively (any event having n likely possible outcome has $1/n$ probability of winning/accuracy). In case of the shuffled bytes, for the initial training process, the training and test accuracy increase gradually and in sync, demonstrating some learning. It is only

Table 1 Top test and training accuracy achieved by VGG (modified), SmallNet, tNet, and convNet on true images, shuffled bytes, and randomly generated noise for binary classification and ten-class classification tasks

Problem	Data variant	Best accuracy VGG (modified) (%)		SmallNet (%)		tNet (%)		convNet (%)		Random guess (%)
		Train	Test	Train	Test	Train	Test	Train	Test	
Binary classification	True image	100	95.90	99.67	97.05	100	94.70	100	97.05	50.00
	Shuffled bytes	100	75.80	92.77	74.80	100	73.95	99.87	75.30	
	Noise	100	50.00	91.49	50.00	100	50.00	50.65	50.00	
10 class classification	True image	100	74.94	79.79	78.28	99.96	68.76	99.90	85.35	10.00
	Shuffled bytes	99.64	20.85	27.68	18.15	99.94	19.27	96.69	20.67	
	Noise	99.77	10.00	20.95	10.00	99.87	10.00	9.98	10.00	

after some initial learning that the model starts to overfit. Even in the absence of semantic and localized relations, the models show some learning, implying that semantics and localized relations are not all that a network looks for while learning. Each model before training produces results that can be considered mere guessing.

The best test and training accuracy for different models on true image data, shuffled bytes, and randomly generated noise for binary and ten-class classification is shown in Fig. 3.

In Fig. 3a, all models achieve high train and test accuracy for true images on binary classification (good generalization, indication of learning). In Fig. 3b, all models achieve near 100% training accuracy on shuffled bytes (memorization) and test accuracy higher than random guessing before memorization (indicating learning). In Fig. 3c, VGG (modified), smallNet, and tNet achieve 100% training accuracy (memorization), but convNet fails to memorize the binary classes of randomly generated noise. All modes achieve test accuracy as good as random guess only (no indication of learning). In Fig. 3d, all models achieve high train and test accuracy for true images on binary classification (Good generalization, indication of learning). In Fig. 3e, smallNet failed to memorize the shuffled bytes; all other models memorized the data, achieving 100% training accuracy. All models including smallNet achieve test accuracy better than random guessing (indicates learning). In Fig. 3f, only VGG (modified) and tNet memorize the ten classes of randomly generated noise.

The order of ease of memorization (100% training accuracy) is true images > shuffled bytes> randomly generated noise; randomly generated noise is the hardest to memorize. All the models achieve 100% training accuracy on true images and shuffled bytes. SmallNet and convNet have fewer parameters and fail to memorize the randomly generated noise. By reducing the explicit regularization and/or increasing the number of trainable parameters in smallNet and convNet, memorization capacity was observed to be improved. The order of ease of learning is similar to the order of ease of memorization, i.e., true images > shuffled bytes> randomly generated noise; true images easiest to learn. There is nothing to be learned in case of randomly generated noise; there is no relation between the training images and labels and the training and validation sets. It is important to note that some information is preserved in shuffled bytes, and neural networks try to capture that relation.

The order of ease of memorization indicates some data are easier to optimize than others. Semantic data was observed to be the easiest to optimize. Here, the order of ease of optimization is based on the number of epochs required to reduce the training error. The training methodology remains the same for true images, shuffled bytes, and noise. The initial generalization in the models for shuffled bytes demonstrates the algorithmic stability of deep convolutional networks.

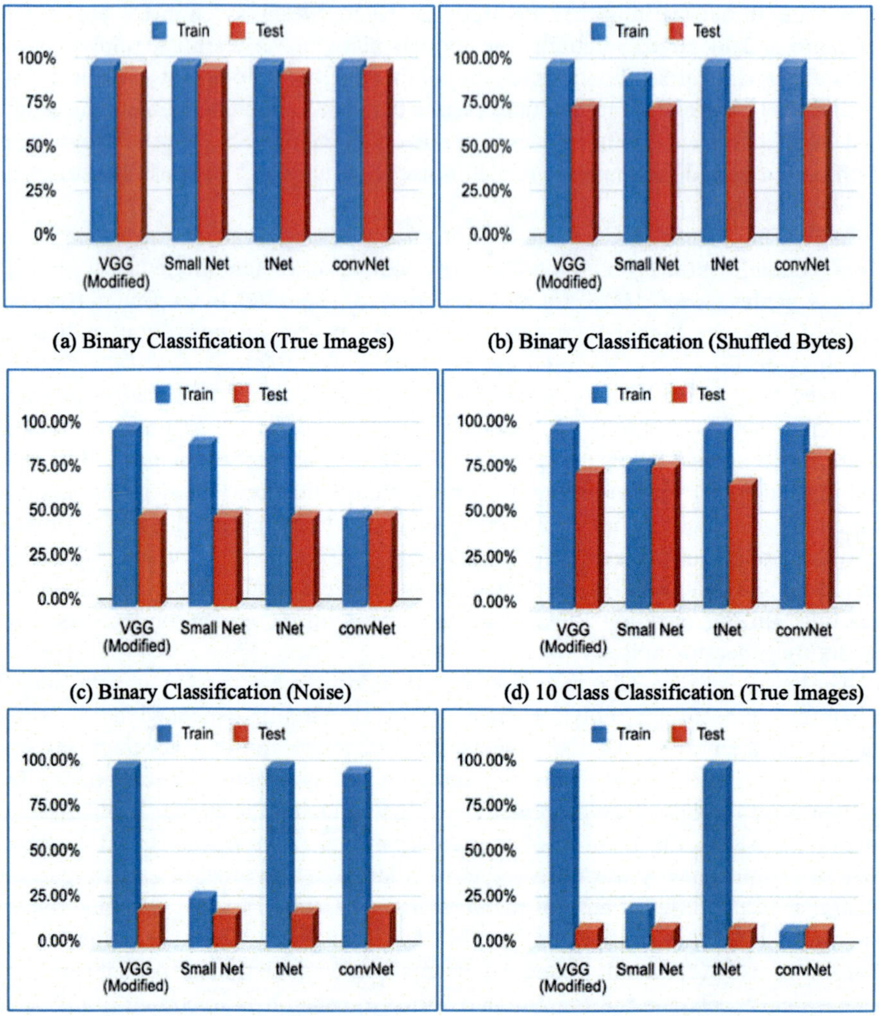

(a) Binary Classification (True Images)
(b) Binary Classification (Shuffled Bytes)
(c) Binary Classification (Noise)
(d) 10 Class Classification (True Images)
(e) 10 Class Classification (Shuffled Bytes)
(f) 10 Class Classification (Noise)

Fig. 3 Top test and training accuracy achieved by VGG (modified), SmallNet, tNet, and convNet on true images, shuffled bytes, and randomly generated noise for binary classification and ten-class classification tasks. (**a**) Binary classification (true images). (**b**) Binary classification (shuffled bytes). (**c**) Binary classification (noise). (**d**) 10 Class classification (true images). (**e**) 10 Class classification (shuffled bytes). (**f**) 10 Class classification (noise)

5 Conclusion

It is observed that over-parameterized deep neural networks have huge model capacities. Most modern deep models can completely shatter the training data. Despite huge model capacity, over-parameterized deep networks can be generalized.

This paper demonstrates huge stability in over-parameterized deep neural networks. A model that achieves generalization on true images, shows initial learning with shuffled bytes with no modification in the algorithm. Neural networks are also great feature extractors and stable learning algorithms. When networks are trained on shuffled bytes of the images, for the binary classification task, VGG (modified), tNet, and convNet achieve 100% training accuracy while smallNet achieved 75% training accuracy. All models achieve at least 74% test accuracy or higher. A similar trend is observed on the ten-class classification task of shuffled pixels. Despite their huge capacity, neural networks tend to learn the features in data before overfitting. It is also demonstrated that over-parameterized deep neural networks also look beyond semantics and localized relation in data.

References

1. Zhang, Chiyuan & Bengio, Samy & Hardt, Moritz & Recht, Benjamin & Vinyals, Oriol. (2016). Understanding deep learning requires rethinking generalization. Communications of the ACM. 64. https://doi.org/10.1145/3446776.
2. Zhang, Chiyuan, Samy Bengio, Moritz Hardt, Benjamin Recht, and Oriol Vinyals. "Understanding deep learning (still) requires rethinking generalization." Communications of the ACM 64, no. 3 (2021): 107–115.
3. Allen-Zhu, Zeyuan, Yuanzhi Li, and Zhao Song. "A convergence theory for deep learning via over-parameterization." In International Conference on Machine Learning, pp. 242–252. PMLR, 2019.
4. H. Salehinejad, S. Valaee, T. Dowdell, E. Colak and J. Barfett, "Generalization of Deep Neural Networks for Chest Pathology Classification in X-Rays Using Generative Adversarial Networks," 2018 IEEE International Conference on Acoustics, Speech and Signal Processing (ICASSP), 2018, pp. 990–994, doi: https://doi.org/10.1109/ICASSP.2018.8461430.
5. Vapnik, Vladimir Naumovich. "Adaptive and learning systems for signal processing communications, and control." Statistical learning theory (1998).
6. Bartlett, Peter L., and Shahar Mendelson. "Rademacher and Gaussian complexities: Risk bounds and structural results." Journal of Machine Learning Research 3, no. Nov (2002): 463–482.
7. Mukherjee, Sayan, Partha Niyogi, Tomaso Poggio, and Ryan Rifkin. "Learning theory: stability is sufficient for generalization and necessary and sufficient for consistency of empirical risk minimization." Advances in Computational Mathematics 25, no. 1 (2006): 161–193.
8. Bousquet, Olivier, and André Elisseeff. "Stability and generalization." The Journal of Machine Learning Research 2 (2002): 499–526.
9. Poggio, Tomaso, Ryan Rifkin, Sayan Mukherjee, and Partha Niyogi. "General conditions for predictivity in learning theory." Nature 428, no. 6981 (2004): 419–422.
10. Wei, Colin, Jason Lee, Qiang Liu, and Tengyu Ma. "Regularization matters: Generalization and optimization of neural nets vs their induced kernel." (2019).
11. Delalleau, Olivier, and Yoshua Bengio. "Shallow vs. deep sum-product networks." Advances in neural information processing systems 24 (2011): 666–674.
12. Hornik, Kurt, Maxwell Stinchcombe, and Halbert White. "Multilayer feedforward networks are universal approximators." Neural networks 2, no. 5 (1989): 359–366.
13. Funahashi, Ken-Ichi. "On the approximate realization of continuous mappings by neural networks." Neural networks 2, no. 3 (1989): 183–192.
14. Barron, Andrew R. "Approximation and estimation bounds for artificial neural networks." Machine learning 14, no. 1 (1994): 115–133.

15. Neyshabur, Behnam, Ryota Tomioka, and Nathan Srebro. “In Search of the Real Inductive Bias: On the Role of Implicit Regularization in Deep Learning.” In ICLR (Workshop). 2015.
16. Allen-Zhu, Zeyuan, Yuanzhi Li, and Zhao Song. “A convergence theory for deep learning via over-parameterization.” In International Conference on Machine Learning, pp. 242–252. PMLR, 2019.
17. Allen-Zhu, Zeyuan, Yuanzhi Li, and Yingyu Liang. “Learning and generalization in over-parameterized neural networks, going beyond two layers.” arXiv preprint arXiv:1811.04918 (2018).
18. Dauphin, Yann N., Razvan Pascanu, Caglar Gulcehre, Kyunghyun Cho, Surya Ganguli, and Yoshua Bengio. “Identifying and attacking the saddle point problem in high-dimensional non-convex optimization.” Advances in neural information processing systems 27 (2014): 2933–2941.
19. Choromanska, Anna, Mikael Henaff, Michael Mathieu, Gérard Ben Arous, and Yann LeCun. “The loss surfaces of multilayer networks.” In Artificial intelligence and statistics, pp. 192–204. 2015.
20. Hochreiter, Sepp, and Jürgen Schmidhuber. “Flat minima.” Neural Computation 9, no. 1 (1997): 1–42.
21. Goyal, Priya, Piotr Dollár, Ross Girshick, Pieter Noordhuis, Lukasz Wesolowski, Aapo Kyrola, Andrew Tulloch, Yangqing Jia, and Kaiming He. “Accurate, large minibatch sgd: Training imagenet in 1 hour.” arXiv preprint arXiv:1706.02677 (2017).
22. Neyshabur, Behnam, Ryota Tomioka, and Nathan Srebro. “In Search of the Real Inductive Bias: On the Role of Implicit Regularization in Deep Learning.” In ICLR (Workshop). 2015.
23. Jaderberg, Max, Karen Simonyan, and Andrew Zisserman. “Spatial transformer networks.” Advances in neural information processing systems 28 (2015): 2017–2025.

Analysis of Depth Sensing and Lane Detection Algorithms for Advanced Driver Assistance Systems

Soumydip Sarkar, Farhan Hai Khan, Srijani Das, Anand Saha, Deepjyoti Misra, Sanjoy Mondal, and Santosh Sonar

1 Introduction

In the twenty-first century, autonomous driving is the need to ensure human comfort. Although autonomous driving is restricted in specific test areas and driving conditions due to safety and other major concerns but in no time, this will expand where we will face highly unpredictable traffic and road conditions. Driverless systems are mainly based no three basic components, Sensors (taking input from environment), perception (what is around the environment) and planning + control (how to drive around). Detection is a part of perception, which helps the car to analyze its surroundings, what is the free space available for driving, and to find obstacles on the driveway. In real world situation like in night-time, foggy weather, snowy weather etc. it is very much important to understand where the car stands and how much area is allotted for it to drive around. There are thousands of variables when driving in a road, a human driver can easily observe and take required decisions, but for a machine to do that, takes a lot more than sensory input and high compute. It starts with taking the sensory input from various sensors like, camera, RADAR, LiDAR, GPS, and others. These inputs get further processed by the Perception module. And based on the analysis of the perception module, planning and control gets executed. These systems are constantly required to analyse their surroundings and understand how far away the other object lies or what is the

S. Sarkar · F. H. Khan · S. Das · A. Saha · D. Misra · S. Mondal (✉)
Department of Electrical Engineering, Institute of Engineering and Management, Kolkata, West Bengal, India

S. Sonar
Department of Electrical and Instrumentation Engineering, Thapar Institute of Engineering and Technology, Patiala, Punjab, India

M. Mishra et al. (eds.), *Applications of Computational Intelligence in Management & Mathematics*, Springer Proceedings in Mathematics & Statistics 417,
https://doi.org/10.1007/978-3-031-25194-8_4

velocity of an object and depth estimation is the computer vision technique that helps to achieve that.

Depth estimation has two parts, one is Monocular depth estimation, which involves estimating depth from a single image of the scene, and the other one is Binocular depth estimation, which involves estimating depth from two images of the scene. Here we only discuss Monocular depth estimation. Previous computer vision work on solving depth estimation was to use these components in a statistical manner. As the time progressed different deep learning-based methods became state-of-the-art for solving the problem. There were a lot of methods revolving around using Convolutional Neural Network (abbr. as CNN hereafter) which generated features without any hand crafter kernels for extracting features. Also works using video sequences as input [1–4], has been published, however, this approach requires more computation and data when compared to monocular depth estimation. And after the introduction of the transformer-based model in 2018, some state-of-the-art papers were also developed such as DPT (Dense Prediction Transformers) [5], and GLPN (Global-Local Path Networks) [6]. The problem of depth estimation can be defined as *pixel-level continuous regression* from a single image. In case of training, the problem is defined as given a training set of I and D, where I belongs to I and D belongs to D, we need to find a mapping function Phi which maps I to D. In modern ADAS systems there are few features like Lane-Departure Warnings, Lane-Keeping Assistance, etc. which solely requires lane detection. For objects with long structure regions that could be occluded, such as lane markings and poles, it still does not function well. In this work we go over a few different CNN-based models for lane detection. Understanding importance is very essential but it is also important to know the underlying problem statement that lane detection implies. We describe lane detection as opposite to depth estimation as it comes under a general category of image segmentation (semantic segmentation or instance segmentation). Image segmentation is defined as *pixel-level classification.* Given a training set of I and S where I belongs to I (the bigger set) and S belongs to S (Bigger set of corresponding segmentation masks) we need to find a mapping function that maps I to S.

1.1 Background

The Conditional GAN, or cGAN, is an extension of the GAN architecture that provides control over the image that is generated, e.g., allowing an image of a given class to be generated. Pix2Pix GAN is an implementation of the cGAN where the generation of an image is conditional on a given image. The generator model is provided with a given image as input and generates a translated version of the image. The discriminator model is given an input image and a real or generated paired image and must determine whether the paired image is real or fake. Finally, the generator model is trained to both fool the discriminator model and to minimize the loss between the generated image and the expected target image. In this paper, we used the stereo vision and as well as monocular depth sensing technique. In stereo vision we have implemented the triangulation theory and in monocular depth sensing we have implemented the methods of deep learning. Previously Hertzmann et al. [5] proposed a supervised stereo vision in their work. Previous work has shown that depth estimation from a stereo pair of images can be formulated as a supervised learning task to be resolved with convolutional neural networks (CNNs). Current architectures depend on patch-based Siamese networks which lacks the means to exploit context information for finding correspondence in ill-posed regions. For solving this issue, they proposed a PSM Net which is a pyramid stereo matching network consisting of two main modules, one is spatial pyramid pooling and another one is 3D CNN. Hinton et al. [6] in their monocular depth based on video proposed a more realistic development. They proposed GeoNet which is a jointly unsupervised learning framework for monocular depth, optical flow, and ego motion estimation from videos. The three components are stuck together by the nature of 3D scene geometry, jointly learned by their framework in an end-to-end manner. The geometric relationships were extracted over the predictions of individual modules and then combined as an image construction loss, reasoning about static and dynamic scene parts separately. They also proposed an adaptive geometric consistency loss to increase robustness towards outliers and non-Lambertian regions, which resolves occlusions and texture ambiguities effectively. In [7], it describes an end-to-end approach to the problem with instance segmentation using CNNs. The proposed method in that paper is an instance segmentation-based method. The model is divided into a semantic segmentation branch that outputs the pixels of the partition line and a branch that outputs the embedding vector used to cluster the pixels for each partition line. Then in [8] which viewed the problem from a semantic segmentation perspective by leveraging the power of CNNs as well as LSTMs. These LSTM units helped the model process the sequence information of the video dataset. They claim the method is real time. One of the interesting works [9], based on the output head part. In general, the output part is a binary map of the lane markings. The value of each greed in the binary map is the probability that the greed is a lane marking. So, a binary map is a vertical * horizontal 2D tensor. The number of channels is 1. In this paper, the number of channels in the final output is not 1. That is no longer binary. Specifically, not only the probability that a certain pixel is a lane marking but also the probability of surrounding pixels is given. The advantage of this method is that there is no need for post-processing, and if you know even one pixel on the lane marking, you can output all the pixels in the Imozuru formula. The rest of the article has been organized as follows. The problem formulation is given in Sect. 2. The simulation results and conclusions are drawn in Sects. 3 and 4 respectively.

2 Proposed Strategy

2.1 Depth Estimation

Taking into consideration the various complications and expenses involved in stereo vision and sensor-based approaches, we propose, develop, and implement two novel monocular depth-sensing strategies in this study: (1) image segmentation and object detection-based hybridized approach and (2) advanced Pix2Pix conditional GAN deep learning model, multiple variants of U-Net.

2.1.1 Image Segmentation and Object Detection-Based Hybridized Approach

Depth has been estimated using properties of similarity of triangles and this has been performed on a multiclass realworld challenge using both instance segmentation & object detection for the same task. After the instance segmentation is performed, we obtain the bounding boxes using translation from the polygons obtained to perform object detection. After the boxes are identified, a simple estimation of the depth from the viewer to the object can be performed using the properties of triangle similarities. The most common and simplest form of measuring the distance between two objects is by placing a scale for measuring tape in between them. This method is best suited for its purpose. The problem arises when this equipment is away from our hands or when the objects are too big/astronomical. For such objects, we can use angular distances between the two objects by extending imaginary lines outward from our eyes. This theory is referred to as the triangulation theory. The key to using this theory is to measure distances and realize that an object's apparent angular size is directly related to its actual size and distance from the observer. The further away an object is, the smaller it appears. This can be shown by a simple example: an astronomical body looks much smaller than any other body present nearby, despite the astronomical body being extremely bigger. A handy tool for measuring such distances just from their distance and angular length is the rule of 57. The rule of 57 states that an object with an angular size of 1 deg. is about 57 times farther away than it is big. The ratio of the angular size of the object to a whole 360° or 2-pi-radian circle should be equal to the ratio of the actual size of the object to the circumference of the circle drawn at that distance from the observer.

Bounding Boxes

For vehicle/obstruction detection, bounding boxes or anchor containers are used to discover the coordinates of the items with reference to the camera or CMOS sensors attached to the driver's car. Now, for a moving item, the picture will increase in size as the object comes in the direction of the CMOS sensor. When it comes closer to the image, it takes up a large region inside the image. Therefore, even though the vehicle type is the same, the sizes can be exclusive inside the photograph. While training for such fashions, exclusive sizes or shapes had distinctive outcomes on the whole model, which led to larger errors for larger-sized objects than small-sized objects. To reduce the impact of the effect, the loss calculation for the width and top of the bounding boxes is improved with the use of normalization. The advanced loss function is,

$$\lambda_{coord}\sum_{i=0}^{S^2}\sum_{j=0}^{B} I_{ij}^{obj}\left[\left(x_i - x\hat{}_i\right)^2 + \left(y_i - y\hat{}_i\right)^2\right] +$$

$$\lambda_{coord}\sum_{i=0}^{S^2}\sum_{j=0}^{B} I_{ij}^{obj}\left[\left(\frac{w_i - w\hat{}_i}{w\hat{}_i}\right)^2 + \left(\frac{h_i - h\hat{}_i}{h\hat{}_i}\right)^2\right]$$

$$+\sum_{i=0}^{S^2}\sum_{j=0}^{B} I_{ij}^{obj}\left(C_i - C\hat{}_i\right)^2$$

$$+\lambda_{noobj}\sum_{i=0}^{S^2}\sum_{j=0}^{B} I_{ij}^{noobj}\left(C_i - C\hat{}_i\right)^2 + \sum_{i=0}^{S^2} I_i^{obj}\sum_{c\in classes}\left(p_i(c) - p\hat{}_i(c)\right)^2 \tag{1}$$

in which xi and yi are the middle coordinates of the field of the ith grid cell, w and h are the widths and the peak of the grid cell, Ci is the confidence of the container, and p.c. is the class probability of the container of the *i*th grid cell, λnoobj denotes the weight of the bounding containers. Additionally, S2 denotes the S × S grid cells, B denotes the boxes, obj i denotes whether the object is positioned in cell i or not, and obj ij denotes that the j-th box predictor in cell i is "responsible" for that prediction. For the bounding containers, the object is detected, and bounding boxes are recognized around the item. The start and give-up coordinates of the bounding field are identified. Then the coordinates of the triangle are recognized, which are as follows: (A) the bottom-most point is 1/9 of the height of the image and below the centre point of the given bounding box of the image, (B) the left point is 1/7 of the height of the bounding box, and (C) the right point is 1/7 of the height of the bounding box. The next step is to identify the angle in the angle of vision in between the two lines. The coordinates will be left (x3,−6 h/7), right (x2,−6 h/7), and bottom (W/2,H/9). So, now for the given figure, let the distance between the extreme left and right points be indicated as w (width).

x_1 = image width/2 = W/2, y_1 = image height/9 = H/9, x_2 = end of boundary box coordinate x, y_2 = −6 h/7 (h = height of boundary box), x_3 = start of boundary box coordinate x, y_3 = −6 h/7 (h = height of boundary box).

Angle between bottom and right points, angle between bottom and left points are

$\Theta = \tan^{-1}(X_1 - X_2/Y_1 - Y_2)$ *180/π, $\phi = \tan^{-1}(X_1 - X_2/Y_1 - Y_2)$ *180/π. Therefore, right angle = 90 + Θ, left angle =90-ϕ, total angle = left angle + right angle, distance = (w*(1/total angle) *57) (by the rule of 57).

Image Segmentation

In the case of pc vision and virtual picture processing, image segmentation is the technique of partitioning a virtual image into a couple of image segments referred to as image regions or objects. The intention of the technique is to simplify and examine the items and their boundaries. More precisely, image segmentation is the technique of naming each pixel in a picture such that the pixels with the same label have comparable traits. Some of the most superior software and designs to put into effect the segmentation are Yolo, unit, and U-Net++. For performing image segmentation, we have used the following SOTA pre-trained models: (i) PointRend, (ii) MobileV3Large, and (iii) Yolo object detection model (you only look once) (Fig. 1).

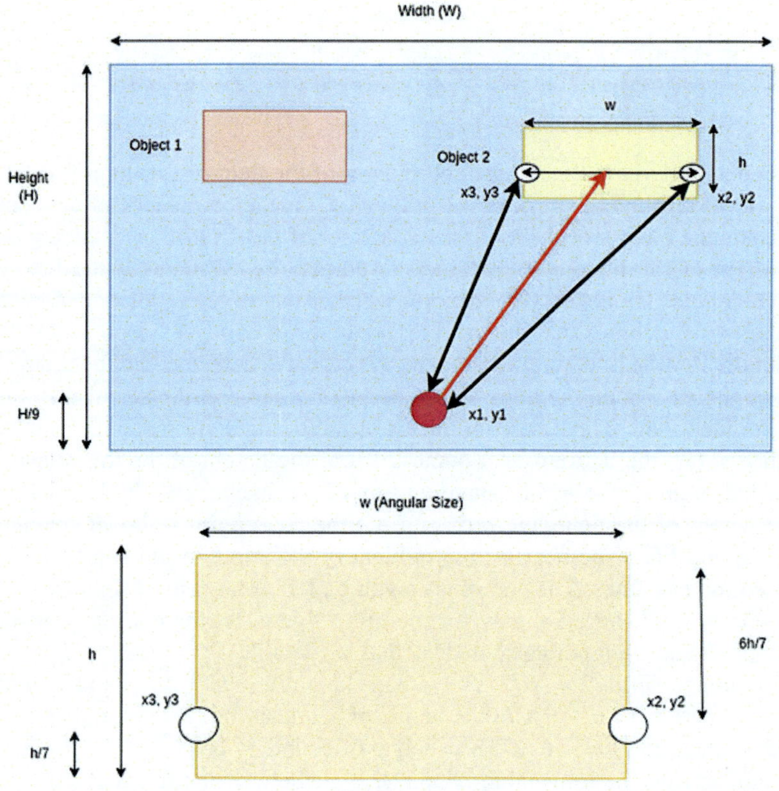

Fig. 1 Depth sensing using the triangulation approach

Fig. 2 Object detection, instance segmentation, and depth sensing results on selected objects of interest using the proposed method of triangulation

Simulation Results (Fig. 2)

2.1.2 Pix2Pix Approach

GAN is considered a neural network trained under unsupervised learning. It is used for the task of deep generative modelling. As an unsupervised learning method, GAN is trained as a method to learn the underlying distribution of the training data. But Pix2Pix GAN is considered a conditional GAN, which comes under supervised learning. In this supervised learning training approach, we provide the input image as well as the corresponding image label. The sole purpose of Pix2Pix GAN is an image-image translation (image-to-image translation is a category of vision and graphics problems in which the goal is to figure out how to transfer an input image to an output image), so here we send the image to be translated as a label to the model. In essence, the generator learns the mapping from the real data as well as the noise for the label, G: $\{x,z\} \rightarrow y$. The discriminator, meanwhile, learns representation from both labels and real data, D(x,y). And at the time of model inference, we are only required to supply one image upon which the translation will take place. U-net and patchGAN are the two major designs in the Pix2Pix, one for the generator and the other for the discriminator, respectively. The architecture of U-Net will be discussed later. But in brief, it is an encoder-decoder-based network which down samples the image, and after it reaches the bottleneck, it up-samples it. The number of up sample and down samples is equal because it helps to concatenate the output

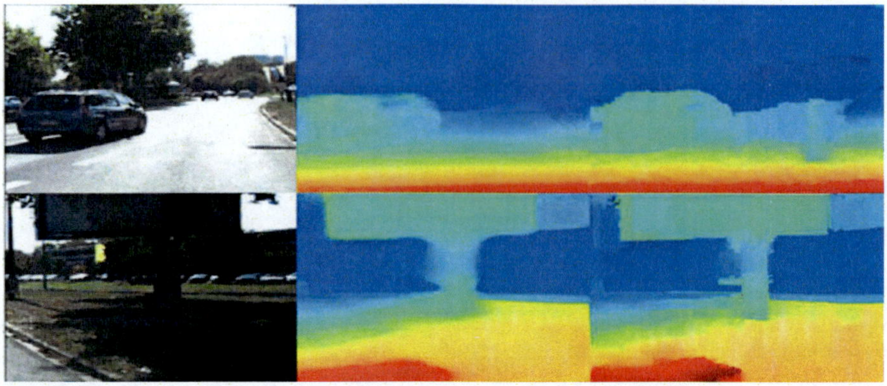

Fig. 3 Pix2Pix output: original image, GAN output, ground truth values. Impressive results even with a low number of EPOCHS chosen due to hardware limitations

of the down samples with the up-sample images via a skip connection, and at the end of the decoder, we get an image the same size as the input image. This architecture is capable of localization, which means it can locate the object of interest pixel by pixel (Fig. 3).

Additionally, U-Net enables the network to transport context information from lower to higher resolution levels. The network can now create high-resolution samples because of this. Patch GAN architecture is used in the discriminator. There are several transposed convolutional blocks in this design. It scans the NxN portion of the image to determine whether it is real or not. The number N can be any size. It can be a fraction of the size of the original image and still generate high-quality results. However, training is a tough procedure since GAN's objective function is more concave-concave than convex-concave. Because of this, finding a saddle point is difficult, which makes training and optimizing GANs challenging. Loss function of vanilla GAN is combined with an L1 loss so that the generator not only fools the discriminator but also produces images that are close to reality. In essence, the generator has an additional L1 loss in the loss function. The generator model is trained using both the adversarial loss for the discriminator model and the L1 or mean absolute pixel difference between the generated translation of the source image and the expected target image.

2.1.3 U-Net

U-Net, one of the deep learning networks with an encoder-decoder architecture, is widely used in scientific image segmentation. U-Net is a U-fashioned encoder-decoder network architecture, which consists of four encoder blocks and four decoder blocks that are linked via a bridge. The encoder network acts as a feature extractor and learns a summary illustration of the input image via a chain of encoder blocks. The encoder network (contracting route) halves the spatial dimensions and doubles the quantity of filters (characteristic channels) at every encoder block. Likewise, the decoder network doubles the spatial dimensions and halves the number of characteristic channels. The decoder network is used to take the summary representation and generate a semantic segmentation mask [10].

After the objects are detected since the touching items are intently positioned next to each other, they are without difficulty merged by using the network. To separate them, a weight map is applied to the output of the network.

$$w(x) = w_c(x) + w_0 \cdot exp\left(-\frac{(d_1(x) + d_2(x))^2}{2\sigma^2}\right) \tag{2}$$

To compute the load map as above, d1(x) is the distance to the nearest detected object at position x, d2(x) is the distance to the second one nearest object. Thus, at the border, the weight is much better than in the figure.

$$p_k(x) = \exp\left(a_k(x)\right) / \left(\sum_{k'=1}^{K} \exp\left(a_{k'}(x)\right)\right) \tag{3}$$

$$E = \sum_{x \in \Omega} w(x) \log\left(p_{l(x)}(x)\right) \tag{4}$$

As a result, the cross-entropy function is penalized at each position through the weight map. And it assists in forcing the network to learn the small separation borders among touching objects [11]. The U-Net model is an encoder and decoder structure with multi-degree cascaded convolutional neural networks. It also takes a feature illustration from the encoder/down sampling route to attention with the aid of concatenating the decoder/up sampling route. And after that, it makes use of a dense layer for prediction, but on the other hand, it is seen that the function vectors from the beginning of the down sampling route are not strong, and the use of that with the same degree of up sampling no longer supplies a great deal of enhancement. So, to cope with this difficulty, we bring interest to the object image. Interest makes use of the same degree down sampling function map and additionally the "characteristic map from one degree beneath" and passes that via an attention gate, which helps to extract a better feature map for concatenating with the corresponding up sampling level. It also facilitates cognizance of the essential part of the picture that is relevant [12]. The attention framework is provided in the figure.

Working: In W_g our edge is 1 and in W_x our edge is 2. However, the number of filters is the same for both. W_g and W_x help to resize g and x. Due to the fact g is from a decreased degree than the present-day stage, it has a smaller form. Now that we have equal sizes, we add them. Aligned weights are extended by means of addition, but unaligned weights are decreased. After that, we send it through it via RELU activation, which makes all weights > = 0. After that, we pass it through a conv block with strobe = 1 and filter = 1. Then we upload a sigmoid addition function to help translate the values between the 0 and 1 barriers. Then we resample

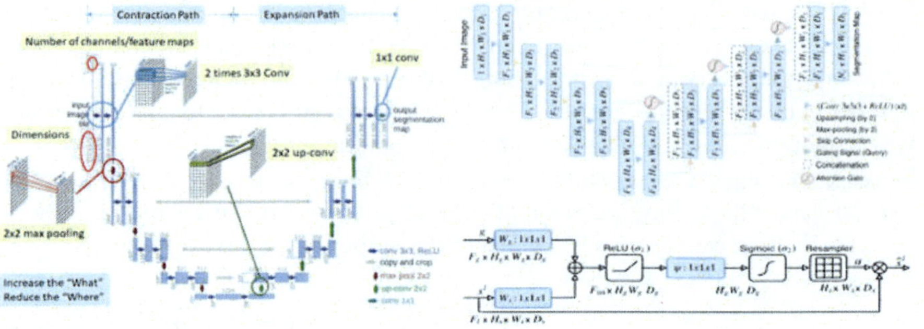

Fig. 4 Layer-wise mathematical model of the U-Net model and U-Net encoder-decoder architecture. Notice that the number of up samples and down samples is the same, resulting in a U-tube-like architecture

them to the scale of x and multiply x by the resampling result. This turns into the final feature map, which is later concatenated with modern-day up sampling [13].

2.1.4 U-Net Backbones

U-Net is a fully convolutional encoder-decoder network extension. The convolutional layers learn low- and high-dimensional features as they iteratively train, using the filters in these layers to extract features. The idea behind U-Net is to encode the image while it is down sampled by sending it through a CNN, then decode or up sample it to retrieve the segmentation mask. The learned weight filters, up sampling and down sampling blocks (which can also be made learnable), and concatenations and skip connections determine which characteristics are identified in the mask. The backbone is an architectural element that determines how these layers are organized in the encoder network and how the decoder network should be constructed [14–16] (Fig. 4 and Table 1).

Vanilla CNNs, such as VGG, ResNet, Inception, EfficientNet, and others, are frequently employed as the backbone since they conduct encoding or down sampling on their own. To create the final U-Net, these networks are extracted, and their counterparts are formed to execute decoding and up sampling. In our experiments, we used ResNet50, ResNext50, and DenseNet121 as the backbone for U-Net.

2.1.5 U-Net ++

The state-of-the-art models for photograph segmentation are the fully convolutional network (FCN) and U-net. U-Net++ is a new segmentation structure based totally on nested and dense pass connections. The primary idea behind U-Net++ is to bridge the semantic hole between the feature maps' ultra-modern encoder and

Table 1 Evaluation metric reports U-Net model variants trained (with hyperparameters)

Model with configuration	Mean squared error (MSE)	Mean absolute error (MAE)	R2 score	Training time taken
Scratch Vanilla (U-Net Adam, bs-16, e40)	0.0024	0.024	0.807	46 m 13 s
Scratch Vanilla U-Net with LR Scheduler (Adam, bs16, e40, Lr-Sched)	0.0025	0.0248	0.8046	59 m 1 s
Attention U-Net Scratch (Adam, bs16, e40, lr1-e4)	0.0025	0.0243	0.8026	1 h 0 m 4 s
U-Net BackBone ResNet50 (Adam, bs-16, e40, Lr: 1e-3)	0.0022	0.0221	0.8231	49 m 33 s
U-Net BackBone ResNext50 (Adam, bs-16, e40, Lr: 1e-3)	0.0022	0.0221	0.8231	47 m 14 s
U-Net BackBone DenseNet 121 (Adam, bs-16, e40, Lr: 1e-3)	0.0022	0.022	0.8312	48 m 26 s
U-Net++ BackBone ResNet50 (Adam, bs-16, e40, Lr: 1e-3)	0.0023	0.0224	0.8202	1 h 9 m 50 s
U-Net ++ BackBone DenseNet 121 (Adam, bs-16, e40, Lr: 1e-3)	0.0022	0.0211	0.8328	1 h 16 m 20 s

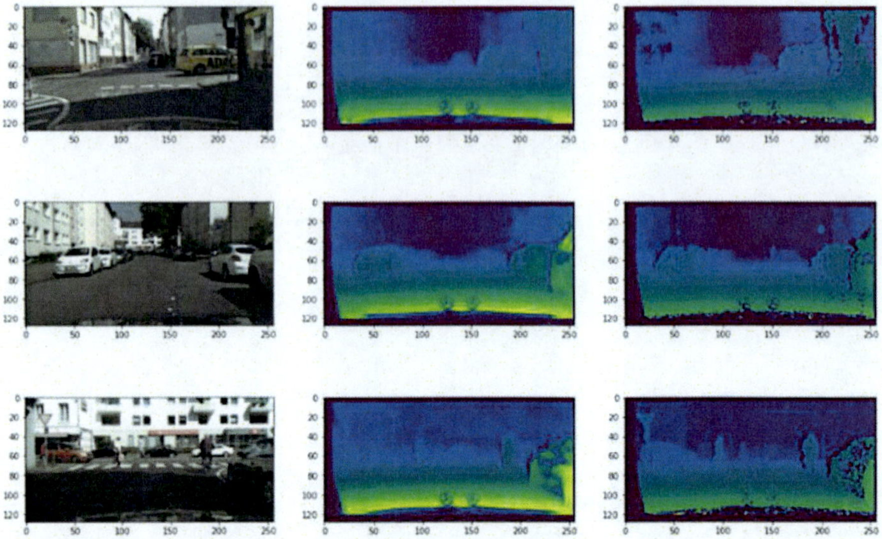

Fig. 5 Output of the Depth Estimation Map for the best-performing U-Net model variant: U-Net ++ BackBone DenseNet 121

decoder. In this paper, we use deep supervision in U-Net++, allowing the model to operate in two modes: (1) correct mode, wherein the outputs from all segmentation branches are averaged, and (2) fast mode, wherein the very last segmentation map is chosen from the best one of the segmentation branches, the choice of which determines the volume of the latest model pruning and the velocity advantage (Fig. 5). The above U-Net++ model suggests an example of the way the characteristic maps tour via the pinnacle bypass pathway of modern-day U-Net++. Owing to the nested skip pathways, U-Net++ generates complete decision feature maps at a couple of semantic ranges. For that reason, the losses are predicted from four semantic ranges, giving a greater targeted knowledge cutting-edge the left regions inside the pix. So, the new loss feature becomes as:

$$L\left(Y, Y^{\hat{}}\right) = -\frac{1}{N}\sum\nolimits_{b=1}^{N}\left(\frac{1}{2}\cdot Y_b\cdot \log Y^{\hat{}}{}_b + \frac{2\cdot Y_b\cdot Y^{\hat{}}{}_b}{Y_b + Y^{\hat{}}{}_b}\right) \tag{5}$$

Lane Detection

Classical Method for Lane Detection

Our used classical approach is a multi-step process which involves many regularly used computer vision algorithms. The method is applied frame by frame to a video input, followed by four steps, pre-processing, bird's-eye view transformation,

Fig. 6 Computer vision approach for lane detection and bird's eye view

lane marking extraction, and finally curve fitting. In pre-processing, we start by converting the frame into a grayscale image, and then again RGB to YCrCb color space transformation takes place on the same image later. We then combine the grey image with the Cr channel. In bird's-eye view transformation, we perform inverse perspective mapping (IPM). The lane marks get extracted using the Gabor filter, or Gaussian modulated sinusoid, which is a Gaussian-like filter with more parameters. Below is the equation:

$$G_c\,[i, j] = Be^{-\frac{(i^2+j^2)}{2\pi^2}} \cos\left(2\pi f\,(icos\theta + jsin\theta)\right) \tag{6}$$

And finally, in curve fitting, the third-order Bezier curve is used. This is the result of the lane detection algorithm we used for the experimentations (Fig. 6).

Deep Learning Approach to Lane Detection

Over the years, people have developed different approaches for lane detection using deep learning; some of them involved pure CNN [3] architecture, some of them used RNNs with CNNs [2], and some current solutions leverage the power of transformer-based architecture also, but at its core, it is a semantic segmentation or instance segmentation problem. As these segmentation models are high-latency and high-computing models, we need to think carefully before designing the architecture so that it can be used in real time. In this work, we benchmarked various state-of-the-art semantic segmentation models and, in the end, came up with a novel approach for performing the same.

2.1.6 U-Net

The vanilla U-Net architecture was trained with 50 epochs and a learning rate of 0.001 along with a combined focal loss + dice loss to give impressive training results, which are displayed at the end of the results. Call-backs included Reduce LR on the plateau for faster convergence and better results [17, 18].

Different Deep Learning Methods Used for Lane Detection

This paper argues that the spatial CNN, a proprietary layer, can be plugged into an existing network of semantic segmentations to improve the accuracy of the model. Their assumption was results of image segmentation depend on the capacity of capturing spatial relationships of pixels across rows and columns of an image. To address this, they created SCNN, which generalizes traditional deep layer-by-layer convolutions to slice-by-slice convolutions within feature maps, thus enabling message passing between pixels across rows and columns in a layer. They used a vgg16 as a backbone model, and they also input additional information about the presence of lane lines, which helped the model to learn better. Our experience with SCNN was very good. It had a very high IoU and F1 score. This is one of the best-performing models with low compute as it was only using VGG16. Except that we also used ERFNet, which is a CNN-based real-time image segmentation framework. This network takes inspiration from a lot of models, like ResNet [3], inception-v3 [4], ENet [5], etc. The most important feature of the model was the residual non-bottleneck-1D block. The down sampler block of its encoder/decoder-based network, inspired by the initial block of ENet, performs down sampling by concatenating the parallel outputs of a single 3×3 convolution with stride 2 and a Max-Pooling module. Dilated convolutions, originally developed in DeepLab [6], are also inserted at certain non-bt-1D layers to gather more context, which led to an improvement in accuracy. Thus, the second pair of 3×1 and 1×3 convolutions are a pair of dilated 1D convolutions. It was also a good performing network, but as it was trained from scratch, it did not have as much IoU and F1 as other models. The results of our previous experience with U-Net motivated us to use this again for lane detection. We treated this as a multiclass semantic segmentation problem. The results were very good, which came with the cost of using pretrained ResNet50 as a backbone, which made the model a bit heavy. We also had a chance to work with self-supervised learning for this problem. We decided to use this as an alternative to transfer learning, which we previously did using VGG16 in SCNN and ResNet50 with U-Net.

SSL-Based Encoder-Decoder Architecture

An ideal dataset is a dataset that has three main qualities: large, clean, and diverse. But getting large diverse annotated data can be very difficult and time-consuming.

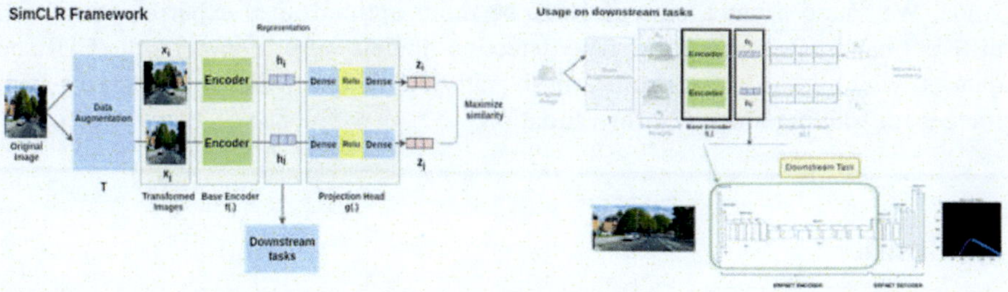

Fig. 7 Sim CLR framework and usage of Sim CLR framework on downstream tasks

So, to solve this problem, SSL has emerged, which handles not only the problem of small datasets but also data without annotation. Because of that, it is a better alternative to transfer learning. In this work, we used SimCLR [7], which is a constructive learning framework. This was the first ground-breaking SSL model which not only outperforms previous self-supervised models but also beats the supervised learning method on ImageNet classification.

The SimCLR framework is based on a very simple concept. To create a pair of augmented pictures xi and xj, an image is obtained, and random transformations are applied to it. To get representations, each picture in the pair is run through an encoder. Then, to get representations z, a non-linear fully linked layer is employed. For the same image, the goal is to maximize the similarity between these two representations, zi and zj (Fig. 7).

$$l_{i,j} = -\log \frac{\exp\left(sim\left(z_i, z_j\right)/\tau\right)}{\sum_{k=1}^{2N} 1_{[k \neq i]} \exp\left(sim\left(z_i, z_k\right)/\tau\right)} \tag{7}$$

τ denotes a temperature parameter. The final loss is computed across all positive pairs, both (i,j) and (j,i), in a mini-batch.

After training, we use the encoder model for the down streaming task, which in our case is semantic segmentation. For doing segmentation, we decided to use ERFNet, which is an encoder-decoder-based network. We used the encoder network in the constructive learning training, and when it is trained, we used that trained encoder as a pretrained backbone to the decoder part. The base model U-Net performed very well, giving an IoU of 0.6911 and an F1 of 0.801. Then the performance of SCNN was a bit better than U-Net IoU of 0.719 and F1 of 0.83. The ERFNet model was the worst, and the performance of the SSL-based encoder-decoder was also less than our base model. This pointed out that the model needs some more hyperparameter tuning, and also, we need to use a better SSL model for this or change the encoder-decoder part of ERFNet to U-Net. While training, we used seeding to keep the experiments reproducible. We used a fixed size of image which is 224 × 224 for training. We found using Adam with weight decay to be better than SGD. For Adam, the learning rate was 0.0001, which is the default best value. We found a batch size of 16 to be more optimal after experimenting with different batch sizes. All of the lane detection models were trained on the CULane dataset, which is a dataset from a dashcam view taken in China. We used ~10 k images for training and ~ 3 k for validation.

3 Results

The inference results for both implementations obtained suitable results based on the testing data and performed accurately with scores above 82%. Some of the outputs obtained from the models are displayed in Figs. 1 and 2. Also, the accuracy and the loss training curve for the Pix2Pix model have been displayed in Figs. 3 and 4. All code and videos are available at https://github.com/khanfarhan10/Advanced-Depth-Sensing-Algorithms/. We attempted to perform depth estimation and lane detection using various methodologies, for better automated vehicle driving. We learned a lot about deep learning, autonomous vehicle, experiment tracking, learned new tools, etc. The future scopes of these works include traffic sign detection, low light obstruction detection and animal detection for accident reduction, path planning, and ultimate ADAS mechanism (Figs. 8 and 9).

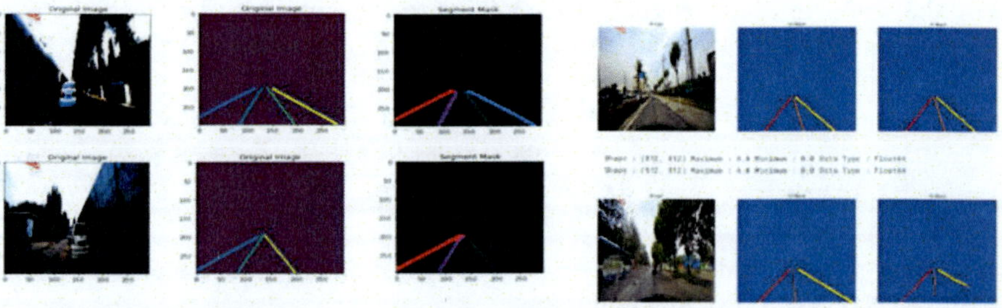

Fig. 8 Output for lane detection from the SCNN model and the U-Net model

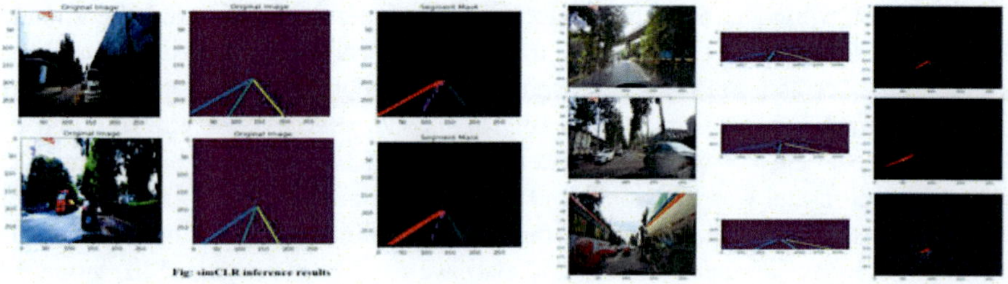

Fig. 9 ERFNet inference results

4 Conclusion

This research mainly focuses on the major methodologies of object detection and summarizes the traditional machine learning methods and the mainstream deep learning technologies for accurate depth sensing. Along with these, the different pros and cons of these methods are discussed. A brief analysis is also given about the two common methods, the Pix2Pix method and the OD and segment method. Though both the methods discussed above are best-in-class designs and detect depths with absolute accuracy, they are used according to the needs of the users. These detection methods are suitable for applications/programs requiring high precision and real-time detection. Depth estimation is a challenging problem that has numerous applications. Through major efforts taken by the research and development community, powerful and rather cheap solutions using machine learning and artificial intelligence are becoming more common. These and many other related solutions have paved the path for innovative applications using depth sensing and estimation in many domains. These technologies are very helpful in the applications of augmented reality and trajectory estimation. More developments are required in the fields concerning fog and haze removal and differentiating between useful and non-useful objects in self-driving vehicles.

References

1. H. Jiang, L. Ding, Z. Sun and R. Huang, "Unsupervised Monocular Depth Perception: Focusing on Moving Objects," in IEEE Sensors Journal, vol. 21, no. 24, pp. 27225–27237, 15 Dec.15, 2021.
2. Kumari, N., Kr. Bhatt, A., Kr. Dwivedi, R. et al. Hybridized approach of image segmentation in classification of fruit mango using BPNN and discriminant analyzer. Multimed Tools Appl 80, 4943–4973 (2021).
3. X. Zou, "A Review of Object Detection Techniques," 2019 International Conference on Smart Grid and Electrical Automation (ICSGEA)", 2019, pp. 251–254.
4. D. Warde-Farley, S. Ozair, A. Courville, and Y. Bengio. Advances in Neural Information Processing Systems. In NIPS, 2014, pp. 201–222.
5. A. Hertzmann, C. E. Jacobs, N. Oliver, B. Curless, and D. H. Salesin. Image analogies. In SIGGRAPH, 2001, pp. 1–4
6. G. E. Hinton and R. R. Salakhutdinov. Reducing the dimensionality of data with neural networks. Science, 313(5786):504–507, 2006.
7. D. Neven, B. D. Brabandere, S. Georgoulis, M. Proesmans and L. V. Gool, "Towards End-to-End Lane Detection: An Instance Segmentation Approach," 2018 IEEE Intelligent Vehicles Symposium (IV), 2018, pp. 286–291.
8. R. Ranftl, A. Bochkovskiy and V. Koltun, "Vision Transformers for Dense Prediction," 2021 IEEE/CVF International Conference on Computer Vision (ICCV), 2021, pp. 12159–12168.
9. S. Hwang, J. Park, N. Kim, Y. Choi, and I. So Kweon. Multispectral Pedestrian Detection: Benchmark Dataset and Baseline. Proceedings of the IEEE Conference on Computer Vision and Pattern Recognition (CVPR), 2015, pp. 1037–1045
10. J. -R. Chang and Y. -S. Chen, "Pyramid Stereo Matching Network," 2018 IEEE/CVF Conference on Computer Vision and Pattern Recognition, 2018, pp. 5410–5418.

11. Z. Yin and J. Shi, "GeoNet: Unsupervised Learning of Dense Depth, Optical Flow and Camera Pose," 2018 IEEE/CVF Conference on Computer Vision and Pattern Recognition, 2018, pp. 1983–1992.
12. M. Hu, S. Wang, B. Li, S. Ning, L. Fan and X. Gong, "PENet: Towards Precise and Efficient Image Guided Depth Completion," 2021 IEEE International Conference on Robotics and Automation (ICRA), 2021, pp. 13656–13662
13. Bar Hillel, A., Lerner, R., Levi, D. et al. Recent progress in road and lane detection: a survey. Machine Vision and Applications 25, 727–745 (2014)
14. Q. Zou, H. Jiang, Q. Dai, Y. Yue, L. Chen and Q. Wang, "Robust Lane Detection From Continuous Driving Scenes Using Deep Neural Networks," in IEEE Transactions on Vehicular Technology, vol. 69, no. 1, pp. 41–54, Jan. 2020
15. K. He, X. Zhang, S. Ren and J. Sun, "Deep Residual Learning for Image Recognition," 2016 IEEE Conference on Computer Vision and Pattern Recognition (CVPR), 2016, pp. 770–778.
16. C. Szegedy, V. Vanhoucke, S. Ioffe, J. Shlens and Z. Wojna, "Rethinking the Inception Architecture for Computer Vision," 2016 IEEE Conference on Computer Vision and Pattern Recognition (CVPR), 2016, pp. 2818–2826.
17. Waleed Alsabhan & Turkey Alotaibai, "Automatic Building Extraction on Satellite Images Using Unet and ResNet50" Hindawi, February 18, 2022. https://doi.org/10.1155/2022/5008854
18. L. -C. Chen, G. Papandreou, I. Kokkinos, K. Murphy and A. L. Yuille, "DeepLab: Semantic Image Segmentation with Deep Convolutional Nets, Atrous Convolution, and Fully Connected CRFs," in IEEE Transactions on Pattern Analysis and Machine Intelligence, vol. 40, no. 4, pp. 834–848, 1 April 2018.

Implementation of E-Commerce Application with Analytics

Rohit Kumar Pattanayak, Vasisht S. Kumar, Kaushik Raman, M. M. Surya, and M. R. Pooja

1 Introduction

The practice of automatically locating meaningful information in massive data warehouses is known as data science. It uses data science approaches to comb through enormous databases in search of novel and relevant patterns that could otherwise go unnoticed. To put it another way, data science is the act of examining data from a variety of angles and integrating it into meaningful knowledge. The associative approach, which looks for patterns of associations between components, is one of the data science methodologies accessible. Affiliation methods are commonly employed to analyze sales transaction data in order to gain a better understanding of market conditions for products that consumers frequently purchase. One method of determining the relationship between time elements is to use a time distribution rule. Clinic is one of the organizations that provides healthcare to the general public.

The difficulty is that these clinics solely use Excel data for drug data summaries, which is a purchase of drug inventory based exclusively on rejected drugs. There are 183 distinct medicines, according to interviews with clinic staff, with up to 630 times more per month. Another unresolved issue is the inability to determine which medicine was administered at any given time. During the monsoon season in October and December, for example, spending on cough suppressants and paracetamol soared by 80 percent compared to normal days. The condition may not be applicable always although required during moderate rainy seasons.

The law of temporal correlation as applied to medical data was used to find the correlation pattern in this investigation, based on aforementioned description. It is hoped that by doing this research, it would be feasible to detect the types and timing

R. K. Pattanayak · V. S. Kumar · K. Raman · M. M. Surya · M. R. Pooja (✉)
Vidyavardhaka College of Engineering, Mysuru, Karnataka, India
e-mail: pooja.mr@vvce.ac.in

M. Mishra et al. (eds.), *Applications of Computational Intelligence in Management & Mathematics*, Springer Proceedings in Mathematics & Statistics 417,
https://doi.org/10.1007/978-3-031-25194-8_5

issues of many pharmaceuticals that are supplied at the same time in accordance with the association's standards.

2 Related Work/Literature Survey

2.1 Purpose

The main goal of this application is to satisfy customers by providing access to various information about prescription drugs and medicines.

2.2 Interactive Web Interface for Association Rule Mining and Medical Applications

The purpose of the application is to create a trouble-free and interactive net interface for association rule mining which represents the act of deriving association rules. Filtrate Association, Aprili, periodical pattern growth, predictive a priori, generalized sequential patterns, HotSpot, and Tertius algorithms are used. The goal is to find non-recurring events and recurring events by implementing data mining to identify time-based patterns in medical data approaches [1].

Clinic is an organization that provides health treatment to a large number of individuals. Patients having medical checks might also get drugs from the facility.

The difficulty is that these clinics solely use Excel data for drug data summaries. Purchase of drug stock based only on data at pharmaceuticals cannot be used Zahrotun et al. [2], Abin et al. [3]. Mining with Causal Effects Adverse Effects of Drugs Induced by Human Allergy are Drug Adverse Events (AEs) for Adverse Identification. Overdosage, chemical interaction between two substances, and so forth. As ADR has become a major issue in recent years, it is vital to reduce these reactions in order to save a patient's life. To avoid negative repercussions, it is critical to discover these adverse impacts as soon as possible. Therefore, it is critical to establish a causal relationship between drug-related events and their findings and shifts.

2.3 Wet Experiment Preparation

The detection of diseases requires a lot of time and labor. Therefore, the development of computational techniques for predicting drug diseases is an urgent task [4]. The overall process is depicted in Fig. 1.

2.4 Methodology

Step 1: In the first step, medical purchase data (drug-based transactions) are collected.

Step 2: Datasets are then pre-processed, and irrelevant data are removed and only relevant data extracted and inputted to the algorithms.

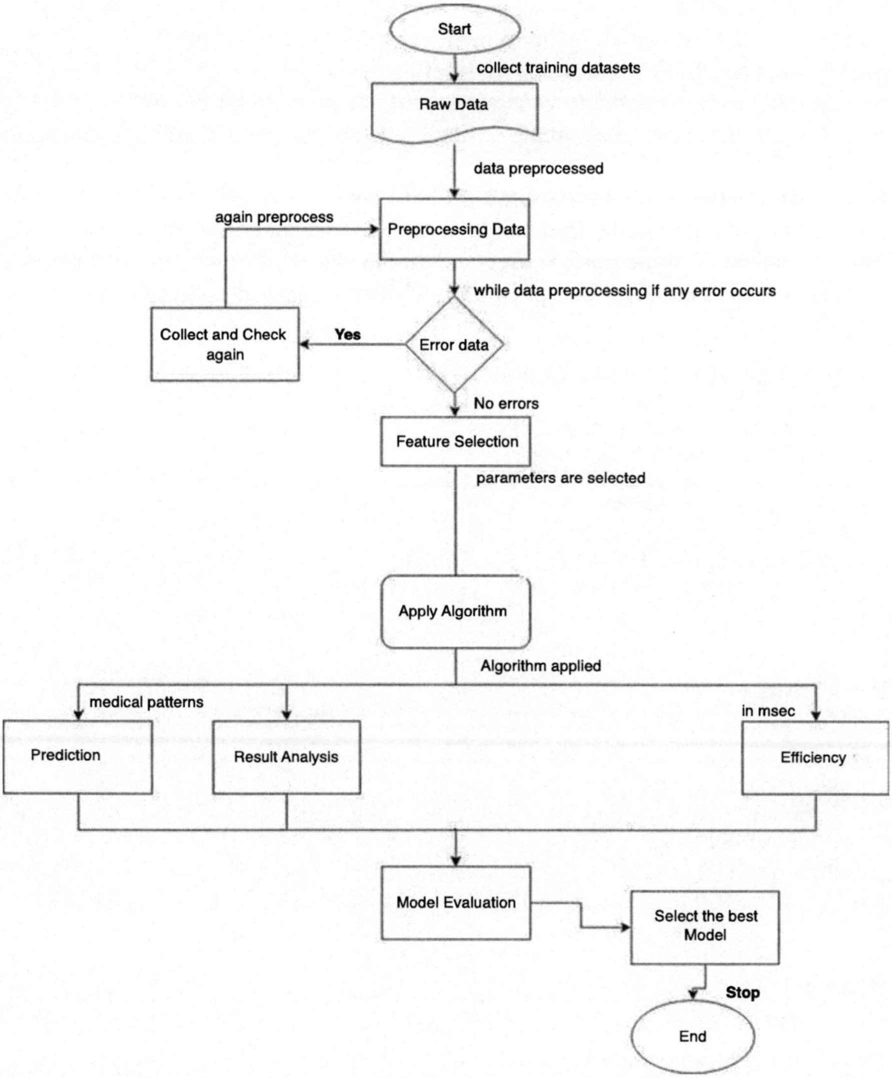

Fig. 1 Process used in the selection of the best model

Step 3: Then, initially the pre-processed data are inputted to Apriori algorithm to discover the medical patterns (purchasing patterns).
Step 4: Apriori algorithm will discover the medical purchasing patterns, which shows the relationship between purchased products and identifies the frequent purchased products.
Step 5: Medical purchasing patterns are displayed on GUI (front end).
Step 6: Results of the data science algorithms are analyzed and represented visually.

Apriori Algorithm
Step 1: Scan the data set and determine the support(s) of each item.
Step 2: Generate L1 (Frequent one item set).
Step 3: Use Lk-1, join Lk-1 to generate the set of candidate k-item set.
Step 4: Scan the candidate k item set and generate the support of each candidate k-item set.
Step 5: Add to frequent item set, until C=Null Set.
Step 6: For each item in the frequent item set, generate all non-empty subsets.
Step 7: For each non-empty subset, determine the confidence. If confidence is greater than or equal to this specified confidence, then add as Strong Association Rule.

Apriori Algorithm (Pseudo-Code)
```
Apriori (T, minSupport)
Cl = {candidate I-itemsets};
L1 = {c∈C1|c.count≥ minsup};
FOR (k = 2; Lk-1ÂŘ˘1; k++) DO BEGIN
Ck=apriori-gen(Lk-1);
FOR all transactions t∈D DO
BEGIN Ct=subset (Ck,t);
FOR all candidates c∈Ct DO
```

3 Count++

```
END
  Lk=c∈Ck
  |c.count •
  {minsup} END
  Answer=*Lk;
```

4 Result

Data Structure – array based

Memory Utilization – depends on the data set [less for small datasets]

No. of. scans – single scan required

Execution time – execution time depends on producing candidates. Details relevant to Table 1 are as follows:

Table 1 Results showing variable execution times

No of instances (records)	Execution time (milliseconds)
Around 500 records	557
Around 300 records	495
Around 200 records	465
100 records	445

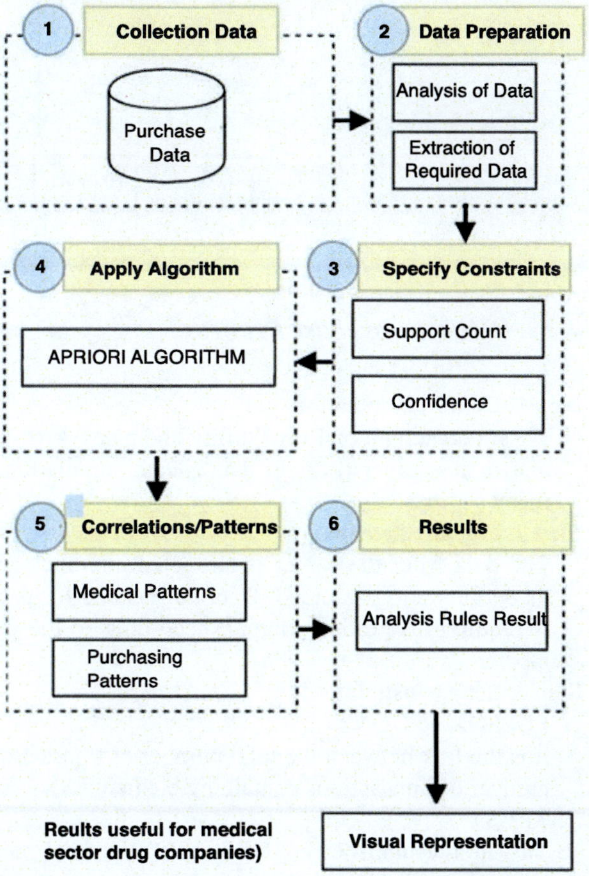

Fig. 2 Implementation methodology

4.1 Unsupervised Learning

There are no predefined labels for understanding the unsupervised gaining technology. The intent is to observe the data and discover the shape in it. The methodology involving the same is depicted in Fig. 2. Unsupervised learning works right with transactional data (Fig. 3).

Fig. 3 Unsupervised learning descriptive model

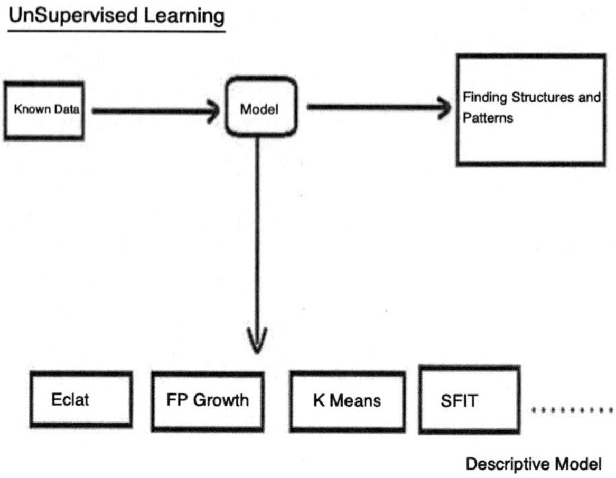

4.2 Pattern Forecast Process

Step 1: Data Accumulation

We are creating a real-time utility and a new software program that incorporates a statistics server (to hold data). Data cumulation entails gathering facts from many sources.

Step 2: Prepare the data

The data from the server is extracted and analyzed here. Data is extracted in its entirety and examined. Data not needed for processing is removed. All that is required to produce output, according to the project, is medicine and only medicine.

Step 3: Set a constraint

SUPPORT COUNT

It is the link between the total number of transactions in the record and the total number of transactions containing element (A).

TRUST

This item establishes confidence. It's the total number of transactions with an item set divided by the total number of transactions with LHS.

Step 4: Association Rule Mining

The most well-known and simplest data mining technique is association (or relation). We make a basic association between two or many objects of the same kind, which is typically used to find a pattern. For example, a shopping cart analysis that tracks a customer's buying habits can determine that the customer always buys cream the next time they buy strawberries, so it's a good idea to buy cream the next time they buy strawberries.

Step 5: Pattern prediction

The system anticipates the interactions between different medications in this case. The positive and unfavorable conditions for the development of pharmaceutical e-commerce in China in the future are quickly examined. Many issues have developed in recent years as the e-commerce of Chinese herbs has grown rapidly. The promise and hazards of Chinese pharmaceutical e-commerce are examined in this section. With its advantages, China's pharmaceutical e-commerce has a big domestic market, mature firms, and excellent technology. However, the solid hospital letter, which is a matter of distribution channels separated by the competition between health insurance and drug sales, remained in a bad state. Therefore, this paper intends to make specific proposals that can theoretically support the national pharmaceutical policy [5]. Prediction and Analysis of Online Shopping Behaviour leverage group learning. With the development of online transaction systems and online shopping forums, there are a growing number of customers. Buy in the online store. However, sellers understand that buyers and sellers cannot encounter face to face. Because our customers' needs are small, we cannot capture our thoughts in a timely manner. Online systems make it possible to record consumer performance, collect data on consumer behavior, and predict consumer preferences. This article explores a variety of e-commerce platform purchase data and uses the Cat Boost model to analyze and predict whether consumers will buy a particular product.

4.2.1 Accuracy

Accuracy and other model conditions are provided to evaluate the performance of the forecast. An accuracy of 88.51 is considered to a best result in predicting the buying behavior on this dataset [6]. Item-based hybrid recommender systems for newly marketed medicines Advertising systems Information systems predict user preferences and offer personal and motivational product / property / service recommendations. Our work aims to introduce such ideas in the field of healthcare. New drugs and their variants continue to be on the market, and tracking each one can be a tedious task. Therefore, the use of such new drugs is biased [7]. Research on e-commerce monitoring systems for prescription drugs is widespread throughout the Chinese industry. However, developing B2Ce commerce was relatively easy. In particular, the development of prescription drugs, which dominate the pharmaceutical market, has been discontinued. Over the past few years, online prescription drug marketing has become unstoppable with the introduction of the Medical Transformation Guidelines series. Making medicine from doctors' e-commerce is becoming a reality, and we rely on outsiders and instructions [8]. Discovering relationships between characteristics like Chinese drug formed upon Data Mining gives technical assistance for upgrading the Chinese pharmaceutical entities. This article focuses on the need to find relationships between the characteristics of traditional herbal medicines [9]. Prediction of Drug Protein Disease Associations and Drug Repositioning have been proposed Using Tensor Decomposition. Many medications work on several targets or diseases rather than just one, according to the classic "one gene, one drug, one disease" paradigm for drug discovery. Drug repositioning, which tries to find new indications for existing medications, is a valuable and cost-effective drug development technique. It is also crucial to figure out how target proteins, medicines, and diseases are functionally clustered, as well as the pathological causes for connections between these clusters and individuals [10]. An analysis of pharmaceutical companies' e-commerce applications shows that 95% of those surveyed reported fewer complaints about their website, including certain features, colors, frames, information updates, website ranking, and more. This shows that few people have requirements. Few people give you tips on how to improve your site. Nearly 90% of users are satisfied with price, 80% with assortment, and 70% with service, reliability, convenience of shopping, and reliability of site information for registered users [11].

4.2.2 Customer Segmentation Using Data Mining Techniques

A case study was performed to explain the segmentation theory of a Turkish supermarket chain. The purpose of this case study is to identify product addiction and shopping habits. Forecasts also determine the promotion and customer profile of product [12]. A study on the empirical effectiveness according to the association rules on herbal medicines has been carried out. Association rule algorithms help find relationships between sets of items in large databases. It is widely used to get rid of the experience of herbal medicine, primarily to study drug combination rules and the relationship between symptoms and syndromes [13].

5 Conclusion

This task collects real-time clinical records from reliable websites and performs data mining to identify exceptional feasible relevance of this information. This helps in discovering and understanding person contributions and perceive useful "patterns" for organizations. This helps docs find the aspect results of many distinctive medications and permits to use them to better prescribe medicinal drugs for other patients with comparable situations. Pharmaceutical groups can observe more than one drug reactions to humans and therefore arrive at n concept of which drugs are popular and which ones must be synthesized.

References

1. Perçin İ, Yağin FH, Güldoğan E, Yoloğlu S. ARM: An Interactive Web Software forAssociation Rules Mining and an Application in Medicine. In 2019 International Artificial Intelligence and Data Processing Symposium (IDAP) 2019 Sep 21 (pp. 1–5). IEEE.
2. Zahrotun L. Soyusiawaty D, Pattihua RS. The Implementation of Data Mining for Association Patterns Determination Using Temporal Association Methods in Medicine Data. In 2018 Inter-

national Seminar on Research of Information Technology and Intelligent Systems (ISRITI) 2018 Nov 21 (pp. 668–673). IEEE.
3. Abin D, Mahajan TC, Bhoj MS, Bagde S, Rajeswari K. Causal association mining for detection of adverse drug reactions. In 2015 International Conference on Computing Communication Control and Automation 2015 Feb 26 (pp. 382–385). IEEE.
4. Zhang W, Yue X, Chen Y, Lin W, Li B, Liu F, Li X. Predicting drug-disease associations based on the known association bipartite network. In 2017 IEEE international conference on bioinformatics and biomedicine (BIBM) 2017 Nov 13 (pp. 503–509). IEEE.
5. Wang Y. A brief analysis of favorable and unfavorable conditions for futuredevelopment of pharmaceutical E-commerce in China. In 2016 International Conference on Logistics, Informatics and Service Sciences (LISS) 2016 Jul 24 (pp. 1–5). IEEE.
6. Dou X. Online purchase behavior prediction and analysis using ensemble learning. In 2020 IEEE 5th International Conference on Cloud Computing and Big Data Analytics (ICCCBDA) 2020 Apr 10 (pp. 532–536). IEEE.
7. Bhat S. Aishwarya K. Item-based hybrid recommender system for newly marketed pharmaceutical drugs. In2013 International Conference on Advances in Computing, Communications and Informatics (ICACCI) 2013 Aug 22 (pp. 2107–2111). IEEE.
8. Qinghua Z, Wang S, Zhang Y, He K. Research on supervision system of prescription drug e-commerce. In2018 13th IEEE Conference on Industrial Electronics and Applications (ICIEA) 2018 May 31 (pp. 411–416). IEEE.
9. Wang R, Li S, Wong MH, Leung KS. Drug-protein-disease association prediction and drug repositioning based on tensor decomposition. In2018 IEEE International Conference on Bioinformatics and Biomedicine (BIBM) 2018 Dec 3 (pp. 305–312). IEEE.
10. Tian L, Wang X, Xue W. Analysis of e-commerce application in pharmaceutical enterprises-A case study of Yaofang. cn. In 2010 International Conference on Future Information Technology and Management Engineering 2010 Oct 9 (Vol. 2, pp. 9699).
11. Güllüoğlu SS. Segmenting customers with data mining techniques. In 2015 Third International Conference on Digital Information, Networking, and Wireless Communications (DINWC) 2015 Feb 3 (pp. 154–159). IEEE.
12. Li H. Jiamin Y. Zhimin Y, Huanyu L, Chunhua H. Research on the drugs of addition and subtraction of empirical formula of traditional Chinese medicine based on association rules. In 2013 IEEE International Conference on Bioinformatics and Biomedicine 2013 Dec 18 (pp. 67–68). IEEE.
13. Pattanayak RK, Kumar VS, Raman K, Surya MM, Pooja MR. E-commerce Application with Analytics for Pharmaceutical Industry. InSoft Computing for Security Applications: Proceedings of ICSCS 2022 2022 Sep 30 (pp. 291–298). Singapore: Springer Nature Singapore.

A Survey on the Latest Intrusions and Their Detection Systems in IoT-Based Network

Partha Jyoti Cheleng, Prince Prayashnu Chetia, Ritapa Das, Bidhan Ch. Singha, and Sudipta Majumder

1 Introduction

We are striving in the era of technology that encompasses every sphere of day-to-day life. The Internet of Things is the new global way of life [1]. In recent years, the growth of the IoT has been a spectacular phenomenon. Devices and objects are embedded with sensors, chips, and software, interconnected with each other like never before [2]. It intends to communicate with each other and other systems worldwide through the Internet. Unlike traditional computers and servers, the IoT is a giant network of devices. The IoT will and is spreading everywhere from industries to healthcare to the household. These IoT devices can transform ordinary objects into "smart" objects capable of adjusting to our needs. For example, today we can create a "smart house" complete with automatic sensing light bulbs and smart locks, allowing for a more convenient and higher standard of living. Also, medical IoT devices allow real-time measurements of pulse rate, blood pressure, or sugar level, a favorable way to a higher chance of patient survival [3].

With the growing requirements, IoT devices need to be implantable, portable, and small while also having ample computing and memory capacity [4] and thus need to be implemented with various lightweight protocols to communicate with other IoT devices. These requirements and the unprecedented growth of IoT networks have resulted in convenient ways for attackers to intrude on the network. As a result, security systems are critical to the successful operation of an IoT paradigm.

Our study mainly focuses on different areas of the IoT such as security, privacy, and robustness of the entire IoT ecosystem. Section 2 of the paper discusses the

P. J. Cheleng · P. P. Chetia · R. Das · B. C. Singha · S. Majumder (✉)
Department of Computer Science and Engineering, DUIET, Dibrugarh University, Dibrugarh, Assam, India
e-mail: sudipta2020@dibru.ac.in

M. Mishra et al. (eds.), *Applications of Computational Intelligence in Management & Mathematics*, Springer Proceedings in Mathematics & Statistics 417,
https://doi.org/10.1007/978-3-031-25194-8_6

various factors which motivated the researchers to develop new technologies to tackle the vulnerabilities present in IoT devices. Section 3 looks into adaptable frameworks and architectural solutions to help IoT devices and systems communicate more effectively. It briefly discusses the layered architecture of the IoT, namely, three-layered, four-layered, and five-layered along with IoT data protocols (MQTT, CoAP, and HTTP) and network protocols (Wi-Fi, Bluetooth, ZigBee). In the later section, i.e., Sect. 4, the numerous vulnerabilities, security threats, and attacks of the IoT paradigm are explained. The main emphasis is on the commonly used attacks that are often implemented by attackers to overpower the systems. These attacks range from simple ones such as eavesdropping and DoS to more sophisticated and advanced ones such as brute force attacks. To tackle these, the researchers have put forward some effective and promising measures using machine learning and other techniques such as blockchain (BC) technology, fog computing (FC), etc. All of these measures are summarized and briefly discussed in Sect. 5 of this paper. We strongly believe that the results of these many surveys can potentially protect the confidentiality and integrity of the system and make IoT networks more reliable and safe. With this, we have concluded our findings and summed up our entire study in Sect. 5 of this paper.

2 Motivation and Contributions

IoT has become a rising field of interest and importance in recent years and is expected to dominate the world in the coming years. While IoT devices are making their dominant presence in every home, office, hospital, etc., the lack of security in IoT may result in numerous problems such as privacy breaches, miscommunication, and even life threats. Hence, security in the IoT has become a major point of concern and discussion. Moreover, finding security solutions for IoT systems is difficult due to the heterogeneous nature of IoT systems, which involve devices with constrained power and computation capabilities. A great deal of research has been done and published in the literature on the IoT and its various components, including architecture, applications, protocols, and standards. However, not enough work has been produced concerning the threats and attacks in the IoT. To illustrate this, we compiled facts and figures from 2010 to 2020 from SCOPUS, one of the world's largest databases.

The number of articles in SCOPUS connected to IoT architecture, IoT architecture and risks, and IoT architecture and assaults is depicted in Fig. 1. The findings clearly show that IoT architecture's danger and attack aspects are understudied. However, the outcomes have steadily improved over the last decade. It is evident by the figures of papers referring to threats and assaults analysis in IoT architecture, which increased from four and zero in 2010 to 128 and 276, respectively, in 2020.

Similarly, the plotted statistics in Fig. 2 show that threats and assaults analysis in IoT protocols has received little attention in the last decade. Nonetheless, these issues have begun to receive the attention they need, as seen by the number of

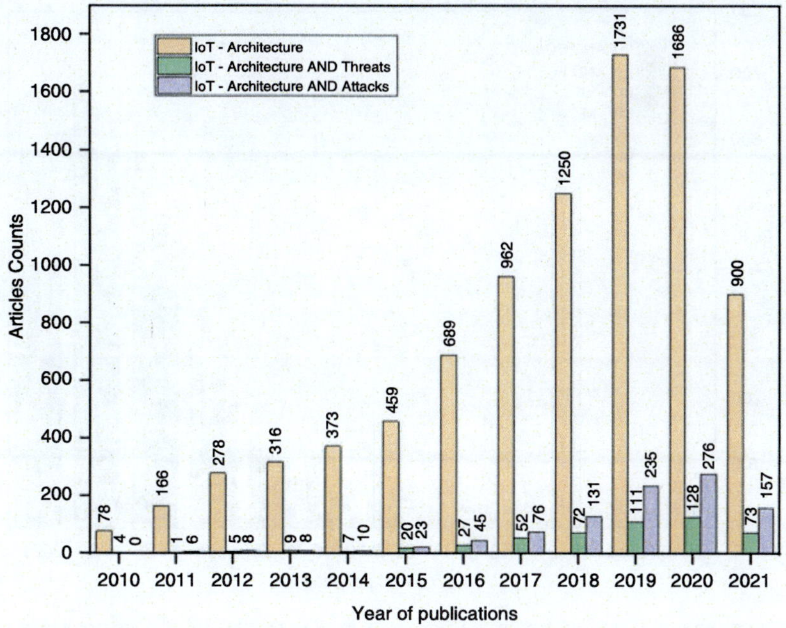

Fig. 1 IoT architecture, IoT architecture and threats, and IoT architecture and attacks statistics from SCOPUS [46]

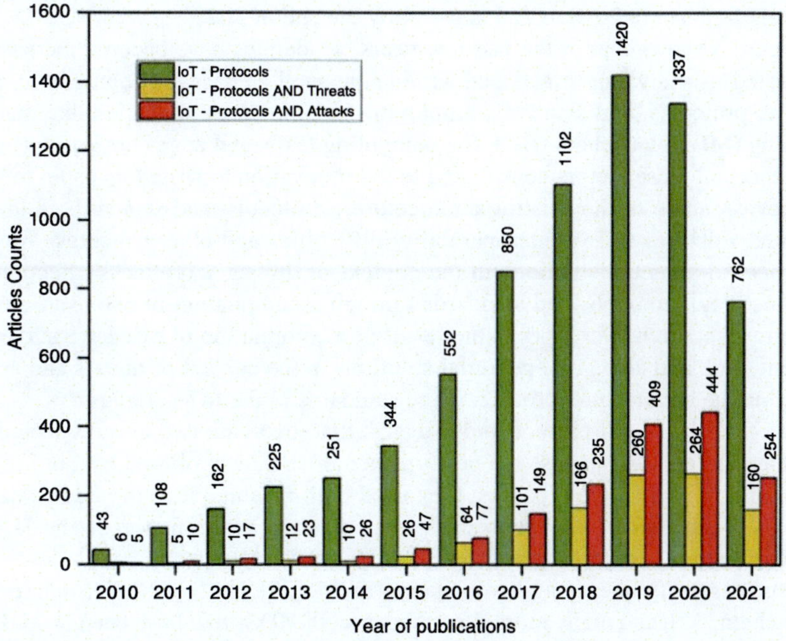

Fig. 2 IoT protocols, IoT protocols and threats, and IoT protocols and attacks statistics from SCOPUS [46]

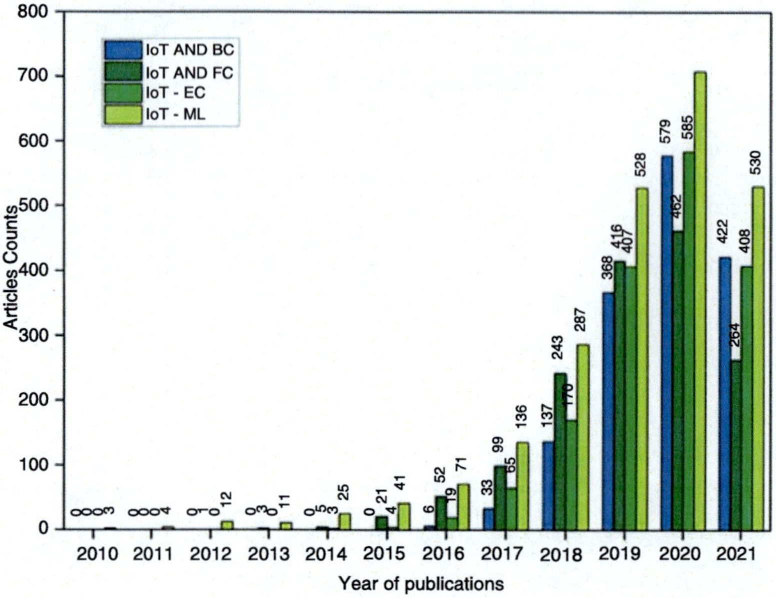

Fig. 3 IoT and BC, IoT and FC, IoT and EC, and IoT and ML statistics from SCOPUS [46]

published works on threats and assaults analysis in IoT protocols, which climbed from six and five in 2010 to 264 and 444 by the end of 2020.

Figure 3 shows that in the last few years, academics have become increasingly interested in analyzing threats and assaults, as well as possible solutions in architecture, protocols, and standards employing ubiquitous technologies like machine learning (ML), blockchain (BC), fog computing (FC), and edge computing (EC).

Hence, all these factors contributed to the motivation to write this paper to study and review some of the existing architectures, protocols, and standards in the IoT and find solutions to IoT interoperability difficulties and other concerns. We have put forward potential solutions in the context of threats and attacks affecting IoT architectures, protocols, and standards through a compilation of extensive surveys of current research. Our work is the result of a compilation of extensive surveys of current work and also gives potential solutions in the context of threats and attacks affecting the architecture, protocols, and standards in the IoT paradigm.

The findings of our present study suggest that involving two or more techniques together can greatly increase the performance of IDS, as proposed by Liu [28] and Garcia-Font [31]. In these papers, they used both machine learning and signature-based techniques for the better and more effective performance of the IDS. In contrast to this, other researchers have applied different approaches to develop an IDS such as the frequency agility manager (FAM) [29], HAN [42], SDN controllers, blockchain [37], and many more. In either case, the IDSs that have been proposed in this paper have successfully detected the intrusions targeted at the system and have opened new scopes for development in the area of IoT network security.

3 Architecture of IoT

The Internet of Things, or IoT, has sparked a surge in interest in adaptable frameworks and architectural solutions to help IoT devices and systems communicate more effectively. This is because IoT devices are designed to be classified across a wide range of application domains and locations. To do this, the system is built with a variety of components, including sensors, actuators, cloud services, and layers in its architecture. This design has distinct layers that track a system's consistency through protocols and gateways. Initially, three layers constituted the architecture. But the latter five-layer model has become well recognized, as shown in Fig. 4. The various layers of the five-layer model are as follows:

1. The perception layer: It is a layer of the IoT that is composed of sensors. These sensors gather information about the environment by sensing physical boundaries or smart objects present in the surrounding [5].
2. The network layer: This layer of the IoT is engaged in connecting smart things, network devices, and servers. The features of the network layer are also useful for transmitting and processing sensor data [5].
3. The application layer: The application layer of the IoT delivers specific application-based services to the users. The layer defines deployable applications for the Internet of Things, for example, smart homes, smart cities, etc. [5]. The original three-layered architecture describes the main functions of the IoT, but it was considered insufficient, so an in-depth analytic architecture composed of

Fig. 4 IoT architecture five-layered model [6]

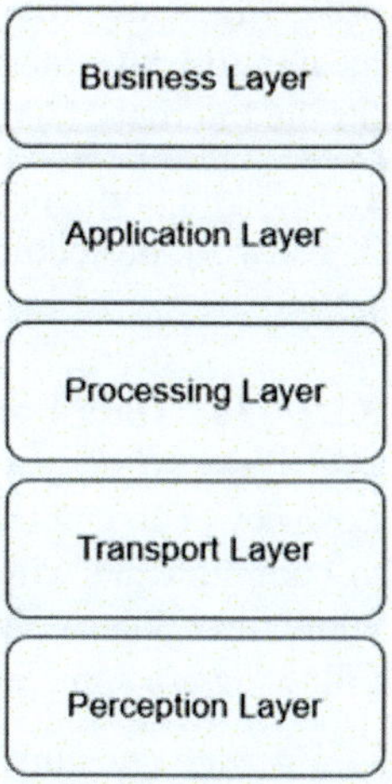

five layers was presented with the additional introduction of the processing layer and the business layer.

4. The transport layer: The transport layer is helpful for transporting sensor data from one layer to another, i.e., the perception layer and processing layer, and vice versa, via wireless networks, 3G networks, LAN, RFID, and NFC [5].
5. The processing layer: This layer of the architecture is known as the middleware layer. The layer processes enormous amounts of data, analyzes them, and stores them as they arrive from the transport layer. It operates databases, cloud computing, etc. It can administer and implement a diverse set of services for other layers [5].
6. The business layer: The business layer regulates the whole of the IoT system, which encompasses applications, business, a profit model, and also the user's privacy [5]. With that being said, the IoT architecture does consist of different layer models

    1. Three-layer model: Perception layer, network layer, and application layer [5].
    2. Four-layer model: Perception layer, support layer, network layer, and application layer [5].
    3. Five-layer model: Perception layer, transport layer, processing layer, application layer, business/physical/data link layer [5]. Comparison of three-layered, four-layered, and five-layered. Models of an IoT network are shown in Fig. 5 above.

3.1 IoT Protocols and Standards

IoT protocols are an integral part of the IoT technology pyramid. Without these protocols and standards, the hardware would be judged unworkable, because the protocols allow the hardware to commute data, and the end-user may extract meaningful information from these pieces of data [7–9]. Protocols and standards for the Internet of Things are divided into two categories. These are:

1. IoT data protocols (presentation/application layers)
2. Network protocols for the IoT (Datalink/physical layers)

3.1.1 IoT Data Protocols

Low-power IoT devices are connected using IoT data protocols. They allow users to communicate with gear without the requirement for an Internet connection. A wired or cellular network is used to connect IoT data protocols and standards. The following are some examples of IoT data protocols:

- *Message Queuing Telemetry Transport (MQTT):* It's a lightweight data protocol with a publisher-subscriber messaging mechanism that allows data to flow

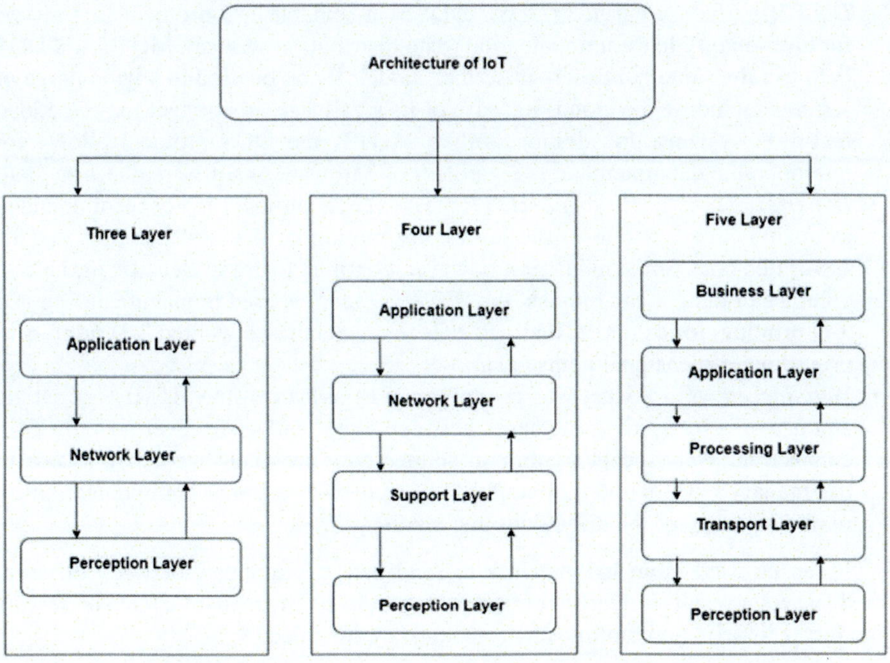

Fig. 5 Layered architecture of IoT, namely, three-layered, four-layered, and five-layered [6]

freely between devices. Its key selling point is its architecture, which features a fundamentally lightweight generic make-up that enables it to implement low power consumption for devices that use the TCP/IP protocol [7].

- *Constrained Application Protocol (CoAP):* The World Wide Web's data transmission infrastructure is based on HTTP-based systems, which stands for Hypertext Transfer Protocol. CoAP is an application layer-based protocol developed to satisfy the needs of HTTP-based systems. By adapting the HTTP concept into limiting devices and network contexts, CoAP has overcome an Internet constraint where IoT apps consume a significant amount of power. While the existing Internet framework is open to all IoT devices and may be used by them, it is typically too heavy and power-hungry for most IoT applications. As a result, many in the IoT community consider HTTP an unsuitable protocol for IoT [7, 8].
- *Advanced Message Queuing Protocol (AMQP):* This protocol is an open standard application layer protocol and is used for managing transactions between servers. The major functions of this IoT protocol are as follows:

1. Collecting and distributing messages in queues
2. Storing messages.
3. Setting up an association between various components employed in the analytical server-based environment due to its level of security and reliability but less widely used elsewhere due to its heaviness [9].

- *Data Distribution Service (DDS):* DDS is a scalable IoT protocol that allows for high-quality IoT communication (data distribution service). MQTT and DDS both use the same publisher-subscriber model. It can be used in a wide range of settings including the cloud. Because of its small size, it's perfect for embedded real-time systems. In addition, unlike MQTT, the DDS protocol allows for interoperable data interchange regardless of hardware or software platform [10].
- *Hypertext Transfer Protocol (HTTP):* The HTTP protocol is not recommended as an IoT standard due to the high consumption of devices' batteries and its associated cost. Some industries, however, continue to use it. Because of the vast volumes of data, it can publish, the HTTP protocol is used in manufacturing and 3-D printing, for example. It allows PCs to connect to networked 3-D printers to print three-dimensional items [11].
- *Web Socket:* Web Socket was created in 2011 as part of the HTML5 initiative, and it works by sending messages between client and server over a single TCP connection. Web Socket's common connectivity protocol, like CoAp's, aids in the management of Internet connections and bi-directional communication by removing many of the difficulties and challenges.

It can be used in an IoT network to send data continuously between different devices. As a result, it's most commonly found in regions where clients or servers interact. Libraries and runtime environments are included [8].

3.1.2 Network Protocols for the IoT

For connecting various IoT equipment and devices in the network, IoT-based network protocols are used. These protocols are often used on the Internet. Examples of IoT network protocols are as follows.

(a) *Wi-Fi*: The most well-known IoT protocol on our list is without a doubt Wi-Fi. However, learning the fundamentals of the most widely used IoT protocol is still worthwhile. To set up a Wi-Fi network, you'll need a device that can send wireless signals. To name a few, there are telephones, computers, and routers. Wi-Fi connects adjacent devices to the Internet within a specific range. Another approach to use Wi-Fi is to create a Wi-Fi hotspot. Mobile phones and PCs can share a wireless or wired Internet connection with other devices by sending a signal. Wi-Fi relies on radio waves to transmit data at certain frequencies, such as 2.4 GHz or 5 GHz. Furthermore, both of these frequency ranges have a variety of channels that various wireless devices can use. This prevents wireless networks from becoming overburdened. A diagram of Wi-Fi architecture is shown in Fig. 6.

The normal range of a Wi-Fi connection is 100 m. On the other hand, the most popular is restricted to a range of 10–35 m. The range and speed of a Wi-Fi connection are largely affected by the environment and whether it delivers interior or external coverage [12].

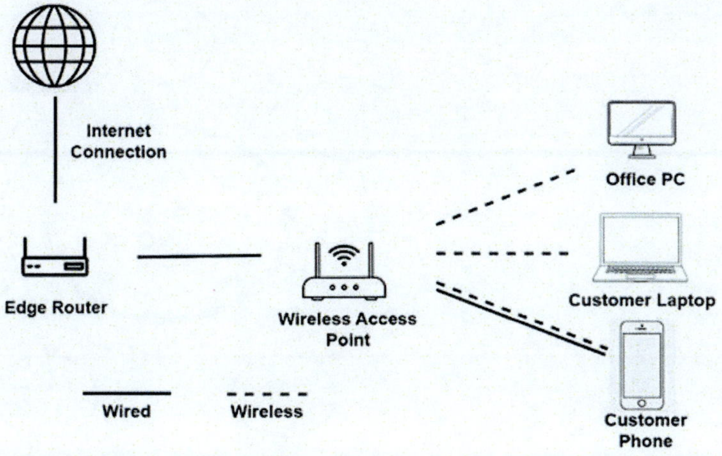

Fig. 6 Wi-Fi network architecture

Fig. 7 Bluetooth network (Piconet) architecture

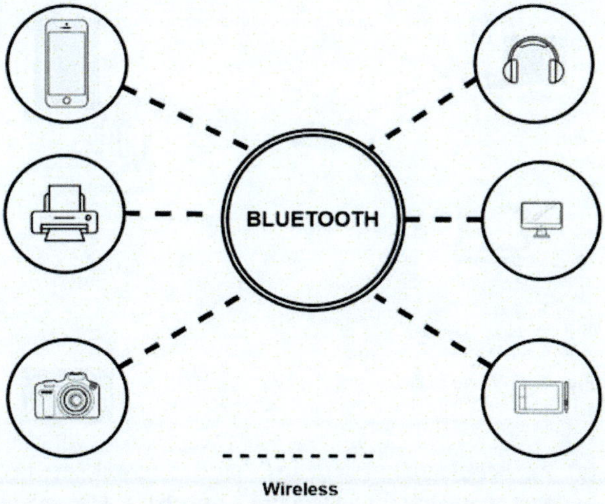

(b) *Bluetooth*: When compared to the other IoT network technologies described below, Bluetooth has a shorter range and has a tendency to frequency hop. Its integration with modern mobile devices, such as smartphones and tablets, as well as wearable technology, such as wireless headphones, has increased its popularity.

Using radio waves in the 2.4 GHz ISM frequency band, Bluetooth standard technology transfers data in the form of packets to one of 79 channels. In contrast, the current Bluetooth 4.0 standard includes 40 channels and a bandwidth of 2 MHz. This allows for a data transfer rate of up to 3 megabits per second [13]. Figure 7 shows the Bluetooth network (Piconet) architecture.

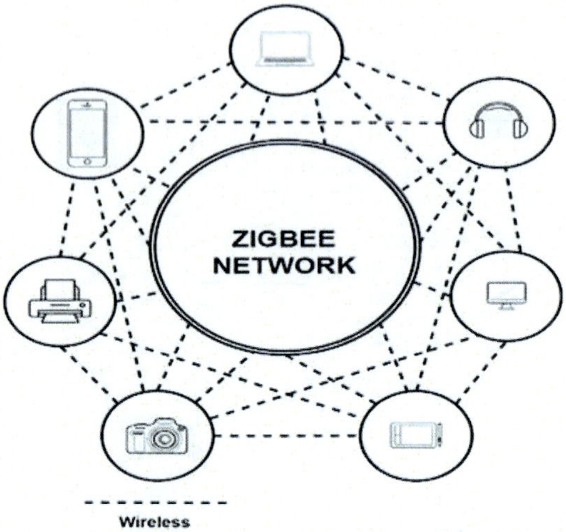

Fig. 8 ZigBee network

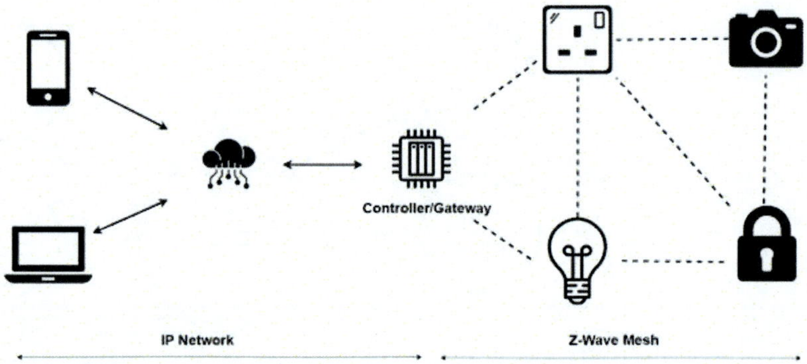

Fig. 9 Z-Wave network

(c) *ZigBee*: In the area of IoT, ZigBee-based networks are similar to Bluetooth in that they already have a huge user base. However, its features far outstrip those of the more widely used Bluetooth. It consumes less energy, has a smaller data range, is more secure, and has a longer communication range (ZigBee can reach 200 m, whereas Bluetooth only reaches 100 m) [7, 8]. The ZigBee network is shown in Fig. 8.

(d) *Z-Wave*: The IoT protocol Z-Wave is gaining popularity. It's a radio frequency (RF)-enabled wireless communication technology used mostly in the Internet of Things (IoT) household applications. It operates on the 800–900 MHz radio frequency. Zigbee, on the other hand, operates at 2.4 GHz, the same frequency as Wi-Fi. Z-Wave rarely faces significant interference because it operates in its own range. The frequency at which Z-Wave devices operate varies by country [8, 9]. Figure 9 shows a Z-Wave network.

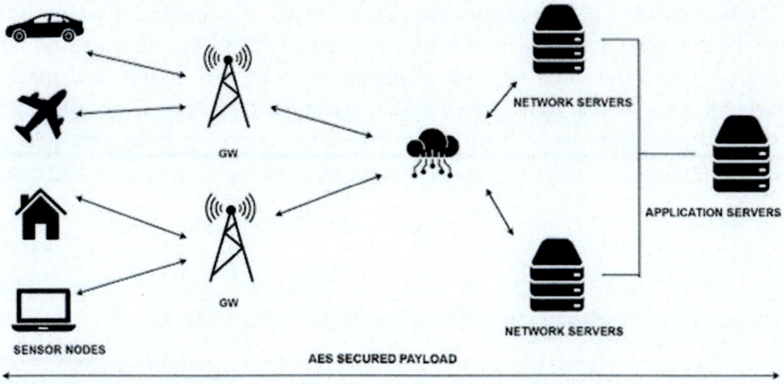

Fig. 10 LoRaWAN network [45]

(e) *LoRaWan*: LoRaWAN is an Internet of Things (IoT) protocol for controlling media access (MAC). Low-powered devices can communicate directly with Internet-connected apps over a long-range wireless link using LoRaWAN. It can also map to the second and third levels of the OSI model [8]. Figure 10 shows a LoRaWAN network.

4 IoT Vulnerabilities, Security Threats, and Attacks

With increasing dependence on IoT technology, the threats to information security systems have also increased exponentially. Every once in a while, a malware attack is initiated on an IoT system or gadgets that are preinstalled with IoT applications. This has given rise to an awareness that the modern generation of electronic devices or systems may be vulnerable to malware and attack. Thus, to safeguard the IoT paradigm, we have to analyze the threats, vulnerabilities, security, and attacks thoroughly.

4.1 Vulnerabilities

Vulnerabilities can be defined as the defects in an IoT network's framework design or functionality that permit an attacker to run commands, access unapproved data, and launch a distributed denial-of-service (DDoS) assault [14]. In order to infiltrate networks, attackers can use IoT devices with known vulnerabilities.

In the IoT framework, both hardware and software systems are prone to flaws in their design. Hardware bugs result from flaws in the design or manufacture and are extremely difficult to detect and repair. On the other hand, software bugs are sourced from human errors and programming complexity. Technical vulnerabilities are produced by human errors, which are caused by a lack of communication between the developer and the client, a lack of resources, a lack of skills, and many other factors [15]. As a result, IoT frameworks are frequently exposed to vulnerabilities. As a result, in the Internet of Things paradigm, vulnerability produces unavoidable threats and attacks. Following next is a summary of security threats and attacks in IoT.

4.2 Security Threats and Attacks in IoT Paradigm

A threat is defined as the act of taking advantage of a system's security vulnerabilities and harming computer systems and organizations. Humans (someone granting access internally) and the environment (earthquakes, floods, fires, etc.) are the two known threats that seek to harm or disrupt a system [16, 17]. Hence, it is a must to formulate backup and contingency plans beforehand to overcome these kinds of threats. Speaking of human threats, it is seen that legal contracts have minimized internal human threats, but still, there are many external entities that continue to pose a threat to security systems. Human threats can also be classified using other parameters which are mentioned as follows:

- Unstructured threats: Unstructured threats are composed primarily of inexpert attackers who install hacking tools that are widely accessible.
- Structured threats: Individuals who are aware of system weaknesses and who can comprehend, develop, and exploit software's code make up this class of attackers.
- Advanced persistent threats (APT): APT can be defined as a highly complex and advanced network attack that targets corporations and governments to steal vital information [18].

From a security standpoint, these threats and attacks offer significant problems for the IoT paradigm. The following section briefly categorizes the security concerns caused by various threats and attacks.

4.3 Security Concern Due to Threats and Attacks in IoT Networks

There are five fundamental layers in IoT architecture, namely, the business layer, application layer, processing layer, transport layer, and perception layer [19]. Each layer serves different functions with specific features and operates uniquely. With the revolution in IoT technology, the connectivity between different Internet-driven smart devices has also increased significantly. Such huge end-to-end connectivity has led to an increase in the risk of breaching the security systems of the network and stealing valuable data from individuals and corporate infrastructures quite easily.

It exponentially increases the number of malicious attacks and threats on different layers of the IoT architecture. Speaking of them, a few of the commonly used attacks that are often implemented by attackers to overpower the systems are mentioned in the following points [20–27]:

- Eavesdropping: Eavesdropping is an attack in the perception layer where, by using comparable IoT devices, the traffic can be sniffed by attackers created by IoT data flows and they can collect sensitive information from users. Eavesdropping can be implemented in the processing layer, transport layer, or perception layer of the IoT networks, thus making it quite difficult to identify.
- Data injection attack: Data injection attack can be defined as an attack where an attacker injects malicious input into the IoT applications and forces them to perform specified commands during an injection attack. This type of sudden injection attack operates on the perception layer and can disclose or harm data, cause service disruptions, or compromise the entire web server instantly.
- Sybil attack: Sybil attack is a class of attack where malware or an attacker node generates a large number of false identities to disrupt the network's overall performance. This attack is sourced at the perception layer and the network layer where the Sybil nodes generate fake reports, spam users with messages, and compromise users' privacy.
- Side-channel attacks: A side-channel attack uses security measures to recover secret data by leveraging execution-related information. While the encryption process is in progress, in order to access the encryption login information of an IoT device, the intruder gathers data and performs reverse engineering. Although it is not possible to retrieve information from plaintext or ciphertext during the execution of the process, it can only be obtained through encryption devices. Such assaults result in timing attacks, power or failure analysis, and electromagnetic attacks. By taking advantage of data leakage, the attacker gathers block cipher keys. To overcome such an attack, an IPS, i.e., intrusion protection system, can be preinstalled in a system. An example of such IPS is Boolean masking.
- Exhaustion attack: It is defined as a subsection of a denial-of-service attack. Here, an attacker devours the memory resources of a system in order to prevent service delivery to authorized users. The exhaustion attack acts on the perception layer or the computing layer, according to the purpose of the attack.
- Denial of service (DoS): In DoS attacks, a specific node is targeted and flooded, causing it to crash. DoS attacks might be active, in which a system's application or task is explicitly rejected, or they might be passive, in which an ongoing task on the device is halted as a result of an attack on one application. Most IoT devices are vulnerable to this type of attack because of their lack of memory capabilities and inadequate computation resources.
- DDoS: A DDoS attack incorporates various remote servers infected with malware to send traffic from multiple locations at the same time in much larger amounts than DoS, thus overloading a server quickly and evading detection. Volumetric attacks, protocol attacks, and application-based attacks are a few

examples of DDoS attacks that increase the damage and may even lead to a catastrophic outcome.

- Hardware Trojans: Introducing hardware Trojans into circuits is the practice of incorporating malicious modifications into elements of the circuits. Many electronic components, including ICs, systems on chips, application-specified programmable circuits, and third-party intellectual property, can contain Trojans. These malicious Trojans can be directly introduced by anyone, including untrustworthy foundries, vendors, and designers into the system hardware for breaching purposes.
- Spoofing attack: When one person or computer successfully impersonates another by misrepresenting data in order to gain an unfair advantage, this is known as a spoofing attack. Domain name spoofing, referrer spoofing, poisoning of file-sharing networks, and e-mail address spoofing are a few instances of spoofing attacks that render IoT networks vulnerable.
- Routing attack: A routing attack primarily destabilizes routing paths. These attacks lower the system's credibility and disrupt its throughput. A few instances of routing attacks are packet mistreatment attacks, hit-and-run attacks, and round table poisoning attacks.
- Blackhole attack: It is defined as the devious behavior of a node attempting to claim to have the shortest path to the destination. The primary or we can say, the main objective of a Blackhole attack is to quickly drop incoming traffic and data packets and to shut away from the nodes or subtrees from the network, thus leading to vulnerabilities in the network layer.
- Wormhole attack: A wormhole attack is a form of internal attack that listens to network activity without modifying it, making it extremely difficult to detect. This class of network attack would detect incoming traffic and reroute it to different locations. Consequently, it clogs channels and reduces efficiency and productivity.
- Selective forwarding attack (SFA): In SFA, the attacker tries to act as a normal node in the routing process, selectively rejecting packets from adjacent nodes.
- Brute force attack: It is a cryptographic algorithm that applies trial and error to guess possible password combinations for logins, encryption keys, or hidden web pages. The brute force attack, or simply known as the exhaustive search, is an efficient and frequently used attack nowadays and is operated at the AL, i.e., the application layer of the IoT framework (Table 1).

5 Solutions

IoT applications require a secure environment and data confidentiality for the smooth functioning of the network. Therefore, ensuring the integrity of the network, data confidentiality, and availability of the network should be the top priorities when setting up an IoT network. Violations in any of these areas can compromise the whole network.

Table 1 Possible consequences and security requirements of various attacks along with their scope on different layers of IoT architecture

Attacks/threats	Possible consequences	Security requirements	Scope on different layers in IoT architecture
Eavesdropping	Loss of data confidentiality	Confidentiality	Perception layer, network layer, transport layer
Data injection	Loss data confidentiality, cause service disruption or compromises the entire system instantly	Authentication	Perception layer
Sybill attack	Compromises users' privacy	Authentication, authorization, and validation	Perception layer, network layer
Side channel attack	Loss of data confidentiality and integrity	Authentication, confidentiality, and integrity	Perception layer
Exhaustion attack	Loss of data availability	Authentication and flexibility	Perception layer
DoS	Loss of data availability	Router control, distributed packet filtering and aggregate congestion control	Transport layer, network layer
DDoS	Affect reliability and data availability	Router control, network segmentation	Transport layer, application layer
Hardware Trojan	Loss of data confidentiality	Integrity	Business layer
Spoofing attack	Loss of data confidentiality and integrity	Authentication, authorization, confidentiality, and integrity	Network layer, application layer
Wormhole attack	Reduces efficiency and productivity	Authenticity and confidentiality	Network layer
Brute force attack	Loss of data confidentiality and privacy	Integrity and confidentiality	Application layer

It is critical to protect the IoT network against hostile threats and attacks, which may be accomplished by the development and deployment of appropriate security measures, one of which is an intrusion detection system. Intrusion Detection Systems (IDS) come in a variety of shapes and sizes. Algorithms are used by an IDS to implement the various levels of intrusion detection. A few of these IDS methods will be briefly explored below.

1. Liu et al. [28] proposed an artificial immune IDS for IoT networks. This model can automatically adapt to the IoT environment and learn new attacks. The technology is built around machine learning and a signature-based paradigm.

Artificial immune system processes are used in the suggested machine learning approach. The system's purpose is to increase the IoT network's security; hence, it's a network IDS. This system's two main characteristics are self-adaptation to new environments and self-learning of new attacks. Simulating and characterizing immune factors in the IoT environment is used to build the immunity theory. The immune detection environment's detection elements develop dynamically to simulate self-adaptation and self-learning mechanisms.

2. Kasinathan et al. [29] proposed an upgraded intrusion detection system based on 6LoWPAN that secures IoT networks by detecting DoS attacks. The key novel features of this system are a frequency agility manager (FAM), a security incident and event management system (SIEM), which are based on the architecture of DoS detection, given in [30]. Together, these components form a monitoring system capable of monitoring massive IoT network systems.
3. Garcia-Font et al. [31] proposed a NIDS for WSNs that implements a signature model and a machine learning technique. To increase the detection rate and FPR, the researchers implemented a signature-based detection engine as well as an anomaly-based detection engine. Using an intrusion detection system and an attack classification schema, NIDS aims to assist smart city managers in detecting intrusions. The system primarily focuses on identifying intrusions happening in WSNs in a variety of smart city contexts. This system's key feature is its capacity to work with large-scale wireless sensor networks (WSNs).

   As the first contribution to this field of study, the researchers have presented a schema for classifying the evidence left by different attacks on smart city WSNs into seven distinct attack models. This paper also demonstrates how combining rule-based detection with OC SVM using basic correlation rules may boost detection results considerably. Thus, the output of the two detection engines combined in a correlation rule outperforms the output of the other approaches running individually. This has been proven in a sophisticated detection situation with a 20% selective forwarding dropping rate (only 10% more than the normal loss rate for the sound network).
4. The suggested RFTrust model by K. Prathapchandran is primarily intended to prevent sinkhole attacks in IoT systems based on the routing protocol for low-power and lossy networks (RPL). The objective function (OF) in RPL outlines the process of calculating routing metrics and constraints to generate the node's rank value. It improves trustworthy routing in the IoT context by employing random forest (RF) and subjective logic (SL).

   Underlying principles involved in the development of the RFTrust model rely on the pure IoT environment having a dynamic topology where IoT nodes are free to move between networks. The presented model works specifically for open and distant IoT applications by reducing unnecessary overhead and energy consumption. The mathematical study demonstrates the model's applicability, which is tested by using a Cooja simulator. The experiment result makes this a promising model with high PDR, high throughput, low average latency, and also consumes less energy. Furthermore, the RFTrust model outperforms the

SoSRPL [44], INTI, and InDReS models in terms of accuracy (85%), false-positive rate (1.4%), and false-negative rate (1.8%) [32].

5. To detect and mitigate intrusions in IoT networks, semi-supervised learning-based security is used. A semi-supervised deep feedforward neural network-repeated random sampling K-means (SDRK) is a new SSML model for intrusion detection. Both supervised and unsupervised deep neural networks, as well as clustering methods, are used in SDRK [33].
6. Mengmeng et al. [34] describe a unique intrusion detection technique for security in IoT systems. The researchers use a recently published IoT dataset to create generic characteristics from field information at the packet level. The model classifies traffic flow using deep learning principles. This paper demonstrates the performance of the proposed scheme by proposing an intelligent binary and multiclass classification mechanism via a FNN model for classifying attacks like DoS, DDoS, reconnaissance, and information against IoT devices, as well as the extraction and preprocessing of field information in individual packets as generic features. The classifier's capacity was proved with a score close to 0.99 in binary classification for DDoS/DoS and reconnaissance attacks across all assessment parameters, including accuracy, precision, recall, and F1 score. The classifier reported a detection accuracy of over 0.99 for DDoS/DoS assaults in the multiclass classification, whereas regular traffic classification provided an accuracy of 0.98.
7. Botnet detection in the IoT using ML. A botnet is a network consisting of multiple bots designed especially to undertake harmful and malicious activities on the target network and controlled by a single unit called the bot-master via a command-and-control protocol. The enormous number and pervasiveness of IoT, which ranges from modest personal gadgets like a smartwatch to a full network of smart grids, smart mining, smart manufacturing, and autonomous driverless cars, has attracted prospective hackers for cyberattacks and theft of data. The primary goal of this research is to offer a novel model that uses a machine learning algorithm to investigate and mitigate botnet-based DDoS assaults in IoT networks. Several machine learning methods, including K-nearest neighbor (KNN), Naive Bayes, and multi-layer perceptron artificial neural network (MLP-ANN), were applied to create a model using data from the BoT-IoT dataset. The optimal algorithm was chosen using a reference point based on accuracy percentage and area under the receiver operating characteristic curve (ROC AUC) score. Machine learning techniques were integrated with feature engineering and the Synthetic Minority Oversampling Technique (SMOTE) (MLAs). The performance of the three algorithms was compared on the class imbalance and class balanced datasets. They proposed the K-nearest neighbor method as an effective method for the determination of a botnet after comparison. The testing of botnet detection algorithms on real-time unbalanced and balanced datasets significantly demonstrated how and why real-time unbalanced datasets were not optimal, how it affects parameters like precision, recall, accuracy, F1 score, and ROC AUC, and how the dataset could be improved. Despite the fact that the unbalanced dataset demonstrated

good accuracy, recall and F1 score were significantly low. This suggests that the accuracy obtained from the unbalanced dataset was maybe illusory. Furthermore, they obtained better consistent accuracy and ROC AUC with the same range of precision, recall, and F1 score after incorporating SMOTE technology [35].

8. Yalin et al. [36] propose novel algorithms based on adversarial machine learning and apply them to three forms of over-the-air (OTA) wireless attacks, including jamming, spectrum poisoning, and priority violation attacks with exploratory, evasion, and causal attacks, to enable IoT systems consisting of heterogeneous devices with varying priorities. The paper evaluates an Internet of Things network in which an IoT transmitter observes the channel and uses deep learning to anticipate the channel condition (idle or busy) based on the latest sensing data. When there are high-priority users, the lower-priority IoT transmitter uses a back-off technique to prevent interfering with high-priority user broadcasts. The experiment shows that deep learning can attain near-optimal throughput. Then, an exploratory attack is used initially by the adversary to anticipate the outcome (ACK or not) of the transmitter's decisions. The adversary then either modifies the evasion attack (test data) or causative attack (data from the retraining process) to lower the performance of the IoT transmitter. The results show that these attacks, with varying levels of energy consumption and stealth capabilities, result in considerable losses in throughput. Hence, the success ratio also drops significantly in wireless communication systems for IoT systems.
9. To address some of the several obstacles and constraints in terms of privacy, security, and computing power that require the research community's immediate attention, the paper offers an IoT network design based on two new developing technologies, SDN controllers and blockchain. Public and private blockchains for peer-to-peer (P2P) communication between IoT devices and SDN controllers are employed in the cluster structure of the proposed model. By including an IoT-specific routing protocol in the SDN controller and removing proof-of-work (POW) from the equation, the proposed solution was able to significantly reduce energy consumption while also improving the security and privacy of communication between IoT devices. In terms of throughput, performance, and energy efficiency, the proposed architecture exceeds the BCF approach, while also providing an enhanced routing protocol that outperforms the EESCFD, SMSN, AODV, AOMDV, and DSDV protocols. For future work, they plan to deliver a high-level P4 architecture with blockchain features in the IoT area and evaluate its efficiency characteristics with the proposed architecture [37].
10. Software-defined networking (SDN) is a method of managing Internet of Things (IoT) networks that is secure, efficient, and autonomous. SDN enables the use of centralized logical control over the network, but due to its centralized architecture, it exposes the network to various potential threats and vulnerabilities. This paper describes one such case, in which a DDoS is started by directing an attack at a computer by overburdening the traffic load to

the target, causing services to fail and valid users unable to access [1]. To quickly mitigate the attack, a DDoS detection algorithm that measures entropy is developed to implement a solution in the central controller that is efficient and lightweight in terms of resource consumption. The suggested system also used Raspberry Pi as OpenvSwitches, demonstrating the viability of using easily accessible hardware for IoT. Thus, the SDN-based system successfully reprimands intrusion network excursions and guards against routinely begun assaults such as DDoS, thereby boosting network security. Although an optimal window size for threshold entropy was established, quantifying the FAR and FRR rates would be useful for enhancing the detection algorithm and avoiding incorrectly detected attacks [38].

11. A hardware Trojan has the capacity to change the functioning of a device. The hardware Trojans are constructed in such a way that they can change the current logic or add new logic, jeopardizing the IC's integrity. Research on such malicious insertions is presented in this work. In addition, the paper describes methods for detecting hardware Trojans and the impact of hardware Trojans such as DoS in a CORDIC processor using a Xilinx Spartan 3A FPGA kit in this study [39].
12. This study underlines the importance of a secure hardware-level framework for the security of devices, as software security alone is insufficient. The sophisticated nature of these HT is discussed in this study, as are their HT taxonomy and insertion techniques, as well as countermeasures. The paper highlights the importance of understanding multiple HT insertion phases during an IC's lifespan in order to detect and avoid HT insertion in various susceptible periods. Moreover, countermeasures are proposed, including detection techniques, design for trust, split manufacturing for trust, and other hardware security solutions such as HSM and TPM [40].
13. This paper proposes a cost-effective technique for establishing IoT system security. Here, the researchers reconstructed the security constraints using graph modelling as the identification of dominant sets where they prioritize the heavier vertices. The dominating set's size is greatly reduced when centralities are used to compute it, whereas graph weighting allows for minor modifications. The researchers modelled compliance (ability to host a requested service) in this case by assigning a priority value to the nodes. The solution has the least effect on the transmission of information while still assuming that it flows optimally via the graph. They also clearly demonstrate that classifying nodes into two groups, high and low values, necessitates a careful selection of the values employed. Increasing the number of categories could lead to better outcomes. However, weighting will be much more difficult. Applying weights while selecting the dominant set optimizes its quality while minimizing the effect on communications and reducing the number of nodes added. Also, adding weighted edges to the graph minimizes the cost of deploying security services by favoring devices with adequate hosting capabilities. Other IoT aspects (such as more comprehensive device capabilities, hazards, or even

bandwidth) might be added to the suggested model to obtain a more realistic model, notably regarding the assignment of weights on the graph [41].

14. HAN is a hybrid encryption technique that combines two algorithms, AES and NTRU, to reduce security risks, improve encryption speed, and reduce the computational complexity of the IoT. The HAN technique uses two algorithms, the first of which is the AES algorithm, which is implemented in hardware and software throughout the world to encrypt sensitive data. The AES algorithm uses three block ciphers, namely, AES-128, which makes use of a 128-bit key length to encrypt and decrypt a block of messages; AES-192, which makes use of a 192-bit key length to encrypt and decrypt a block of messages, and lastly AES-256, which makes use of a 256-bit key length to encrypt and decrypt a block of messages. All these ciphers encrypt and decrypt data in blocks of 128 bits using cryptographic keys of 128, 192, and 256 bits, respectively [42]. The other algorithm that the technique uses is the NTRU encryption algorithm. This algorithm is based on the shortest vector problem in the lattice. It relies on the presumed difficulty of factoring certain polynomials in a truncated polynomial ring into a quotient of two polynomials having very small coefficients. Specifically, NTRU operations are based on objects in a truncated polynomial ring R = Z [X]/(X^N-1) with convolution multiplication, and all polynomials in the ring have integer coefficients and degrees to at most N-1:

$$a = a0 + a1X + a2X^2 + \cdots + aN - 2X^{N-2} + aN - 1X^{N-1}.$$

15. The integration of cloud computing with IoT is enhancing the productivity of a vast variety of applications in industries such as supply chains, engineering, manufacturing, and so on. Presently, privacy and security are major concerns in cloud computing technology and the Internet of Things. This research presents a unique Chinese remainder theorem (CRT)-based safe storage and privacy-preserving paradigm for securely storing cloud data and accessing the data by authorized cloud users. Furthermore, a new and unique CRT-based group key management scheme is provided for accessing the encrypted cloud data stored in the cloud database from the cloud server. The suggested safe storage and privacy-preserving paradigm employs new algorithms for executing encryption and decryption operations. Also, the suggested key generation technique includes a Caesar cipher encryption scheme for encrypting the key. Various experiments have been performed in order to evaluate the suggested privacy-preserving data security model. The results show that the suggested security model outperformed current cryptographic algorithms, and the model has attained a high degree of security when compared to existing cryptographic algorithms such as AES, DES, BF, CRGK, and the cloud with the IoT model. For future work, they are planning to introduce new lightweight encryption and decryption algorithms that can reduce the computational complexity in cloud- and IoT-based applications [43] (Table 2).

Table 2 Detection methods and the implementation strategy of various research papers along with their scope of attacks/threats in IoT architecture

Papers	Detection	Implementation strategy	Attacks/threats
Liu C et al. [28]	Signature	AIS artificial immune system	Harmful antigens
Garcia-Font V et al. [31]	Hybrid	A security information and event management (SIEM)	Data tampering, eavesdropping
Kannimuthu et al. [32]	Anomaly	RFTrust model	Rank attack, sinkhole attack
Nagarathna Ravi et al. [33]	Anomaly	SDRK model	Data Deluge Attack
Mengmeng Ge et al. [34]	Anomaly	Feed-forward neural network model	DoS, DDoS, Service scanning, data exfiltration
Satish Pokhrel et al. [35]	Anomaly	KNN, Naive Bayes method MLP ANN and SMOTE	Botnet-based DoS
Narmadha Samban dam et al. [38]	Anomaly	Blockchain-based SDN controller	DoS
Tanguy Godquin et al. [41]	Signature	Weighted graph	

6 Conclusion

The Internet of Things' diversified design increases the attack surface and introduces new issues to an already vulnerable IoT network. To ensure that critical vulnerabilities are minimized, the IoT ecosystem's entire security must be considered. To counter the threats and attacks, policies and protocols must be strictly enforced. In this paper, we have presented a thorough survey on IoT security threats and attacks on the expanding IoT infrastructure, as well as security weaknesses and solutions. Any resolution would be ineffectual and antiquated owing to the varied presentations and limitations of IoT devices. More countermeasures and vulnerabilities are predicted to be discovered in the near future as a result of the evolving nature of technology. In the future, it is stated that the authors will focus on ML and IoT integration to increase the security of IoT-based systems in constantly changing environments.

References

1. Jha, A.V., Appasani, B., Ghazali, A.N. Performance Evaluation of Network Layer Routing Protocols on Wireless Sensor Networks. In Proceedings of the 2019 International Conference on Communication and Electronics Systems (ICCES), Coimbatore, India, 17–19 July 2019, 1862–1865.

2. Tiwary, A., Mahato, M., Chidar, A., Chandrol, M.K., Shrivastava, M., Tripathi, M. Internet of Things (IoT): Research, architectures and applications. Int. J. Future Revolut. Comput. Sci. Commun. Eng. 2018, 23–27.
3. Internet of Things in Healthcare: Applications, Benefits, and Challenges. Available online: https://www.peerbits.com/blog/internet-of-things-healthcare-applications-benefits-and-challenges.html (accessed on 1 May 2022).
4. Ryan, P.J., Watson, R.B. Research Challenges for the Internet of Things: What Role Can OR Play? Systems. 2017, 24.
5. Sarah A. Al-Qaseemi., Hajer A. Almulhim., Maria F. Almulhim., and Saqib Rasool Chaudhry. IoT Architecture Challenges and Issues: Lack of Standardization. 2016. 7–8.
6. Muhammad AminuLawal., Raizahmed Shaikh., and Syed Raheel Hassan.Security Analysis of Network Anomalies Mitigation Schemes in IoT Networks. 2020. 15–20.
7. Dae-Hyeok Mun., Minh Le Dinh., and Young-Woo Kwon. An Assessment of Internet of Things Protocols for Resource-Constrained Applications. 2016. 5–6.
8. Heđi., I. Špeh., and A. Šarabok. IoT network protocols comparison for the purpose of IoT Constrained networks. 2017. 4–5.
9. Vinoski, S. (2006). Advanced message queuing protocol. IEEE Internet Computing, 10(6), 87–89.
10. Paolo Bellavista., Antonio Corradi., Luca Foschini., and Alessandro Pernafini. Data Distribution Service (DDS): A Performance Comparison of Open Splice and RTI Implementations. 2013. 5–6.
11. Lu Hou., Shaohang Zhao., Xiong Xiong., Kan Zheng., Periklis Chatzimisios., M. Shamim Hossain., and Wei Xiang. Internet of Things Cloud: Architecture and Implementation. 2016. 32–39.
12. Jianfei Yang., Han Zou., Hao Jiang., and LihuaXie. Device-free Occupant Activity Sensing using WiFi-enabled IoT Devices for Smart Homes. 2018. 3991–4002.
13. Angela M. Lonzetta., Peter Cope., Joseph Campbell., Bassam J. Mohd., and Thaier Hayajneh. Security Vulnerabilities in Bluetooth Technology as used in IoT. 2018. 25–26.
14. Pipkin, D.L. Halting the Hacker: A Practical Guide to Computer Security, 2nd ed.; Prentice Hall Professional: Hoboken, NJ, USA, 2003.
15. Kizza, J.M. Guide to Computer Network Security, 1st ed.; Springer: Heidelberg, Germany, 2009. 387–411.
16. Dahbur, K., Mohammad, B., Tarakji, A.B. A survey of risks, threats, and vulnerabilities in cloud computing. In Proceedings of the 2011 International Conference on Intelligent Semantic Web-Services and Applications, New York, NY, USA, 18–20 April 2011. 1–6.
17. Rainer, R.K., Cegielski, C.G. Ethics, privacy, and information security. An Introduction to Information Systems: Supporting and Transforming Business; John Wiley & Sons: Hoboken, NJ, USA, 2010. Volume 3, 70–121.
18. Tankard, C. Advanced persistent threats and how to monitor and deter them. Netw. Secur. 2011. 16–19.
19. Swamy, S.N., Kota, S.R. An Empirical Study on System Level Aspects of Internet of Things (IoT). Access 2020, 8, 188082–188134.
20. Butun, I., Österberg, P., Song, H. Security of the Internet of Things: Vulnerabilities, attacks, and countermeasures. Commun. Surv. Tutor. 2019. 616–644.
21. Makhdoom, I., Abolhasan, M., Lipman, J., Liu, R.P., Ni, W. Anatomy of threats to the internet of things. Commun. Surv. Tutor. 2018. 1636–1675.
22. Restuccia, F., D'Oro, S., Melodia, T. Securing the internet of things in the age of machine learning and software-defined networking. Internet Things J. 2018. 4829–4842.
23. Khanam, S., Ahmedy, I.B., Idris, M.Y.I., Jaward, M.H., Sabri, A.Q.B.M. A survey of security challenges, attacks taxonomy and advanced countermeasures in the internet of things. Access 2020, 8, 219709–219743.
24. Cherian, M., Chatterjee, M. Survey of security threats in IoT and emerging countermeasures. In Proceedings of the International Symposium on Security in Computing and Communication, Bangalore, India, 19–22 September 2018. 591– 604.

25. Sicari, S., Rizzardi, A., Grieco, L.A., Coen-Porisini, A. Security, privacy and trust in Internet of Things: The road ahead. Comput. Netw. 2015, 146–164.
26. Sepulveda, J., Willgerodt, F., Pehl, M. SEPUFSoC: Using PUFs for memory integrity and authentication in multi-processors system-on-chip. In Proceedings of the GLSVLSI'18: Proceedings of the 2018 on Great Lakes Symposium on VLSI, Chicago, IL, USA, 23–25 May 2018. 39–44.
27. Bîrleanu, F.G., Bizon, N. Reconfigurable computing in hardware security–a brief review and application. J. Electr. Eng. Electron. Control Comput. Sci. 2016, 1–12.
28. Liu C, Yang J., Chen R, Zhang Y., Zeng J. (2011) Research on immunity-based intrusion detection technology for the internet of things In: 2011 Seventh International Conference on Natural Computation, vol. 1. 212–216, Shanghai.
29. Kasinathan P., Costamagna G., Khaleel H., Pastrone C., Spirito MA (2013) DEMO: An IDS framework for internet of things empowered by 6LoWPAN In: Proceedings of the 2013 ACM SIGSAC Conference on Computer; Communications Security, CCS'13, 1337–1340, Berlin.
30. Kasinathan P., Pastrone C., Spirito MA., Vinkovits M. Denial-of-service detection in 6LoW-PAN based internet of things In: 2013 9th International Conference on Wireless and Mobile Computing, Networking and Communications (WiMob), 2013. 600–607, Lyon
31. Garcia-Font V., Garrigues C., Rifà-Pous H. Attack classification schema for smart city WSNs. Sensors. 2017, 1–24.
32. Prathapchandran, K., & Janani, T. A trust-aware security mechanism to detect sinkhole attack in RPL-based IoT environment using random forest – RFTRUST. 2021. 1–18.
33. Nagarathna Ravi, and S. Mercy Shalinie. Semi-Supervised Learning-based Security to Detect and Mitigate Intrusions in IoT Network. 2020. 2–11.
34. Mengmeng Ge, Xiping Fu, Naeem Syed, ZubairBaig, Gideon Teo, Antonio RoblesKelly. Deep Learning-based Intrusion Detection for IoT Networks. 2019. 1–9.
35. Satish Pokhrel and Robert Abbas and Bhulok Aryal. 2021. IoT Security: Botnet detection in IoT using Machine learning. CoRR. 2021. 2–4.
36. Yalin E. Sagduyu, Yi Shi, and TugbaErpek. IoT Network Security from the Perspective of Adversarial Deep Learning. 2019. 1–9.
37. Abbas Yazdinejad, Reza M. Parizi, Ali Dehghantanha, Qi Zhang, Kim-Kwang Raymond Choo. An Energy-efficient SDN Controller Architecture for IoT Networks with Blockchain-based Security. 2020. 1–13.
38. NarmadhaSambandam, Mourad Hussein, Noor Siddiqi, Chung-Horng Lung, "Network Security for IoT using SDN: Timely DDoS Detection". (2018), 1–2.
39. Azhar Syed, Dr. Mary Lourde R. Hardware Security Threats to DSP Applications in an IoT network. 2016. 1–4.
40. Sidhu, S.; Mohd, B.J.; Hayajneh, T. Hardware Security in IoT Devices with Emphasis on Hardware Trojans. J. Sens. Actuator Netw. 2019. 2–17.
41. Tanguy Godquin, Morgan Barbier, Chrystel Gaber, Jean-Luc Grimault, Jean-Marie Le Bars. Applied graph theory to security: A qualitative placement of security solutions within IoT networks. 2020. 2–9.
42. AfsoonYousefi, Seyed Mahdi Jameii. Improving the Security of Internet of Things using Encryption Algorithms. 2017. 1–5.
43. Balasubramanian Prabhu kavin, Sannasi Ganapathy, "A secured storage and privacy preserving model using CRT for providing security on cloud and IoT-based applications". 2019. 181–190.
44. Alejandro Molina Zarca, Jorge Bernal Bernabe, Ivan Farris, YacineKhettab, Tarik Taleb, Antonio Skarmeta. Enhancing IoT security through network softwarization and virtual security appliances. 2018. 2–16.
45. Elijona Zanaj, Giuseppe Caso, Luca De Nardis, Alireza Mohammadpour, Ozgu Alay, Maria-Gabriella Di Benedetto. Energy Efficiency in Short and Wide-Area IoT Technologies-A Survey. 2021, 9(1), 22.
46. Krishna, R. R., Priyadarshini, A., Jha, A. V., Appasani, B., Srinivasulu, A., & Bizon, N. (2021). State-of-the-art review on IoT threats and attacks: Taxonomy, challenges and solutions. *Sustainability*, *13*(16), 9463.

Design and Analysis of Z-Source Inverter with Maximum Constant Boost Control Method

Subha Maiti, Santosh Sonar, Saima Ashraf, Sanjoy Mondal, and Piyali Das

1 Introduction

There are two classic inverters: one is a voltage source inverter and another is a current source inverter [1]. Talking about the voltage source inverters, they are widely utilized but have a number of drawbacks [2], which include the following: (i) the output voltage is less than the input voltage [3–5]. That's why it can also be termed a buck inverter [5–10]. (ii) Turning on the top and bottom switches for the same phase simultaneously is not feasible [2]. Otherwise, a short circuit will occur, which will damage the switches. That's why there is a dead time for both the upper and lower switches of the inverter [1], and that is again causing a wave distortion.

Below given points are some of the conceptual and theoretical challenges and limitations faced by basic current-source inverters (CSI) [15]: (i) The output ac voltage is greater than the dc inductor's primary supply voltage [13]. That's why CSI basically refers to it as a boost inverter. (ii) At least one upper and one bottom switch must be gated on all the time [2]. Otherwise, there will be an open circuit, and the switches will be damaged by the generated high voltage. So, for safe current

S. Maiti · S. Sonar (✉) · S. Ashraf
Department of Electrical and Instrumentation Engineering, Thapar Institute of Engineering and Technology, Patiala, Punjab, India
e-mail: santosh.sonar@thapar.edu

S. Mondal
Department of Electrical Engineering, Institute of Engineering and Management, Kolkata, West Bengal, India

P. Das
Department of Electrical Engineering, North Eastern Regional Institute of Science and Technology, Nirjuli, Arunachal Pradesh, India

M. Mishra et al. (eds.), *Applications of Computational Intelligence in Management & Mathematics*, Springer Proceedings in Mathematics & Statistics 417,
https://doi.org/10.1007/978-3-031-25194-8_7

commutation, the I-source converter requires an overlap time [15], which again causes waveform distortion.

So, both the V-source converter and the I-source inverter experience the following problems [10–12]: (i) They can only be a buck or a boost converter [3], not a buck–boost converter. That is, they can attain an output voltage range [6] that is either higher or lower than the input voltage [12–14]. (ii) Their basic circuits cannot be interchanged [16, 17]. In other words, it is not possible to use the V-source converter's main circuit for the I-source converter, and vice versa [15]. The impedance source inverter can be used in renewable energy applications, such as electrical vehicles, PV applications, wind energy [10] power conversion, etc.

To eliminate these issues, a Z-Source converter [19] has been developed.

2 Z-Source Inverter

2.1 Schematic Diagram

The basic diagram of a Z-source inverter (ZSI) is shown in Fig. 1. It is a two-port network that comprises two identical inductors L1 and L2 and two identical capacitors C1 and C2, respectively, in a cross form to provide an impedance source. This network can be coupled with any power electronics converter.

2.2 Operating Principle

The ZSI is a modified three-phase inverter that eliminates some of the limitations that were faced with traditional three-phase inverters. In terms of ZSI's theoretical characteristics, the output voltage can be as low as zero and as high as infinity. Unlike the traditional three-phase inverter, the output voltage is independent of the input voltage.

There are nine switching states in a three-phase ZSI. It has six active states just like the traditional three-phase inverter, i.e., there is a power transfer between the DC side and AC side. And also, there are two zero states, i.e., when all the upper

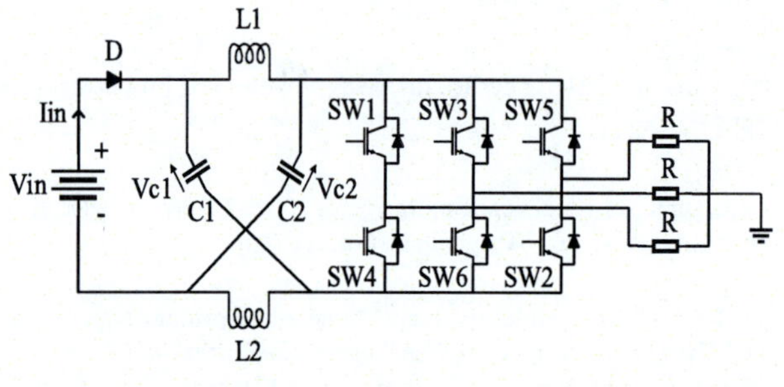

Fig. 1 General circuit diagram of 3-*phase* ZSI

switches are gated on or all the lower switches are gated on. By eliminating those eight states, there is also one additional state that cannot be seen in the traditional zero state, i.e., in upper and lower switches, any single phase is gated on, any two-phase switches are gated on, or all six switches are gated on. That is termed the shoot-through (ST) zero state.

2.3 *Circuit Diagram for Different Operations and Analysis*

Figure 2 shows the equivalent circuit diagram of ZSI.

When there is an ST zero state at that time, the ZSI is comparable to being short-circuited as shown in Fig. 4. As shown in Fig. 3, in any of the one active states, the circuit will function as an equivalent current source.

Both the inductors L1 and L2 are assumed to be identical and have the same values in the circuit analysis. Similarly, the capacitors C1 and C2 are assumed to be the same and have the same values. As a result of the identical values of inductors and capacitors as stated above, the ZSI network can be analyzed in terms of a symmetrical network.

$$V_{C1} = V_{C2} = V_C \tag{1}$$

$$V_{l1} = V_{l2} = V_l \tag{2}$$

Considering Fig. 4, T_{st} is the duration at which ST will occur, and the total time is T. From this, it can be stated that

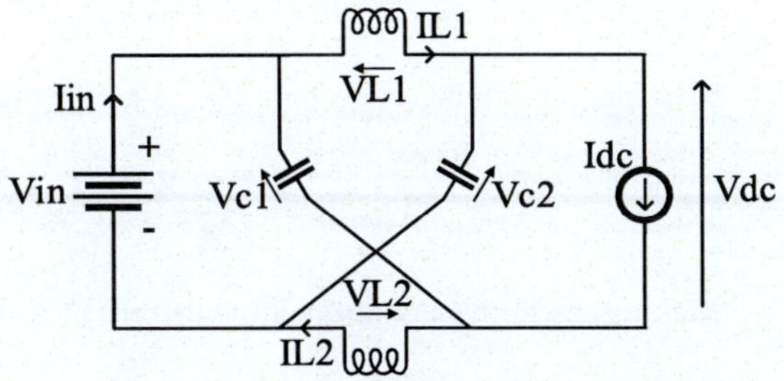

Fig. 2 Equivalent circuit diagram ZSI

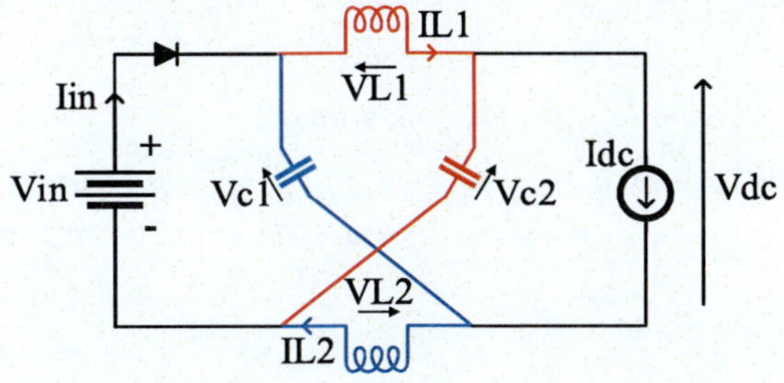

Fig. 3 Circuit diagram of ZSI at non-shoot through (NST)

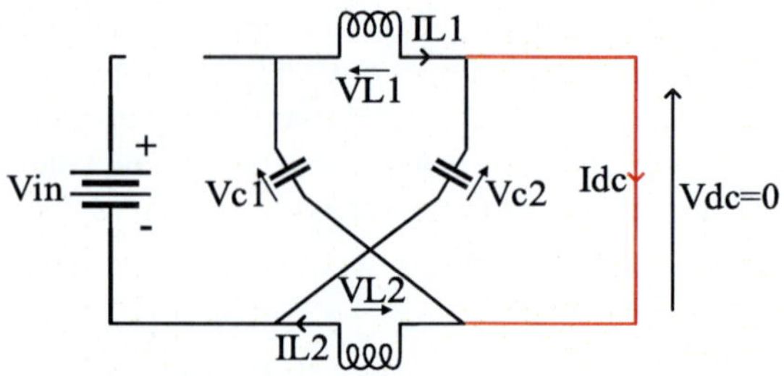

Fig. 4 Circuit diagram of ZSI at shoot-through (ST) state

$$V_l = V_C \tag{3}$$

$$V_{dc} = 0 \tag{4}$$

Now consider Fig. 3, i.e., one of the non-shoot through (NST) states and time periods of NST is represented by T_a and the total time for one cycle is T. From the equivalent circuit in Fig. 2, we have

$$V_l = V_{in} - V_C \tag{5}$$

$$V_{dc} = V_C - V_l \tag{6}$$

Now, putting (5) into Eq. (6), one has

$$V_{dc} = 2V_C - V_{in} \tag{7}$$

Now, average inductor voltage $V_{l(avg)}$ and average DC link voltage $V_{dc(avg)}$

$$V_{l(avg)} = \frac{T_{st} \cdot V_C + T_a \cdot (V_{in} - V_C)}{T} \tag{8}$$

Making the above equation equal to zero at a steady state, we get

$$V_C = \frac{T_a}{T_a - T_{st}} \cdot V_{in} \tag{9}$$

Similarly, for DC link voltage,

$$V_{dc(avg)} = \frac{T_{st} \cdot 0 + T_a \cdot (2V_C - V_{in})}{T}$$

$$= \frac{T_a}{T_a - T_{st}} V_{in} = V_C \tag{10}$$

By substituting (9) in Eq. (7), the peak DC link voltage can be found out.

$$V_{dc} = \frac{T}{T_a - T_{st}} V_{in} \tag{11}$$

$$V_{dc(Peak)} = B.V_{in} \tag{12}$$

where B is the boost factor, i.e.,

$$B = \frac{T}{T_a - T_{st}} \tag{13}$$

$$B = \frac{1}{1 - 2\frac{T_{st}}{T}} \tag{14}$$

The output ac peak voltage can be written from a traditional voltage source inverter as,

$$V_{ac(Peak)} = 0.5 \times V_{dc(peak)} \times \frac{V_{ref}}{V_{car}} \tag{15}$$

$$V_{ac(Peak)} = M \times \frac{V_{dc(peak)}}{2} \tag{16}$$

From Eqs. (12) and (16), it can be stated that,

$$V_{ac(Peak)} = M \times B \times \frac{V_{in}}{2} \tag{17}$$

From (1), (9) and (14), the capacitor voltage can be obtained

$$V_{C1} = V_{C2} = V = \frac{\frac{T_a}{T}}{\frac{T_a}{T} - \frac{T_{st}}{T}} \cdot V_{in} \tag{18}$$

$$\text{As, } T = T_a + T_{st} \tag{19}$$

Putting (19) in (18), we have,

$$V_C = \frac{1 - \frac{T_{st}}{T}}{1 - \frac{2 \cdot T_{st}}{T}} \cdot V_{in} \tag{20}$$

Here, T_{st} is the time duration of the shoot-through state, and T is the total time in one cycle. Talking about boost factor B, it is crucial because the right boost factor value will be required to obtain the required output voltage. And this boost factor can be controlled by the ST zero-state interval. If T_{st} rises, the boost factor will rise, and as a result, a higher output AC voltage can be obtained and vice versa.

3 Maximum Constant Boost Control

With this control technique, the maximum obtainable voltage gain (G) is achieved while keeping the duty ratio (D_{st}) constant. Now, as with the simple boost control method (SBC), it has a total of six modulation curves with a slight difference. Figure 5 shows the proposed modulation waveform of the Maximum Constant Boost Control (MCBC) strategy.

Here the three reference voltage sources (V_a, V_b, V_c) are the same as in the case of a simple boost control method (SBC)—one carrier triangular wave with a positive peak (+1) and a negative peak (−1). Another two envelopes are added over here, which are different from those that have been used in SBC. Those two envelopes have been represented in Fig. 5 as V+ and V−.

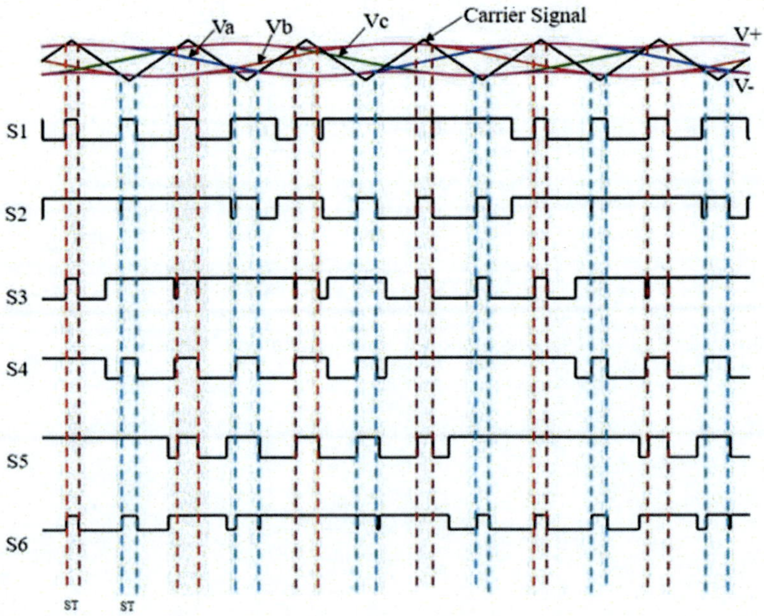

Fig. 5 Sketch Diagram of MCBC method

Whenever the carrier signal is more positive than the **V+** or more negative than the **V−**, then shoot through will occur. And for the rest of the cases, it will act as traditional PWM inverters.

According to Eq. (14), the shoot-through duty ratio determines the boost factor (B). So, it is very obvious that to maintain a constant boost, the shoot-through duty cycle must be kept constant.

In Fig. 5, the mentioned two shoot-through envelopes, i.e., V+ and V-, must be periodic and three times the output frequency.

Taking the first half period as 0 to $\frac{\pi}{3}$ so, during this period, the upper and lower envelope curves can be expressed as

$$V_{+1} = \sqrt{3}M + Msin\left(\theta - \frac{2\pi}{3}\right) \tag{21}$$

and

$$V_{-1} = Msin\left(\theta - \frac{2\pi}{3}\right) \tag{22}$$

And for next $\frac{\pi}{3}$ to $\frac{2\pi}{3}$, the upper and lower envolpe curves can be expressed in equations

$$V_{+2} = Msin\,(\theta) \tag{23}$$

and

$$V_{-2} = Msin\,(\theta) - \sqrt{3}M \tag{24}$$

Now it can be observed from both of these equations that if Eq. (22) is substracted from (21), we get $\sqrt{3}M$, i.e., the distance between two curves is constant. And the shoot-through duty ratio can be calculated as Eq. (28).

Gain of any system can be find out by,

$$Gain = \frac{V_{out}}{V_{in}} \tag{25}$$

$$G = \frac{V_{ac(peak)}}{V_{in}/2} \tag{26}$$

From Eq. (17), one has

$$\frac{V_{ac(Peak)}}{V_{in}/2} = M \times B = G \tag{27}$$

$$\frac{T_{st}}{T} = D_{st} = \frac{2 - \sqrt{3}M}{2} = 1 - \frac{\sqrt{3}M}{2} \tag{28}$$

The relationship between gain (*G*), modulation index (*M*), and shoot-through duty ratio can be obtained by putting Eq. (14) in (27)

$$G = M \times \frac{1}{1 - D_{st}} \tag{29}$$

$$G = \frac{M}{\sqrt{3}M - 1} \tag{30}$$

Now here in Eq. (30), as soon as the modulation index reaches $\frac{\sqrt{3}}{3}$, the voltage gain (G) becomes infinity.

DC link voltage Vdc is actually the voltage stress across the switch during OFF condition of the switch. So, from Eq. (12),

$$V_s = B \cdot V_{in} \tag{31}$$

From Eqs. (30) and (34)

$$V_s = G\frac{V_{in}}{M} \tag{32}$$

Putting (28) in (14)

$$B = \frac{1}{\sqrt{3}M - 1} \tag{33}$$

After putting (33) in (31), we get the relation of voltage stress in terms of modulation index and DC input voltage.

$$V_s = \frac{V_{in}}{\sqrt{3}M - 1} \tag{34}$$

So, for any required voltage gain *G*, the possible maximum modulation index can be taken from (30), i.e.,

$$M = \frac{G}{\sqrt{3}G - 1} \tag{35}$$

From (31), (32), and (35), we can get the voltage stress in this MCBC modulation technique

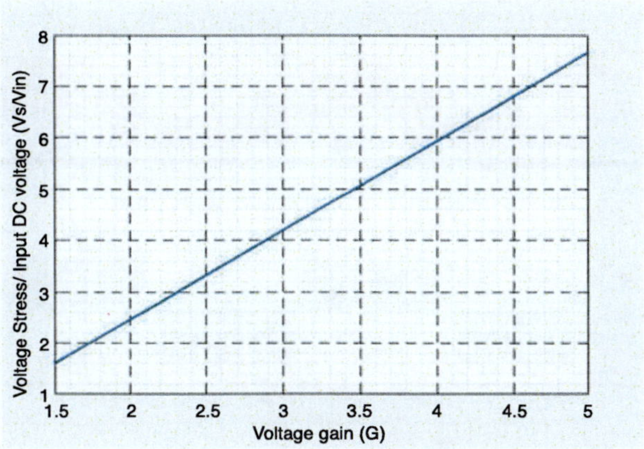

Fig. 6 Voltage stress of switches vs voltage gain of the MCBC method

Table 1 MATLAB Simulink Parameters

Parameters taken	Value
DC input voltage (V_{in})	130 V
Inductors ($L_1 = L_2$)	2 mH
Capacitor ($C_1 = C_2$)	330μH
Three-phase loads	25 Ω
Fundamental frequency (f)	50 Hz
Carrier frequency (f_s)	2.1kHZ
Modulation index (M)	0.95

$$V_s = \left(\sqrt{3}G - 1\right) V_{in} \tag{36}$$

In Fig. 6, the (V_s/V_{in}) vs the voltage gain (MB) is shown. It has been observed that when using the MCBC method, the voltage stress Vs is slightly high.

4 Simulation Results

To verify the validity of the above-proposed method, MATLAB SIMULINK has been used. The system has been simulated at a 0.09 sec time frame.

4.1 Figures and Tables

Certain parameters have been taken for verifying the results that are shown in Table 1.

Here, the simulation has been carried out with a modulation index of 0.95, and as per relation between the modulation index and the shoot-through duty ratio (D_{st}),

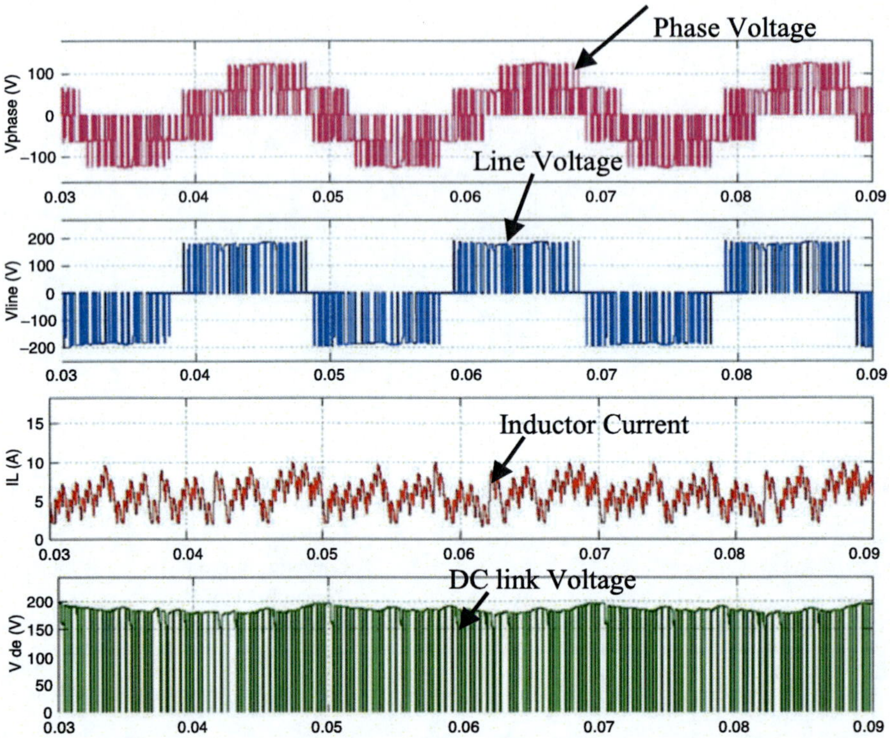

Fig. 7 Simulation results with $M = 0.95$ and input voltage 130 V dc for MCBC

D_{st} becomes 0.1772, which is slightly small. As a result, a medium gain has been observed in the output.

In Fig. 7, for the MCBC method, different waveforms have been analyzed, i.e., the DC link voltage, the current across the inductor current L1, and also the phase and line voltage across the load. Here the DC link voltage is nothing but the voltage stress across the switch, i.e., when the switch is in OFF condition, the reverse voltage is applied across the switch. From Fig. 8, it is clear that both capacitor voltages are the same. That is true according to Eq. (1).

Figures 9 and 10 represent the FFT analysis of load current and line voltage. A suitable filter is to be designed to obtain sinusoidal output voltage and current.

5 Conclusion

This paper discussed one of the most popular PWM techniques suitable for high voltage gain with reduced voltage stress compared to a simple boost control technique, known as the maximum constant boost control technique for three-phase ZSI. The shoot-through insertion process is discussed based on a sine-triangle comparison-based approach. Mathematical derivations in terms of boost factor,

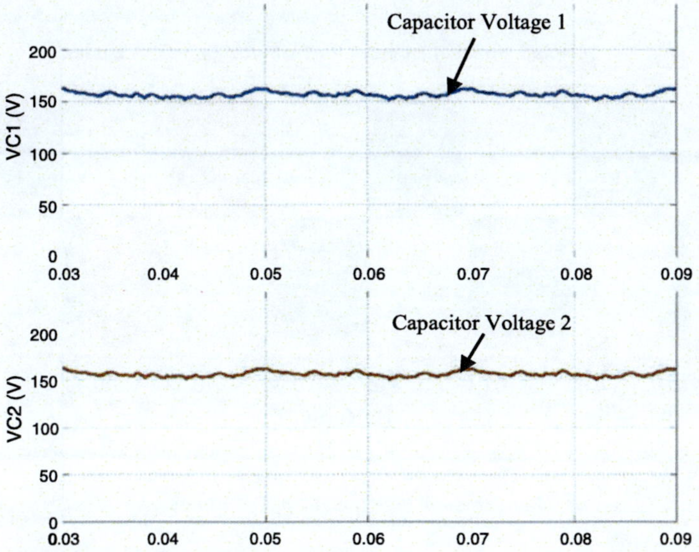

Fig. 8 Capacitor C1 and C2 waveforms

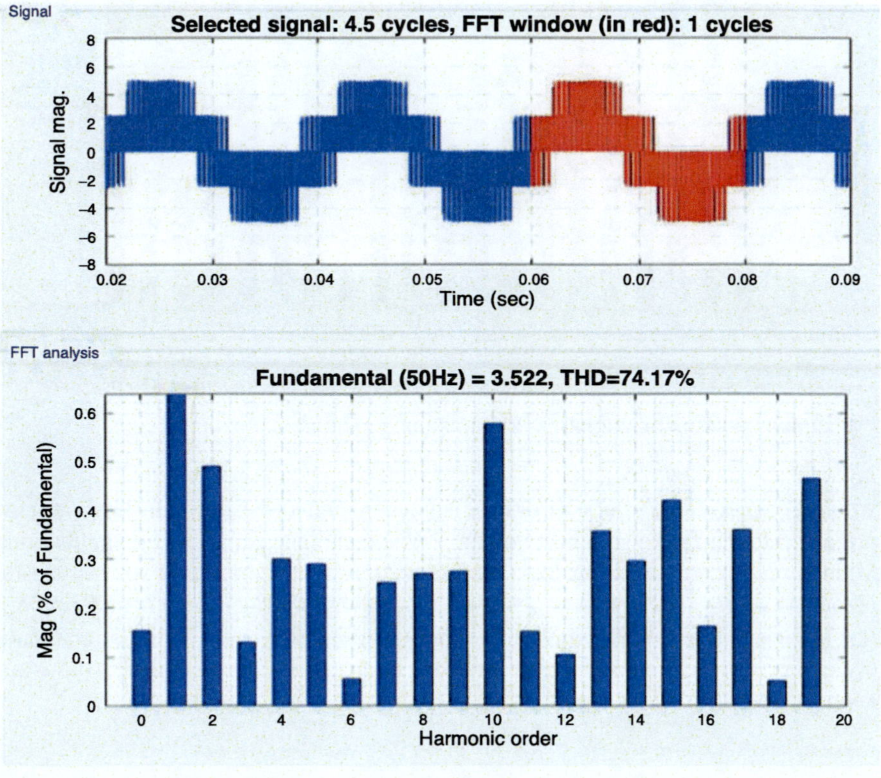

Fig. 9 Load current FFT analysis of the proposed method

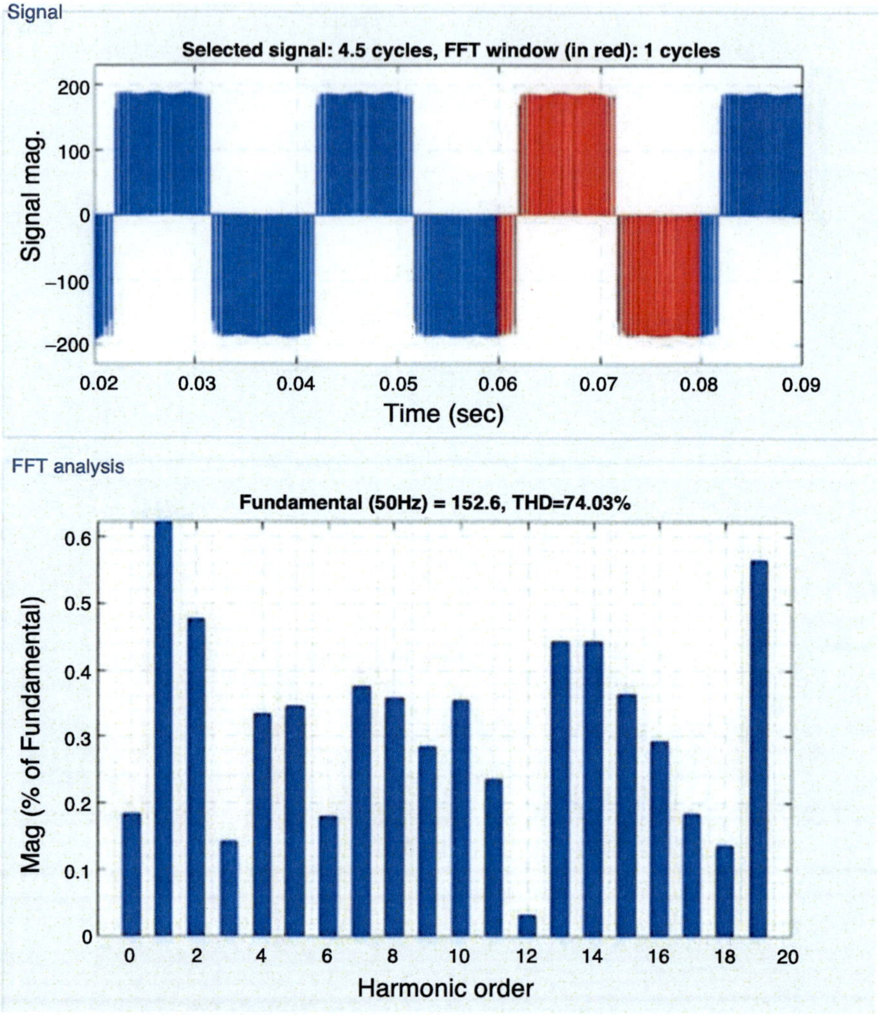

Fig. 10 Line voltage FFT analysis

voltage gain, voltage stress across the switches, and shoot-through duty ratio are presented to verify the suitability of the PWM technique for boosting applications. Simulation is carried out in MATLAB, and results are presented to analyze the output waveform quality. It is observed that though the inductor current ripple is low because of the low variation in the shoot-through duty ratio in one fundamental cycle, the voltage stress across the switches is very high, which limits its industrial applications.

References

1. Power Electronics, P.S. Bhimra, Khanna Publishers, 1991
2. M. H. Rashid, Power Electronics Handbook, Academic Press, New York, 2001.
3. S. Ghosh, K. Sarkar, D. Maiti and S. K. Biswas, "A single-phase isolated Z-source inverter," 2016 2nd International Conference on Control, Instrumentation, Energy & Communication (CIEC), 2016, pp. 339–342, https://doi.org/10.1109/CIEC.2016.7513738.
4. S. Singh and S. Sonar, "Space vector based PWM sequences for Z-source principal derived inverter topologies," 2018 IEEMA Engineer Infinite Conference (eTechNxT), 2018, pp. 1–6, https://doi.org/10.1109/ETECHNXT.2018.8385336.
5. S. Sonar and S. Singh, "Improved Space Vector PWM Techniques of the Three level ZSI," 2019 International Conference on Computing, Power and Communication Technologies (GUCON), 2019, pp. 508–513.
6. S. Sonar and T. Maity, "Design and simulation of a novel single phase to three phase wind power converter," 2012 1st International Conference on Recent Advances in Information Technology (RAIT), 2012, pp. 725–730, https://doi.org/10.1109/RAIT.2012.6194585.
7. X. P. Fang, Ji Min Cui, Jie Liu and Mao Yong Cao, "Detail research on the traditional inverter and Z-source inverter," 2011 International Conference on Applied Superconductivity and Electromagnetic Devices, 2011, pp. 318–321, https://doi.org/10.1109/ASEMD.2011.6145133.
8. S. Singh and S. Sonar, "Improved Maximum Boost Control and Reduced Common-Mode Voltage Switching Patterns of Three-Level Z-Source Inverter," in IEEE Transactions on Power Electronics, vol. 36, no. 6, pp. 6557–6571, June 2021, https://doi.org/10.1109/TPEL.2020.3040908.
9. M. K. Islam, M. M. Zaved, A. M. Siddiky and K. A. Al Mamun, "A comparative analysis among PWM control Z-Source Inverter with conventional PWM Inverter for induction motor drive," 2016 International Conference on Innovations in Science, Engineering and Technology (ICISET), 2016, pp. 1–6, https://doi.org/10.1109/ICISET.2016.7856496.
10. S. Sonar and T. Maity, "Z-source inverter based control of wind power," 2011 International Conference on Energy, Automation and Signal, 2011, pp. 1–6, https://doi.org/10.1109/ICEAS.2011.6147087.
11. R. R. Patil, S. P. Patil, S. D. Patil and A. M. Mulla, "Designing Of Z-source inverter for photovoltaic system using MATLAB/SIMULINK," 2017 International Conference on Circuit ,Power and Computing Technologies (ICCPCT), 2017, pp. 1–5, https://doi.org/10.1109/ICCPCT.2017.8074331.
12. Fan Zhang, Xupeng Fang, F. Z. Peng and Zhaoming Qian, "A new three-phase ac-ac Z-source converter," Twenty-First Annual IEEE Applied Power Electronics Conference and Exposition, 2006. APEC '06., 2006, pp. 4 pp, https://doi.org/10.1109/APEC.2006.1620526.
13. Fang Zheng Peng, Miaosen Shen and Zhaoming Qian, "Maximum boost control of the Z-source inverter," in IEEE Transactions on Power Electronics, vol. 20, no. 4, pp. 833–838, July 2005, https://doi.org/10.1109/TPEL.2005.850927.
14. S. Singh and S. Sonar, "A New SVPWM Technique to Reduce the Inductor Current Ripple of Three-Phase Z-Source Inverter," in IEEE Transactions on Industrial Electronics, vol. 67, no. 5, pp. 3540–3550, May 2020, https://doi.org/10.1109/TIE.2019.2916298.
15. Fang Zheng Peng, "Z-source inverter," in IEEE Transactions on Industry Applications, vol. 39, no. 2, pp. 504–510, March-April 2003, https://doi.org/10.1109/TIA.2003.808920.
16. M. Shen, Jin Wang, A. Joseph, F. Z. Peng, L. M. Tolbert and D. J. Adams, "MCBC of the Z-source inverter," Conference Record of the 2004 IEEE Industry Applications Conference, 2004. 39th IAS Annual Meeting., 2004, pp. 147, https://doi.org/10.1109/IAS.2004.1348400.
17. M. A. Kumar and M. Barai, "Performance analysis of control and modulation methods of z-source inverter," 2015 IEEE International Conference on Signal Processing, Informatics, Communication and Energy Systems (SPICES), 2015, pp. 1–5, https://doi.org/10.1109/SPICES.2015.7091395.

18. P. Kumar, S. Sonar and P. Shaw, "Comparative analysis of three phase ac-ac Z-source converter topologies," 2016 IEEE 1st International Conference on Power Electronics, Intelligent Control and Energy Systems (ICPEICES), 2016, pp. 1–6, https://doi.org/10.1109/ICPEICES.2016.7853427.
19. M. Li, R. Iijima, T. Mannen, T. Isobe and H. Tadano, "New Modulation for Z-Source Inverters With Optimized Arrangement of Shoot-Through State for Inductor Volume Reduction," in IEEE Transactions on Power Electronics, vol. 37, no. 3, pp. 2573–2582, March 2022, https://doi.org/10.1109/TPEL.2021.3109672.

SPV/Wind Energy-Based Hybrid Grid-Connected System with Optimum Power Flow Operation

Sheshadri Shekhar Rauth, Venkatesh Vakamullu, Madhusudhan Mishra, and Preetisudha Meher

1 Introduction

In recent years, with the quantitative rise in the demand for energy and its associated factors, viz. rise in energy costs, limited volume of energy reserves, and contemporary environmental pollution, there has been a drive towards the usage of renewable energy [1]. In fact, these sources are abundantly available in nature, have easy access, and do not impart any environmental adversities [2]. Though various renewable energy sources are available in nature, solar and wind energies are the most prominent among them. At present, generation of energy is the primary objective; in addition, efficient conversion of the generated power into usable format, preservation and management of the energy, and supply of reliable power to the appliances are also essential objectives. At present, generating electrical energy from renewable sources is manifested as a primary objective [3–7].

The biggest dream project of India is to establish a more sustainable, reliable, and pollution-free energy setup to aid the infrastructure in Indian smart cities. However, this can be accomplished by probing the ample energy generation by renewable sources. This deal imparts win-win situations in India by offering clean energy to

S. S. Rauth (✉) · V. Vakamullu
Department of Electrical Engineering, Indian Institute of Technology Kharagpur, Kharagpur, West Bengal, India
e-mail: sheshadri@iitkgp.ac.in

M. Mishra
Electronics & Communication Engineering, North Eastern Regional Institute of Science and Technology (NERIST), Itanagar, Nirjuli, Arunachal Pradesh, India

P. Meher
Department of Electronics and Communication Engineering, National Institute of Technology, Jote, Arunachal Pradesh, India

M. Mishra et al. (eds.), *Applications of Computational Intelligence in Management & Mathematics*, Springer Proceedings in Mathematics & Statistics 417,
https://doi.org/10.1007/978-3-031-25194-8_8

smart cities while downsampling carbon emissions, which are targeted to 33–35% of global emissions by 2030 [8].

Since the availability of SPV and wind sources in the universe is random, the ensemble of the energies emanated from SPV and wind sources is a much more intricate process than their individual operation. However, this problem can be eliminated by incorporating the best combinations of power electronics controllers and converters. Therefore, the resultant system can impart highly consistent power and maintain efficiency in controlling power to the loads. It is essential to take into consideration the isolation of solar sources, variable temperature profiles, and wind velocity throughout the day. Further, in order to identify the optimal operating point for the operation of the SPV and WT systems under certain weather conditions and switch the system to facilitate active support between them, MPPT is probed into the system [9]. MPPT sets the reference point for SPV and WT switches for tracking the maximum electrical power and feeding it to the inverter for AC output generation.

Typically, two categories of SPV and wind energy systems exist in the literature: off-grid systems and on-grid systems. However, the government of India targeted generating 100 GW for SPV and 60 GW for wind power by 2022 to facilitate off-grid or on-grid systems [10].

The significant rise in the number of distributed generators (DGs) that are penetrated by the grid may cause serious problems on the grid side. Grid synchronization is one of the biggest problems of all [11], and irregularity at the grid side is another crucial problem [33, 34]. Further, panel costs are the predominant issue in photovoltaic systems. Usually this cost accounts for around 30–50% of the total cost of the system. However, this cost is diminishing remarkably, which in turn puts the burden on the inverter cost [12]. However, the overall system can be made to have a minimum cost by adopting transformer-less topologies that facilitate the resolution of leakage current, which in turn leads to higher efficiency [13]. Multilevel inverter topologies can fulfil this requirement [14]. Inverters aided with soft switching techniques can find notable efficiency [15]. In addition, optimization of the overall system is essential objective [32]. There are profound systems that have been delineated in the literature, viz., Institute of Electrical and Electronics Engineers (IEEE), International Electrotechnical Commission (IEC), National Electrical Code (NEC), etc. The preceding section describes the various standards of grid-connected systems associated with renewable power generation.

2 Various Standards

Standardization of grid-connected systems is a highly anticipated phenomenon for universal acceptance of norms set by the organization. IEEE 1547 depicts the information about the configuration pertaining to distributed connection to the grid [16], DIN (German institute for standardization) EN 50530 imparts the measure of efficiency [19], IEC 61683 [20], IEEE 921 set the standards for designing and operating converters and controllers by UL (Underwriters Laboratories) 741 [18],

IEEE 929 formulates the standards for grid-connected solar system [19], IEEE 512 facilitates the harmonics measurement [20], IEEE 1373 set the standards for testing and islanding detection [17], IEC 62116, IEC 60364 sets the standards for safety related issues VDE (association for electrical, electronic, and information technologies), 0126, IEC 60269, IEC 62109 [21] and IEC 61173 [17], etc. standardize the over voltage protection.

3 Grid-Connected Hybrid System Topology

Figure 1 depicts the energy system representing the solar-wind ensemble grid-tied system. The proposed system consists of various components, viz., MPP tracker, control circuit, DC/AC converter, filter, step-up transformer, battery, and anti-islanding block. SPV system is directly coupled to DC common taping point with a diode; the wind turbine system is tied after AC to DC conversion block governed by rectifier circuit to DC common taping point using a dedicated switch. Perturbation and Observation (P&O)-based MPP tracker is employed in the system for tracking maximum value of electrical power. Further, the MPP tracker for wind energy is probed to ensure the voltage level is the same as the solar panel voltage for the flexible coupling value. However, the output voltage emanating from wind turbines varies above the standard voltage or below the preset threshold based on the wind velocity, and hence, self-isolation is not needed for maintaining DC voltage at a common value. However, after the wind speed falls between the preset threshold points, the system resumes normal operation. In this system, SPV and WES produce DC voltages, which are further converted into AC by using a bridge converter so that the generated power can be used for domestic and household purposes. However, the harmonics that are in the AC can be discarded by employing an efficient low-pass LC filter in the SPV and WES. As the AC output emanating from the inverter is lower than the standard voltage of 230 V, an efficient step-up transformer is incorporated into the system to achieve the required level of output

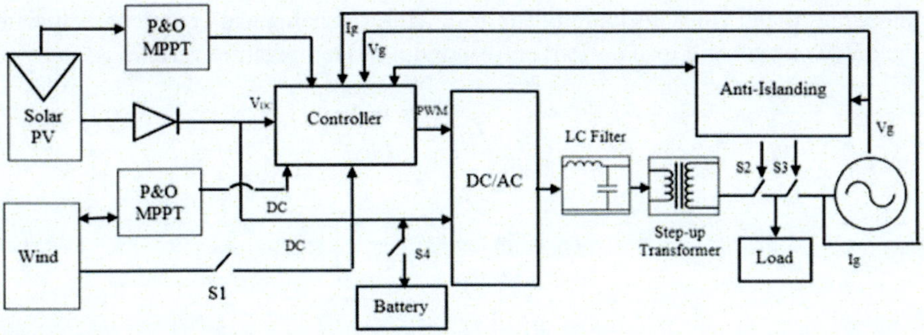

Fig. 1 Typical view of the grid-connected solar-wind hybrid system

voltage. However, throughout the implementation process, interfacing the generated voltage to grid is a hassle task as the wide fluctuation occurs in the voltage and frequencies of grid components. Hence, to mitigate the flux and achieve perfect synchronization, it is essential to establish the standard for various key parameters, viz., voltage, phase, frequency, reactive power, harmonics, etc. The PR controller establishes the control mechanism by using the current reference value of solar and wind power. Further, the controlled output guided by PWM generation using MOSFET switches enables the inverter to produce AC power from DC power. Anti-islanding protection is employed in the system when power is unavailable at the grid side, and at the same time, the system also gets isolated when low-power generation takes place on a step-up transformer. Hence, either generated power or grid power is seamlessly supplied to the load. However, if the power is unavailable at the grid side, switch S3 remains open, and hence, the entire system has to be detached from the grid to safeguard the system from malfunctionality. Nevertheless, during this phase, the installed system supplies power to the load. Further, if the power is unavailable at the generator side, switch S2 remains open, and hence, the system will be detached from the load. In this situation, the grid directly supplies power to the load. However, batteries supply the power in a special case where grid power and solar-wind power are unavailable. The charging of the battery takes place only when the system doesn't have the need to supply power to the grid and due to the availability of extra power after providing sufficient power for domestic use. Switch S4 controls the charging and discharging phenomena.

3.1 Solar Model

Solar PV cells are manifested as the basic building blocks of a solar PV system. PV cells are static devices made up of semiconductors containing multiple P-N junction configurations, and they transfer the irradiance obtained from the Sun into DC voltage by emanating a large quantity of freely movable electrons to the driving circuit. Figure 2 depicts the equivalent electric model of the conventional PV cell [22]. The current generated by the photovoltaic effect depends on the atmospheric temperature and solar insolation. The following mathematical equations delineate the diode current and load current corresponding to temperature variations.

$$I = I_L - I_D - \frac{V + IR_s}{IR_{sh}} \tag{1}$$

$$I_D = I_0 \left[e^{\frac{V+IR_s}{nV_T}} - 1 \right] \tag{2}$$

$$I_L = I_{L(T=298\ K)} \left[1 + \left(T - 298\ K \left(5 \times 10^{-4} \right) \right) \right] \tag{3}$$

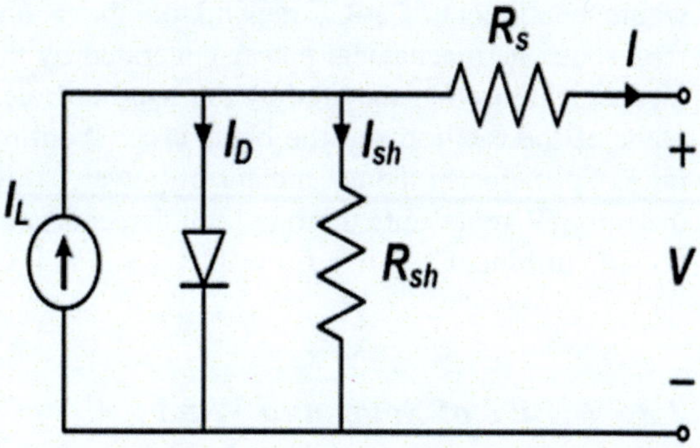

Fig. 2 Equivalent circuit of PV module

where I is the current ejected from the solar panel, I_L is the photovoltaic current, V is the output voltage of the solar panel, I_D is the diode current, I_O is the reverse saturation current at 25 °C, k is Boltzmann's constant, R_S is the series resistance, R_{sh} is the shunt resistance, V_T is the thermal equivalent voltage, and T is the temperature.

3.2 Wind Model

Typically, a wind turbine converts the energy contained in the kinetic form of the wind into mechanical form. Further, this mechanical form of energy is converted into an electrical form with the aid of a generator. Hence, the conversion of electrical energy primarily relies on wind resources. According to the variations in wind velocity, the production of electrical energy varies in a positive or negative direction. The proposed model presented in this article is developed on the basis of the steady-state characteristics of the wind turbine as explained in Eqs. 4–7 [23].

$$P_m = C_P\,(\lambda, \beta)\,\frac{\rho A}{2} V_{wind}^3 \tag{4}$$

$$T_m = \frac{C_P\,(\lambda, \beta)\,\frac{\rho A}{2}\,V_{wind}^3}{\omega_r} \tag{5}$$

$$C_P\,(\lambda, \beta) = C_1\left(\left(\frac{C_2}{\lambda_i}\right) - C_3\beta - C_4\right) e^{-C_5/\lambda_i} + C_6\lambda \tag{6}$$

$$\frac{1}{\lambda_i} = \frac{1}{\lambda + 0.08\beta} - \frac{0.035}{\beta^3 + 1} \tag{7}$$

where coefficients C_1–C_6 depend on the geometric details of the turbine. P_m represents the mechanical power generated by the wind turbine, T_m represents the mechanical torque generated by the wind turbine, β refers to the angle between the plane of the rotation and the blade cross section chord, λ represents the tip speed ratio, Cp indicates the coefficient of power, A indicates the surface area swept by the rotor, V represents the rated wind speed, and ρ refers to the air density. For a specific turbine, $C_1 = 0.5$, $C_2 = 116$, $C_3 = 0.4$, $C_4 = 0$, $C_5 = 5$, and $C_6 = 21$.

3.3 MPPT of Solar and Wind

The cumulative efficiency imparted by the SPV and WT systems can be enhanced by employing the MPP tracker in the proposed system. The P&O MPPT mechanism is probed in an SPV system where the perturbation parameter is the DC voltage [34]. On the other hand, in wind system, dc current is manifested in improved P&O MPPT algorithm as perturbation parameter. A modified version of the perturbation and observation algorithm is found to be stable and able to track abrupt speed changes in wind quickly. The combination of maximum power point tracking (MPPT) and a boost converter derives the maximum power from the sources based on their availability.

The maximum power P(t) can be computed from instant voltages and currents using P&O MPPT and compared with latest computed value of power P(t-1). The MPPT algorithm persistently changes the parameters of the system if a positive variation is seen in the operating point; else, the orientation of perturbation is changed for a positive variation in the operating point [21]. The flow diagram representation of the P&O algorithm is depicted in Fig. 3. It is unavoidable fact that the wind velocity doesn't remain constant and it alters its direction and magnitude all day. Change in the detection wind velocity is computed by the slop in the voltage. Sample voltage and current are considered for the power computation as depicted in the flow diagram represented by Fig. 4 [24].

3.4 Inverter Bridge Model

An anti-parallel coupled MOSFET-diode-based inverter circuit using the H-bridge topology is depicted in Fig. 5. A typical H-bridge inverter owns four switches (MOSFET OR IGBT). During the closing position of switches M2 and M3 (switch M1 and switch M4 are open), a positive voltage is to be probed across C and D terminals. Similarly, during the opening positions of M2 and M3 (closing positions of M1 and M4), the voltage will be reversed across the C and D terminals. The H-bridge inverter yields an output that holds the harmonics pertaining to the switching

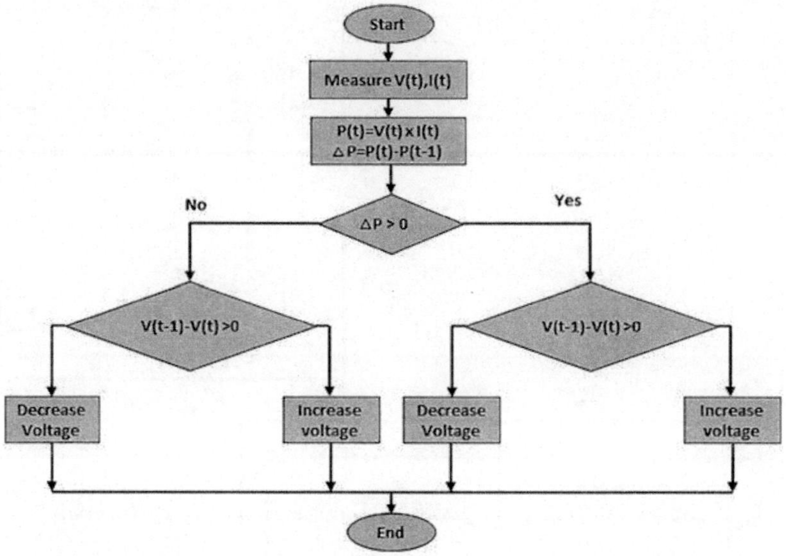

Fig. 3 Flowchart of P&O MPPT algorithm

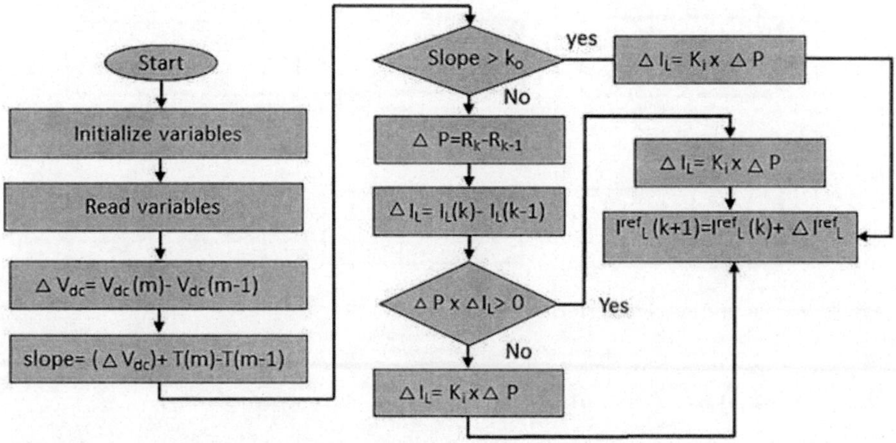

Fig. 4 Modified P&O MPPT algorithm for wind energy system

frequency of the PWM waveform; hence, it is desirable to mitigate the harmonics with the aid of a selective harmonics compensator using sinusoidal PWM (SPWM) and a passive component constructed filter at the output side.

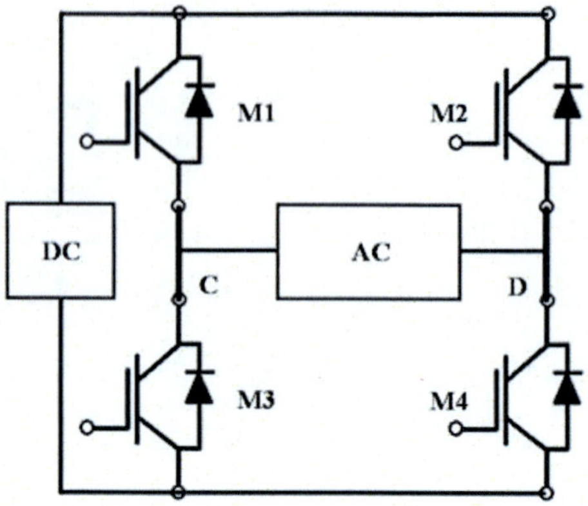

Fig. 5 Typical view of single-phase H-bridge inverter topology

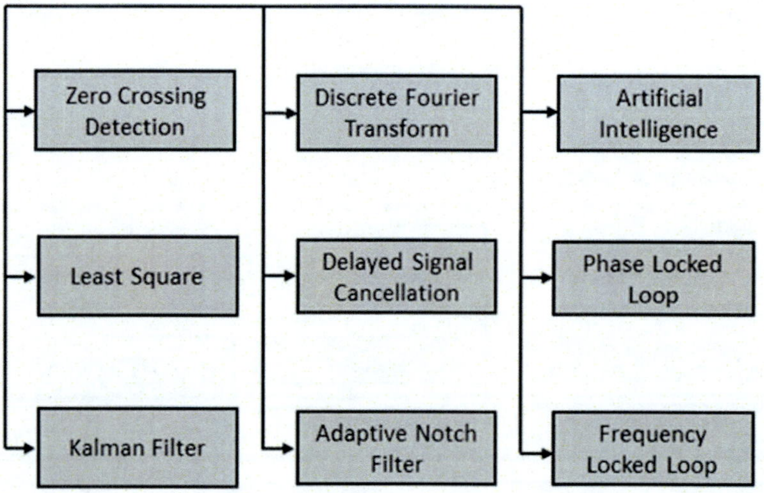

Fig. 6 Various grid synchronization techniques

3.5 Grid Synchronization

Figure 6 depicts the information about various synchronized techniques outlined [25]. presented a work explaining how adaptive notch filters (ANF) aided grid synchronization. The results of a single-phase grid-tied system yield reasonably good performance. Works presented in [26] discussed the modified ANF by considering the magnitude parameter of the ANF which is known as the amplitude adaptive notch filter (AANF). Further, phase-locked loop (PLL) using a second-order generalized integrator (SOGI) is quite common for single phase systems as it offers a low computational burden and greater filtering capacity [30].

3.5.1 AANF

The ANF-based methods are as simple as the PLL-based method. In fact, it addresses the limitations of PLL systems [26]. The following equations delineate the mathematical foundation of AANF.

$$\ddot{x} + \theta^2 x = 2\zeta\theta e(t) \tag{8}$$

$$\dot{\theta} = -\gamma x \theta e(t) \tag{9}$$

$$e(t) = u(t) - \dot{x} \tag{10}$$

$$\gamma = \frac{\varepsilon}{\left(\frac{A^2 N}{4\zeta^2} + 1\right)\left(\theta^2 \mu + 1\right)} \tag{11}$$

where θ refers to the frequency, A indicates amplitude, and ζ, ε, μ, and N represent real positive parameters. ζ and γ compute the estimated value of accuracy and the convergence velocity.

3.5.2 SOGI-PLL

The first preliminary model of PLL was reported in 1923, and its first application was to lock the frequency ranges of radio signals. However, over the years, the detailed refinements and advancements in PLL have found applications in numerous fields, like grid interfacing [28]. Figure 7 depicts the basic PLL circuit facilitated for grid interfacing. It constitutes three primary units: a phase detector (PD), a loop filter (LF), and a voltage-controlled oscillator (VCO). PD is one of the prominent blocks of PLL. PD inputs the error signal to LF, which eliminates high-frequency components and transmits the resultant signal to the VCO for accurate frequency alignment. Based on the categories and functionality of PD, PLL is categorized in multiple ways. SOGI supported Quadrate Signal Generator (QSG) is the primary element in the PD block. The transfer functions pertaining to SOGI-PLL aided PD are explained in Eqs. 12 and 13 [27].

$$\frac{v_\alpha}{v_i} = \frac{kws}{s^2 + kws + w^2} \tag{12}$$

$$\frac{v_\beta}{v_i} = \frac{kw^2}{s^2 + kws + w^2} \tag{13}$$

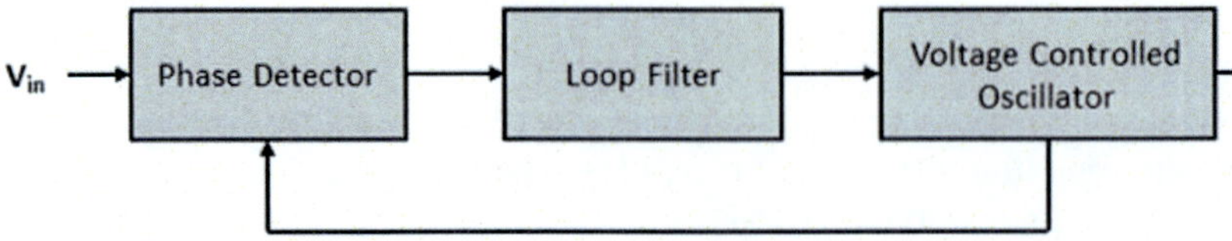

Fig. 7 Basic block diagram of PLL

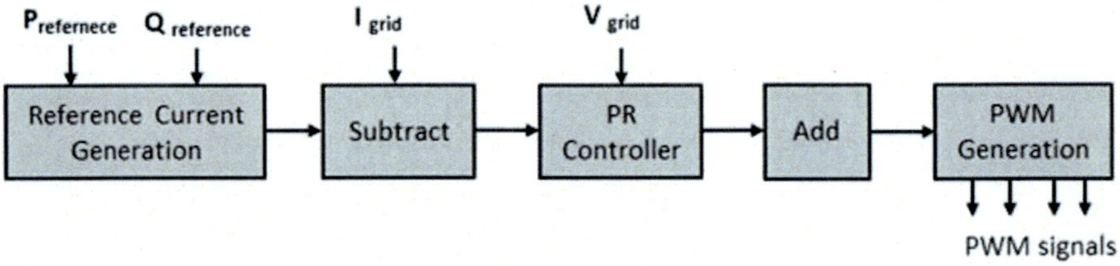

Fig. 8 Reference current generation, PR controller, and PWM generation

where ω represents frequency, k represents gain; V_α and V_β refer to the outputs seen in quadrature signals which are grabbed from the PD detector in SOGI-PLL. A modified AANF-PLL was discussed in [29]. Here, in the present work, the same AANF-PLL was employed for obtaining the grid voltage phase information.

3.6 *Current Control Topology*

A PR controller cannot impart the steady state error, unlike conventional PI controller, while tracking the sinusoidal waveform since it doesn't hold the integral term as seen in PI controllers. The PR controller offers infinitely large gain at the resonance frequency, and all other magnitudes are limited to zero gain and zero phase sift [31].

In a grid-connected PV-wind system, the output voltage imparted from the inverter remains stable. Hence, the power flow to the grid can be controlled. However, mere current needs to maintain a stable value. Different current control network topologies are available in the state of the art [29]. This PR-based current control delivers great response among others, particularly for this application. In the present work, the PR-based controller is designed and implemented to comply with the maximum value of power reference observed from hybrid renewable energy resources with the aid of the P&O MPPT method. The maximum value of power reference obtained from the MPP tracker is used for the generation of reference current needed for the controller (Fig. 8). The comparison between the actual grid current and the error signal feeds the PR controller. The reference voltage signal,

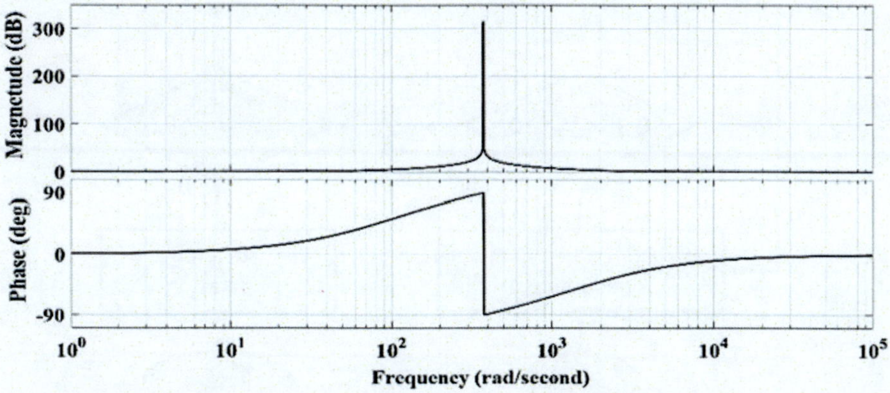

Fig. 9 Magnitude and phase response of PR controller with frequency

along with the PR controller output, is incorporated to instigate the SPWM signal. Figure 9 depicts the magnitude and phase response of the controller. The preliminary transfer function PR controller is explained in Eq. 14:

$$G_{PR} = K_P + K_I \frac{s}{s^2 + \omega_0^2} \tag{14}$$

where K_P and K_I represent the proportional and integral constant, respectively.

3.7 *Passive Islanding*

Figure 10 shows the passive islanding detection algorithm facilitated for this work. Passive islanding detection operates on the grounds of voltage and frequency measurement. Particularly, anti-islanding protection is employed either in the absence of power at the grid side or when a fault occurs at the grid side. Either generated power or grid power continuously drives the load. However, during the absence of power at the grid side, switch S3 should be kept open, and the entire system has to be detached from the grid to safeguard it from malfunction. Hence, during this period, power can be supplied to the load by generating power. On the other hand, if the power is unavailable at the generation side, switch S2 remains open, and the system has to be detached from both load and grid.

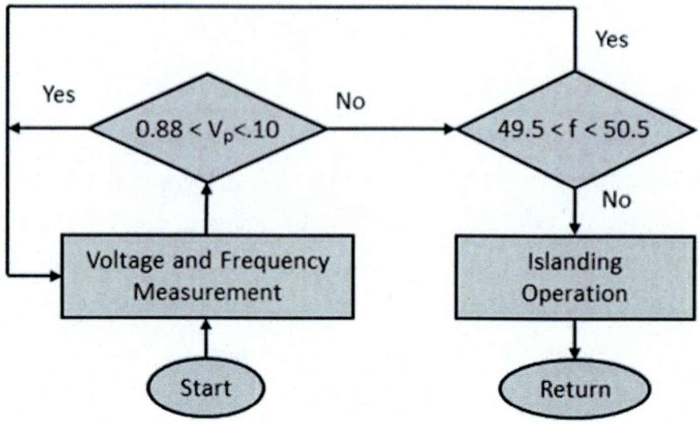

Fig. 10 Passive islanding detection algorithm

4 Results and Discussion

In this work, the proposed MPPT-based grid-tied SPV and WT system is simulated using MATLAB/Simulink. 1 kW system was considered for simulating the results in MATLAB/Simulink environments. The bifurcation of PV and WT was done w.r.t. the availability of the solar and wind data at Chennai, India. The data consists of approximately 67.5% solar data, and the rest (32.5%) is wind data. This combination is often identified as the best combination for modelling hybrid systems [30, 31]. The typical changes in the values of solar and wind power were considered to provide more reliable and stable power, which can further be implemented to execute all operations, viz., grid interfacing, anti-islanding, and the charging and discharging phenomena of batteries. Table 1 displays the information about the changes in solar and wind power as key inputs in percentages. In its first mode, solar input is seen as 0 watts, which means 0%, while the wind power input is seen as 162.5 watt, i.e., 50%. Similarly, in the second mode, solar input seen as 0% and wind input seen as 100%, in the third mode, solar input seen as 50%; wind input seen as 0%, and so on. Figures 11 and 12 represent the MATLAB simulation results of the proposed SPV-WP generation. Figure 11 depicts the information about active and reactive powers generated at the output. Each mode of operation shown in the result indicates the time duration of that mode, i.e., time duration 0.1 *sec* to 0.2 *sec* represents the first mode; duration 0.2 *sec* to 0.3 *sec* represents second mode and so on.

For each mode of operation, the reactive power is maintained at values ranging from zero to a value which imparts a unity power factor (UPF) for the corresponding load. However, as per the demand, it is often needed to supply reactive power after fixing the active power supply at a low value or during the grid recovery period before proceeding to re-synchronization or an islanding event.

Table 1 Variation of solar and wind power as input

Modes of operation	Solar power share (in %)	Wind power share (in %)
1	0	50
2	0	100
3	50	0
4	50	50
5	50	100
6	100	0
7	100	50
8	100	100

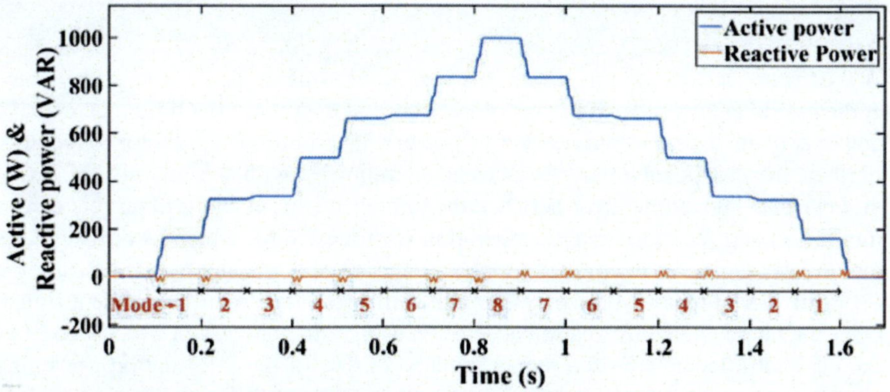

Fig. 11 Real and reactive power supplied to the grid in all modes

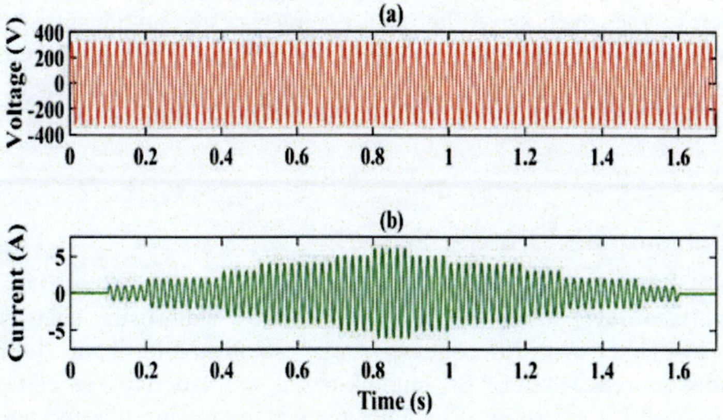

Fig. 12 Voltage and current output of the inverter side in all modes

Figure 12 shows the plots pertaining to the output voltage that remains constant during the operation of the grid itself. The supplied power can be controlled by continuing the changes in the current magnitude. The combination of SPV and

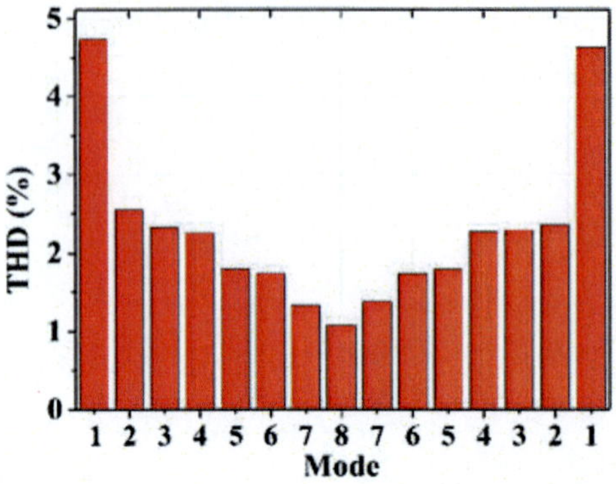

Fig. 13 THD of inverter current in all modes

WT delivers the maximum power that can be used as an active power reference at the hybrid energy side. However, during normal operation, the reference reactive power is fixed at zero to provide active power with a unity power factor to the grid. Further, the computed reference current is compared with the quantity of actual grid current. The output error between the reference current and grid current can be minimized and facilitate inverter operation with the PWM, which in turn supports the inverter to impart maximum supply, which is accomplished by configuring the PR controller to manifest the output of the compared signal with the current signal. Once the phase angle information is detected using grid voltage and a modified PLL, the PR controller imparts the output, which consists of the information pertaining to maximum power and is combined to produce the reference signal, which is compared to the analog triangular pulse to produce PWM waves.

Figure 12 depicts the waveforms representing the output voltage and current of the inverter 1. Once the passive filtering is complete, a step-up transformer is probed to raise the voltage level to around 230 V. From Fig. 13, it is observed that total harmonic distortion (THD) is significantly reduced with the increment in power level.

5 Conclusion and Future Scope

In this work, a novel framework for hybrid energy stem using solar-wind combinations associated with PR controllers was established to obtain the maximum power from sources. MATLAB/Simulink-based simulations were carried out on the design of a 1 kW system. The MPP tracker for finding the maximum power point was simulated to aid the experimentation with solar and wind turbine systems. The proposed model generates a reference value of current to work with a grid-tied inverter. Further, a modified AANF-PLL was incorporated to determine the phase angle information of the grid voltage. Further, the PR controller was instigated to design the control algorithm to manifest the safety of the inverter; in addition, the controlled output power is tested in various conditions. Depending on the value of the reference current, all modes were implemented. The results obtained from the model proved that the grid voltage and current are in phase, and hence, UPF is maintained in the currents and voltages. The performance of the controller and the dynamic behavior of the proposed system have been experimented in a MATLAB/SIMULINK environment.

The future implementation of this work leads to implementation of a hardware model with the aid of an appropriate analog or digital controller and other circuitry. However, the present work doesn't deal with the instability of the PR controller due to infinitely large gain at the resonance frequency. In the future, this can be perfectly tuned and minimized by selecting the appropriate value of damping. Further, the losses incurred by the inverter can be reduced by maintaining a high voltage level at the DC bus. Further, this work can be improved by establishing a multilevel inverter topology accompanied with soft switching, viz., silicon carbide (SiC)-based power switches to enhance the efficiency.

References

1. Monfared, Mohammad, and Saeed Golestan. "Control strategies for single-phase grid integration of small-scale renewable energy sources: A review." *Renewable and Sustainable Energy Reviews* 16.7 (2012): 4982–4993.
2. Quaschning, V., 2016. *Understanding renewable energy systems*. Routledge.
3. Ackermann, T. ed., 2005. *Wind power in power systems*. John Wiley & Sons.
4. Nema, S., Nema, R.K. and Agnihotri, G., 2010. Matlab/simulink based study of photovoltaic cells/modules/array and their experimental verification. *International journal of Energy and Environment*, *1*(3), pp.487–500.
5. Rauth SS, Srinivas K, Kumar M. Grid connected PV/wind single stage converter using PR based maximum power flow control. In 2017 International Conference on Technological Advancements in Power and Energy (TAP Energy) 2017 Dec 21 (pp. 1–7). IEEE.
6. Rauth SS, Kumar M, Srinivas K. A proportional resonant power controller and a combined amplitude adaptive notch filter with pll for better power control and synchronization of single phase on grid solar photovoltaic system. In 2018 International Conference on Smart Systems and Inventive Technology (ICSSIT) 2018 Dec 13 (pp. 378–384). IEEE.
7. Rauth SS, Samanta B. A Grid-Connected Solar Photovoltaic Hybrid System for Reliable Power and Water Supply in Modern Irrigation Application. In Advances in Smart Grid Automation and Industry 4.0 2021 (pp. 493–506). Springer, Singapore.
8. IEA, O., 2015. Energy and climate change, world energy outlook special report.
9. Keyrouz, F., Hamad, M. and Georges, S., 2013, March. A novel unified maximum power point tracker for controlling a hybrid wind-solar and fuel-cell system. In *Ecological Vehicles and Renewable Energies (EVER), 2013 8th International Conference and Exhibition on* (pp. 1–6). IEEE.
10. www.renewableenergyworld.com [Online]
11. http://www.electronics-tutorials.ws/transistor/tran_7.html [Online]
12. Lewis, N.S., 2007. Toward cost-effective solar energy use. *Science*, *315*(5813), pp. 798–801.
13. Islam, M., Mekhilef, S. and Hasan, M., 2015. Single phase transformerless inverter topologies for grid-tied photovoltaic system: A review. *Renewable and Sustainable Energy Reviews*, *45*, pp. 69–86.

14. Rodríguez, J., Bernet, S., Wu, B., Pontt, J.O. and Kouro, S., 2007. Multilevel voltage-source-converter topologies for industrial medium-voltage drives. *IEEE Transactions on industrial electronics, 54*(6), pp. 2930–2945.
15. Rodriguez, J., Lai, J.S. and Peng, F.Z., 2002. Multilevel inverters: a survey of topologies, controls, and applications. *IEEE Transactions on industrial electronics*, *49*(4), pp. 724–738.
16. Basso, T.S. and Deblasio, R.D., 2003, September. IEEE P1547-series of standards for interconnection. In *Transmission and Distribution Conference and Exposition, 2003 IEEE PES* (Vol. 2, pp. 556–561). IEEE.
17. Jana, J., Saha, H. and Bhattacharya, K.D., 2017. A review of inverter topologies for single-phase grid-connected photovoltaic systems. *Renewable and Sustainable Energy Reviews, 72*, pp. 1256–1270.
18. Inverters, C., 2005. Controllers and interconnection system equipment for use with distributed energy resources. *UL Std, 1741*.
19. IEEE. IEEE Recommended Practice for Utility Interface of Photovoltaic (PV) Systems. IEEE; 2000.
20. I F II. Ieee recommended practices and requirements for harmonic control in electrical power systems; 1993.
21. IEC. Iec 62109-1: Safety of power converters for use in photovoltaic power systems-part i: General requirements; 2010.
22. Solanki, C.S., 2015. *Solar photovoltaics: fundamentals, technologies and applications*. PHI Learning Pvt. Ltd..
23. Eggleston, D.M. and Stoddard, F., 1987. Wind turbine engineering design.
24. Femia, N., Petrone, G., Spagnuolo, G. and Vitelli, M., 2005. Optimization of perturb and observe maximum power point tracking method. *IEEE transactions on power electronics, 20*(4), pp. 963–973.
25. Jaalam, N., Rahim, N.A., Bakar, A.H.A., Tan, C. and Haidar, A.M., 2016. A comprehensive review of synchronization methods for grid-connected converters of renewable energy source. *Renewable and Sustainable Energy Reviews*, *59*, pp. 1471–1481.
26. Yin, G., Guo, L. and Li, X., 2013. An amplitude adaptive notch filter for grid signal processing. *IEEE Transactions on Power Electronics*, *28*(6), pp. 2638–2641.
27. Xiao, F., Dong, L., Li, L. and Liao, X., 2017. A frequency-fixed SOGI-based PLL for single-phase grid-connected converters. *IEEE Transactions on Power Electronics*, *32*(3), pp. 1713–1719.
28. Guo, X.Q., Wu, W.Y. and Gu, H.R., 2011. Phase locked loop and synchronization methods for grid-interfaced converters: a review. *Przeglad Elektrotechniczny*, *87*(4), pp. 182–187.
29. Rauth S.S., Kumar M., Srinivas K., "A Proportional Resonant Controller with Adaptive Notch Filter Based Phase Locked Loop for Single Phase Grid Connected Solar Photovoltaic System Using a Diode Based Isolation" unpublished.
30. https://www.researchgate.net/deref/http%3A%2F%2Fwww.nrel.gov%2Frredc%2Fsmarts%2F [Online].
31. https://www.researchgate.net/deref/http%3A%2F%2Fwww.wunderground.com%2Fhistory%2F [Online].
32. Long, H., Eghlimi, M. and Zhang, Z., 2017. Configuration Optimization and Analysis of a Large-Scale PV/Wind System. *IEEE Transactions on Sustainable Energy*, *8*(1), pp. 84–93.
33. Busada, C.A., Jorge, S.G. and Solsona, J.A., 2017. Resonant current controller with enhanced transient response for grid-tied inverters. *IEEE Transactions on Industrial Electronics*.
34. Yang, Z., Duan, Q., Zhong, J., Mao, M. and Xun, Z., 2017, May. Analysis of improved PSO and perturb & observe global MPPT algorithm for PV array under partial shading condition. In *Control And Decision Conference (CCDC), 2017 29th Chinese* (pp. 549–553). IEEE.

A Study on Benchmarks for Ectopic Pregnancy Classification Using Deep Learning Based on Risk Criteria

Lakshmi R. Suresh and L. S. Sathish Kumar

1 Introduction

In the past few decennia, there has been a drastic increase in the number of EPs. Implantation along with the zygote's subsequent development at a place other than the regular intrauterine cavity is termed EP, or extrauterine pregnancy [1]. Mostly, women aged more than 35 years and those of non-white races are detected with EPs [2]. Postoperative implantation is engendered by viable sperm or a pre-implanted fertilized ovum that is considered to be available in the fallopian tube (FT) [3]. 1.3% of EPs emerge in abdominal locations, and the leftover is seen in the ovary, cervix, or abdominal cavity, whilst the majority of the EPs (95.5%) arise in the FT [4]. One of the major usual gynecologic surgical emergencies is EPs [5]. Complexities like extreme hemorrhage, shock, or renal failure are brought about by undiagnosed or untreated EP [6]. A history of EP, a history of pelvic inflammatory disease (PID), the utilization of an intrauterine device, and a history of tubal surgery are all high-risk factors for EP. Over the past 20 decennia, the occurrence of EP has augmented quickly; however, at the same time, mortality has decreased. It is largely due to ameliorated diagnostic gauges [7]. The difference between Normal Pregnancy (NP) and EP is portrayed in Fig. 1.

It is unusual to diagnose tubal rupture based on its clinical signs (hypovolaemic shock). Screening women at risk has resulted EP's earlier detection with progres-

L. R. Suresh (✉)
School of Computing Science and Engineering, VIT Bhopal University, Bhopal, Madhya Pradesh, India

L. S. S. Kumar
Department of Gaming Division, School of Computing Science and Engineering, VIT Bhopal University, Bhopal, Madhya Pradesh, India
e-mail: sathish.kumar@vitbhopal.ac.in

M. Mishra et al. (eds.), *Applications of Computational Intelligence in Management & Mathematics*, Springer Proceedings in Mathematics & Statistics 417,
https://doi.org/10.1007/978-3-031-25194-8_9

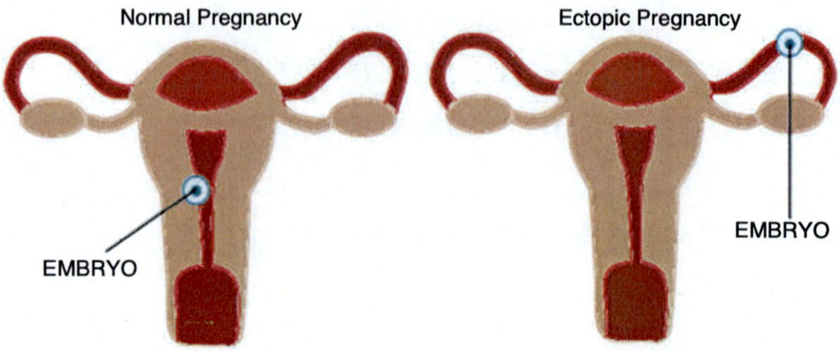

Fig. 1 Normal and ectopic pregnancies with implanted eggs

sions in imaging technology together with protocols [8]. Blood human chorionic gonadotropin (hCG) determination and laparoscopy, the existing diagnosis, along with EP's treatment has been significantly enhanced with the extensive application of transvaginal ultrasound (TVS) [9]. The chief diagnostic tool for medically stable women with an alleged EP is TVS [10]. Women devoid of evidence of an intrauterine pregnancy (IUP) or an EP subsequent to assessment with TVS are said to have a PUL. Out of all the initial pregnancies, about 8–31% are categorized as PUL. Neither IUP nor EP is confirmed or discarded in the event of PUL [11]. Nevertheless, when analogizing the hCG values, an augmented number of visits may augment the accuracy of the diagnosis. Particularly during pregnancy, the utilization of AI to ameliorate women's health has restricted its clinical utilization [12]. Treatment options encompass medical, surgical, or expectant management (EM) subsequent to a definite diagnosis [13]. Elimination of ectopic tissue may be performed by employing medication, laparoscopic surgery, or abdominal surgery when the EP is found. It still sustains to be a grave situation despite advances in diagnostic techniques. Owing to EP, there are about 75% of fatalities in the first trimester, along with 9% of all pregnancy-associated fatalities [14]. Whilst quick diagnosis and management are critical, several diagnostic drawbacks can engender negative results, which all are brought about by an accurately carried out and interpreted pelvic ultrasound [15].

The leftover work is arranged as follows: The related work is exhibited in Sect. 2. The current research is elucidated in Sect. 3. The outcomes and discussion of the current research are explicated in Sect. 4. Lastly, the conclusion is deduced in Sect. 5.

2 Literature Review

Heliza Rahmania Hatta et al. [16] Experts were intended for constructing web-centered expert systems (ES), namely, a doctor or a patient to identify pregnancy from anywhere; hence, it could assist women to be aware of the pregnancy disorder's symptoms. The system can't exist in the clinical decision support system (DSS) group.

ES was constructed by Dian Sa'adillah Maylawati et al. [17] utilizing an artificial neural network and back propagation algorithm to forecast pregnancy with illnesses earlier. An accuracy of about 78.248% was obtained when the ANN was deployed to forecast the pregnancy disorders, as per the training and testing process.

Maethaphan Kitporntheranunt & Watcharachai Wiriyasuttiwong [18] created a software-centered medical ES (MES) to bolster the diagnosis of EP. It surveyed 32 women's medical records that were diagnosed with EP. For assisting the decision-making in the EP's diagnosis, this MES was discovered to be pre-eminent.

Gudu et al. [19] Until now, the D&T ES for hypertension in pregnancy has been at the testing stage of its life cycle and has not been carried out. As the number of knowledge collections increased, the detection of knowledge quickly turned out to be a very intricate challenge.

Alberto De Ramón Fernández et al. [20] constructed a DSS centered on a three-stage classifier. The classifiers had been executed and tested centered on "four" diverse help algorithms: multilayer perceptron, DL, support vector machine (SVM), and Naive Bayes.

Ploywarong Rueangket et al. [21] For EP result forecasting, in a conservative cohort analysis on 377 pregnancies of women from unspecified locations, training along with verification was performed utilizing a group of 22 features. The investigation was done by "three" diverse ML models like neural networks (NNs), DTs, and SVMs.

3 Benchmarks for Ectopic Pregnancy Classification Using Deep Learning

An evaluation of an amalgamation of features encompassing the woman's symptoms and hCG levels along with ultrasound findings is frequently mandated for the EP's identification. The diagnosis is contingent on an amalgamation of ultrasound scanning and serial serum beta-hCG (β-hCG) measurements in developed countries in recent practice. One of some medical situations that can be handled hopefully, medically or surgically is EP.

3.1 Risk Factor in Ectopic Pregnancy

For EP, the circumstances that harm the tube's integrity and damage these tasks are termed RF. Earlier EP, history of infertility, pregnancy conceived with IVF, or intraoperative recognition of contralateral tubal pathology (TP) are the chief RF that are noticed in the research. To distinguish betwixt women who are in a low-risk, moderate-risk, or high-risk category, most of the study on EP has concentrated on these RF for assessment [22]. The RF for EP is enumerated in Table 1.

Table 1 Risk factors for EP

Category of risk	Risk factors	Odd ratio
High Risk	Earlier EP	9.3–4.7
	Earlier tubal surgery	6.0–11.5
	Tubal ligation	3.0–13.9
	TP	3.5–25
	In utero DES exposure	2.4–13
	Present IUD utilization	1.1–45
Moderate	Infertility	1.1–28
	Earlier cervicitis	2.8–37
	History of PID	2.1–3.0
	Manifold sexual partners	1.4–4.8
	Smoking	2.3–3.9
Low Risk	Earlier pelvic surgery (PS)/abdominal surgery	0.93–3.8
	Vaginal douching	1.1–3.1
	Early age of intercourse	1.1–2.5

The risk of EPs is unavoidably augmented by these few factors. Due to the augmentation in RF, the EP's frequency has augmented. The occurrence is 3% in pregnancies with medically assisted procreation along with 10% in patients with earlier EPs in subpopulations at risk. EP is responsible for 4–10% of the entire maternal deaths.

3.2 *Types of Ectopic Pregnancy and Their Locations*

Centered on the regions where the embryo is implanted, there are several kinds of EP. This could happen in any of the reproductive system's structures and even outside them, as per a few adepts. It is categorized as per the location wherein the embryo is implanted [23]. The FT is the most usual location for an EP. However, it may also be available in other locations. The diverse EP locations are explicated in Table 2.

Regarding the position of EP, there are nine kinds of EP that are taken. The rate of tubal EP still remains greater than the rate of other EP with impulsive pregnancy amongst all, wherein 75–80% of EPs happen in the ampullary portion, 10–15% of EPs happen in the isthmic portion, and around 5% of EP is in the FT's fimbrial end. The rarest variants of all EP are cervical EP, ovarian EP, and scar EP. Furthermore, another form of EP is the PUL, which is categorized when there is a positive pregnancy test; however, no droop is discovered even with transvaginal ultrasonography.

Table 2 Ectopic pregnancy locations

Types of EP	Locations	Possibility of occurrence (%)
Tubal pregnancy	Exists in the FT, ambullary, isthmic, or fimbrial regions of the tube	90–95
Interstitial pregnancy	Implants occur in the FT's interstitial portion	2.5
Abdominal pregnancy	Growth of implants in peritoneal surface, tubal lumen	1.3
Cervical pregnancy	Happens inside the cervical canal	0.15
Scar pregnancy	Gestational sac (GS) implanted in the myometrium at the place of an earlier cesarean section	6
Heterotopic pregnancy	EPs conjuction with IUP	1–3
Ovarian pregnancy	It is seen inside the ovarian cortex	0.15–3

3.3 *Symptoms Associated with Ectopic Pregnancy*

EP's symptoms are quite the same as those of a uterine pregnancy. The risk of the pregnancy being ectopic during an earlier pregnancy is very high. The following are the EP's symptoms.

Frequently, on one side of the body, there is less stomach pain

- When analogized to the normal period, vaginal bleeding (VB) may be dark, watery, heavier, lighter, or highly continuous.
- Pregnancy symptoms, namely, a missed menstrual period, breast tenderness, recurrent urination, or nausea.
- While defecating or urinating, bowel and bladder issues like diarrhea and pain occur.
- When lying down, a feeling of fullness that is not related to eating, especially in a person who has already had a child.
- Back pain.

Until the rupture of the FT or closed organs, it is feasible to possess an EP devoid of undergoing symptoms. Tube rupture is engendered when the fertilized egg keeps on developing. Severe stomach ache, VB, light-headedness, fainting, or shoulder pain are the symptoms that are entailed.

3.4 *Diagnosis*

A pregnancy test and a pelvic exam are frequently encompassed in the EP tests and diagnosis, which may include an ultrasound to view the uterus and FTs. Ultrasound, hCG test, and laparoscopy (keyhole surgery) are the most generally used diagnostic tests. The beta sub-unit of β-hCG discriminatory level may be helpful if transvaginal

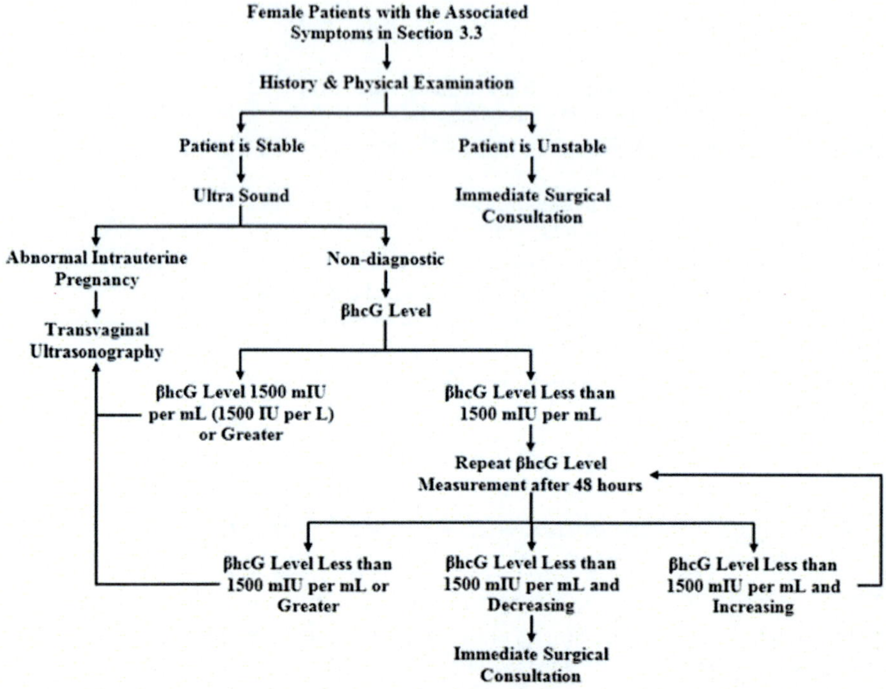

Fig. 2 Levels of diagnosing the suspected ectopic pregnancy

ultrasonography is nondiagnostic. The levels of diagnosis for suspected EP using ultrasound and β-hCG levels are elucidated in Fig. 2.

The first-trimester hCG concentration augments, doubling about every "two" days in an NP. In the event of EP, serial measurement is most helpful for corroborating fetal viability. Laparoscopy is another alternative utilized to confirm the diagnosis if hCG and ultrasound results are unclear.

3.5 Treatment

The EP's pregnancy can be disintegrated naturally, devoid of any interference. Treatment for EP includes methotrexate (MTX) therapy, open or laparoscopic surgery, or EM. Instant surgical treatment is designated for patients who are medically unsteady or undergoing hemorrhage (Table 3).

The growth of the embryo would be stopped by injecting a dose of MTX [24]. Salpingectomy or salpingostomy are encompassed in the surgical options. Only when the other FT looks healthy, the whole tube is eliminated, or else an attempt is made to eliminate the pregnancy.

Table 3 Methotrexate treatment protocols

Protocol	Single dose	Multiple dose
Medication	50 mg per square meter of body surface MTX Intramuscular (IM)	Alternate each other day: 1 mg per kg MTX IM along with 0.1 mg per kg leucovorin
Laboratory values	Liver function tests (LFTs), Complete blood count (CBC), along with renal function at baseline, day 4, and day 7	LFTs, CBC, along with renal function at baseline β-hCG at baseline, day 1, day 3, day 5, and day 7 until levels reduce
Repeat Medication	If the β-hCG level doesn't reduce by 15% betwixt day 4 and day 7, Recur regimen	If the β-hCG level doesn't reduce by 15% with every measurement, recur regimen (for up to '4' doses of every medication)
Follow-up	β-hCG level weekly as well as continue regimen until no longer identified.	β-hCG level weekly, as well as continue regimen until no longer identified.

3.6 Deep Learning in EP

ML has a huge impact on the medical sector owing to its capability to perform intricate tasks automatically. To attain the goals of EP classification and access to care, ML could assist clinicians. The computational models that attempt to resolve issues that can't be resolved with statistical approaches are called ML algorithms.

4 Result and Discussion

By utilizing univariate analysis, the data of 377 patients were examined. Also, specific factors like ultrasound findings, symptoms, and RP of EP were studied [21]. The final diagnosis was 200 EPs along with 177 non-EPs amongst 377 patients whose initial visit was recognized as PUL.

A total of 347 pregnant women with supposed PUL at early diagnosis are scrutinized in Table 4. Centered on the features of ultrasound findings, namely, intrauterine anechoic content, endometrial thickness, which should be below 14 mm, the adnexal mass of intricate echogenicity, and FF in CDS, the investigation was performed. 53% were EP, and 47% were non-EP amongst the 377 patients. Figure 3a portrays the pictorial depiction of ultrasound findings.

Intricate adnexal mass along with FF in the CDS by early ultrasound findings was highly close in EP patients, which was 46.41% (199 patients) vs. 7.69% (177 patients) and 33.15% (195 patients) vs. 8.22% (175 patients), correspondingly, when analogized to patients with no EP. The intrauterine anechoic content was noticed to be extremely less in EP and non-EP patients, other than the entire features.

For 270 (71.61%) patients, abnormal VB was present, of which 39.78% were diagnosed as EP and 38.46% were diagnosed as non-EP. Only 24–43 (6.36–11.40%)

Table 4 Ultrasound findings

Characteristics	Total ($n = 377$)	
	EP	Non-EP
Intrauterine anechoic content	197	177
Endometrial thickness >14 mm	183	172
Adnexal mass of complex echogenicity	199	177
Free fluid (FF) in cul-de-sac (CDS)	195	175

patients have been noticed with the conventional triad of nausea, faintness, and cervical motion tenderness. The major symptom is abdominal pain, which happened in 303 patients, of which 46.15% of EP and 34.21% of non-EP patients have been diagnosed.

Centered on the symptoms, like abdominal pain, abnormal VB, nausea, vomiting, faintness, abdominal tenderness, and cervical motion tenderness, the patient's data have been scrutinized in Table 5. The abdominal ache was the most usual exhibiting complaint noticed in 303 (80.37%) patients. Figure 3b displays the diagnosed percentage of EP and non-EP. Table 5 examines the EP's occurrence in patients with the RF's features. History of PS, smoking, EP's history, PID's history, current utilization of the emergency pill, and assisted reproductive technology are the features that are regarded. In the majority of the patients, the PS's history is noticed, which is 283. Figure 4 displays the pictorial depiction of this investigation.

The RF like an earlier history of EP, a history of EP, along with a history of PID were noticed in 17 patients (4.50%), 15 patients (3.97%), and 13 patients (3.44%), respectively. 71.85% of patients with EP were nonsmokers, as well as only 8.75% were in the present use of the emergency pill. The performance of machine algorithms utilized in diagnosing EP and non-EP is scrutinized in Fig. 4b. The metrics like sensitivity and specificity were employed for this investigation. The logistic regression (LR) sensitivity and the DT techniques' specificity are analogized to the other methodologies in Fig. 4b. ML exhibits superior performance in the prediction process of EP, as per the investigation.

5 Conclusion

For diagnosing the EP when carried out and elucidated by skilled sonographers and radiologists, TVS is more sensitive. The incidence of EP when ultrasound assessment doesn't detect an intrauterine GS along with β-hCG level is more than 1500 mIU per mL is suggested by the serum hCG level in women with PUL. The benefit of utilizing the DL in the medical field includes finding EP and these patient states (i.e., pregnancy phenotypes), and it also could aid in ameliorating unfavourable findings and proffer further insight into diseases during pregnancy.

Table 5 Symptoms and risk of EP

Symptoms of EP			Risk factors for EP		
Characteristics	Total ($n = 377$)		Characteristics	Total ($n = 377$)	
	EP	Non-EP		EP	Non-EP
Abdominal pain	195	177	History of PS	198	163
Abnormal VB	200	177	Smoking	199	161
Nausea, vomiting	165	116	History of EP	199	175
Faint	167	120	History of PID	193	140
Abdominal tenderness	200	177	Present utilization of an emergency pill	196	154
Cervical motion tenderness	197	177	Assisted reproductive technology	199	174

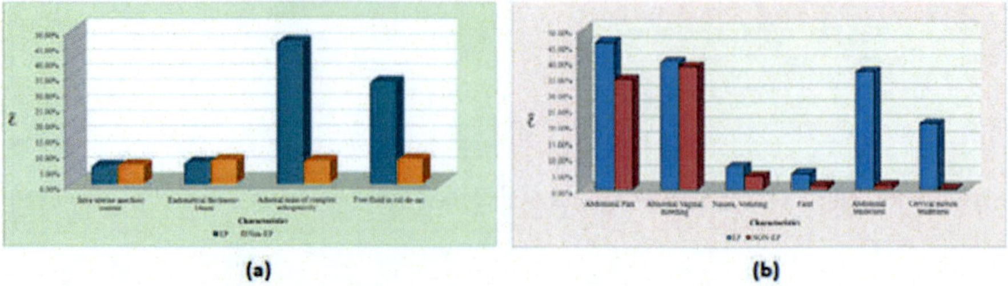

Fig. 3 (**a**) Ultrasound findings. (**b**) Symptoms of EP

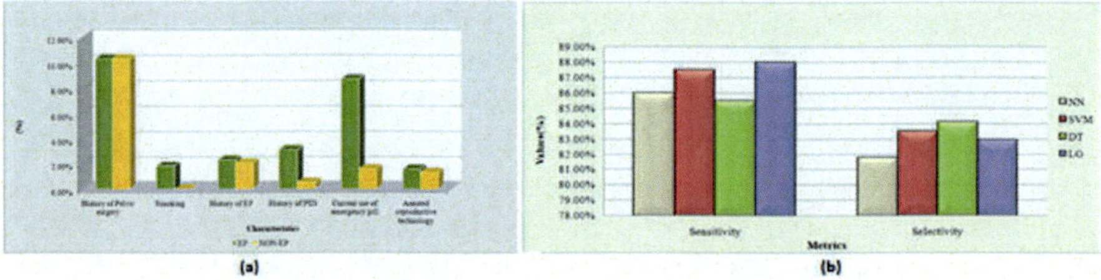

Fig. 4 (**a**) RF of EP (**b**) Performance analysis of machine learning techniques

References

1. Kathpalia, S. K., D. Arora, Namrita Sandhu, Pooja Sinha,: Ectopic pregnancy: Review of 80 cases. Medical Journal Armed Forces India 74(2), 172-176 (2018).
2. Kang, Ok Ju, Ji Hye Koh, Ji Eun Yoo, So Yeon Park, Jeong-Ik Park, Songsoo Yang, Sang-Hun Lee et al,: Ruptured Hemorrhagic Ectopic Pregnancy Implanted in the Diaphragm: A Rare Case Report and Brief Literature Review. Diagnostics 11(12), 2342 (2021).
3. Shao, Emily X., Kendra Hopper, Matthew McKnoulty, Alka Kothari,: A systematic review of ectopic pregnancy after hysterectomy. International Journal of Gynecology & Obstetrics 141(2), 159-165 (2018).
4. Kirk, Emma, Cecilia Bottomley, Thomas Bourne,: Diagnosing ectopic pregnancy and current concepts in the management of pregnancy of unknown location. Human reproduction update 20(2), 250-261 (2014).
5. Ozcan, Meghan CH, Jeffrey R. Wilson, Gary N. Frishman,: A systematic review and meta-analysis of surgical treatment of ectopic pregnancy with salpingectomy versus salpingostomy. Journal of Minimally Invasive Gynecology 28(3), 656-667 (2021).
6. Robertson, Jennifer J., Brit Long, Alex Koyfman,: Emergency medicine myths: ectopic pregnancy evaluation, risk factors, and presentation. The Journal of emergency medicine 53(6), 819-828 (2017).
7. Cacciatore, B., Ylostalo, P., Seppali, M.: Early diagnosis of ectopic pregnancy. 1st Edition. Springer London, ISBN: 978-1-4471-1987-6 (1994).
8. Alur-Gupta, Snigdha, Laura G. Cooney, Suneeta Senapati, Mary D. Sammel, Kurt T. Barnhart,: Two-dose versus single-dose methotrexate for treatment of ectopic pregnancy: a meta-analysis. American journal of obstetrics and gynecology 221(2), 95-108 (2019).
9. OuYang, Zhenbo, Shiyuan Wei, Jiawen Wu, Zixian Wan, Min Zhang, Biting Zhong,: Retroperitoneal ectopic pregnancy: A literature review of reported cases. European Journal of Obstetrics & Gynecology and Reproductive Biology 259 ,113-118 (2021).

10. Ghaneie, Ashkan, Joseph R. Grajo, Charlotte Derr, Todd R. Kumm,: Unusual ectopic pregnancies: sonographic findings and implications for management. Journal of Ultrasound in Medicine 34(6), 951-962 (2015).
11. Fields, Loren, Alison Hathaway,: Key Concepts in Pregnancy of Unknown Location: Identifying Ectopic Pregnancy and Providing Patient-Centered Care. Journal of Midwifery & Women's Health 62(2), 172-179 (2017).
12. Davidson, Lena, Mary Regina Boland,: Towards deep phenotyping pregnancy: a systematic review on artificial intelligence and machine learning methods to improve pregnancy outcomes. Briefings in bioinformatics 22(5) ,1-29 (2021).
13. Joshua H Barash, Edward M Buchananand Christina Hillson,: Diagnosis and managementof ectopic pregnancy. American Family Physician 90(1), 34-40 (2014).
14. Poonam Rana, Imran Kazmi, Rajbala Singh, Muhammad Afzal, Fahad A Al-Abbasi, Ali Aseeri, Rajbir Singh, Ruqaiyah Khan, Firoz Anwar,: Ectopic pregnancy a review. Archives of Gynecology and Obstetrics 288(4), 747-757 (2013).
15. Mausner Geffen E, Slywotzky C, Bennett G,: Pitfalls and tips in the diagnosis of ectopicpregnancy. Abdominal Radiology 42(5), 1524-1542 (2017).
16. Heliza Rahmania Hatta, Fadhilah Ulfah, Khairina D. M, Hamdani Hamdani, Santy Maharani,: Web-expert system for the detection of early symptoms of the disorder of pregnancy using a forward chaining and bayesian method. Journal of Theoretical and Applied Information Technology 95(11), 2589-2599 (2017).
17. Dian Sa'adillah Maylawati, Muhammad Ali Ramdhani, Wildan Budiawan Zulfikar, Ichsan Taufik, Wahyudin Darmalaksana,: Expert system for predicting the early pregnancywith disorders using artificial neural network. 5th International Conference on Cyber and IT Service Management (CITSM), pp. 8–10. IEEE Conference, Denpasar, Indonesia, (2017).
18. Maethaphan Kitporntheranunt, Watcharachai Wiriyasuttiwong,: Development of a medical expert system for thediagnosis of ectopic pregnancy. Journal of the Medical Association of Thailand 93,43-49 (2010).
19. Gudu J, Gichoya D, Nyongesa P, Muumbo A,: Development of a medical expert system as an expertknowledge sharing tool on diagnosis and treatment of hypertension in pregnancy. International Journal of Bioscience, Biochemistry and Bioinformatics 2(5), 297-300 (2012).
20. Alberto De Ramon Fernandez, Daniel Ruiz Fernandez, Maria Teresa Prieto Sanchez,: A decision support system for predicting the treatment of ectopicpregnancies. International Journal of Medical Informatics 129, 198-204 (2019).
21. Ploywarong Rueangket, Kristsanamon Rittiluechai,: Predictive analytical model for ectopic pregnancy diagnosis statistics vs machine learning methods. Peer Reviewed Journal (Preprint), (2022).
22. Antonio Ragusa, Alessandro Svelato, Mariarosaria Di Tommaso, Sara D'Avino, Denise Rinaldo and Isabella Maini,: Updates in the management of Ob-Gynemergencies. 1stEdition, Springer Cham, ISBN: 978-3-319-95113-3 (2019).
23. Ioannis Tsakiridis, Sonia Giouleka, Apostolos Mamopoulos, Apostolos Athanasiadis, Themistoklis Dagklis,: Diagnosis and management of ectopic pregnancy a comparative review ofmajor national guidelines. Obstetrical and Gynecological Survey 75(10), 611-623 (2020).
24. Anne-Marie Lozeau, Beth Potter,: Diagnosis and managementof ectopic pregnancy. American Family Physician 72(9),1707-1714 (2005).

An Ontology-Based Approach for Making Smart Suggestions Based on Sequence-Based Context Modeling and Deep Learning Classifications

Sunitha Cheriyan and K. Chitra

1 Introduction

Sustainability goals (SDGs) appear to concern themselves with encouraging sustainable models of production and consumption patterns. The notion of sustainable tourism, which supports sustainable development and promotes local culture, is, nevertheless, also a key component of this global aim when one digs further into its linked targets. Many ideas, such as smart cities and smart structures, have been given the label "smart." ICT-related travel-related services and platforms have recently entered the tourist industry, paving the way for simpler travel in terms of trip planning and administration as well as local environment monitoring. ICT use in the tourism sector has spawned the concept of "smart tourism" [1], which emphasizes human-computer interactions to enhance consumer decision-making and information processing for effective service delivery across the industry. ICT integration has long been considered logical for advancement from traditional travel ideas coupled with advances of technology in several tourist-related activities. Smart tourism technologies (STTs) may gather and utilize data from physical infrastructures and portable devices to enhance travel experiences by providing a helpful feedback mechanism and promoting destinations via a variety of channels, including social media. Travel agency, destination, and mobile app websites and applications are the most widely used STTs. These STTs can facilitate planning and, consequently, the decision-making process by interactively providing essential information about the site. If users are knowledgeable about the people, places, and leisure activities that

S. Cheriyan (✉)
Department of Computer Science, Madurai Kamaraj University, Madurai, India

K. Chitra
Department of Computer Science, Government Arts College, Melur, India

M. Mishra et al. (eds.), *Applications of Computational Intelligence in Management & Mathematics*, Springer Proceedings in Mathematics & Statistics 417,
https://doi.org/10.1007/978-3-031-25194-8_10

are offered where they wish to use the application, they will have a better experience. By estimating the whole cost of travel and sightseeing using STTs, customers may reduce the probability of accruing unforeseen supplemental costs while travelling.

Recommender systems are systems that can forecast a consumer's future preferences for a group of things and suggest the best among them. The massive quantity of information that can be maintained in recommender systems, such as user information, preferences, situational factors, points of interest (POIs), and so on, necessitates the use of increasingly complicated knowledge. To date, a few strategies have been suggested for accomplishing this goal. People nowadays utilize recommender systems more often in numerous aspects of daily life. Recommender systems make it easier to decide which items to buy, which movies to watch, and which books to read, among other things. They assist users in saving time while choosing while also increasing satisfaction. People are beginning to utilize numerous applications and websites that assist them locate destinations (POI) such as restaurants, bars, sports clubs, and others as the mobile Web develops. In this case, a good POI recommender system might help with the research by providing personalized recommendations of the best POIs based on the user's past behavior. When a user moves to a new city or country and has minimal understanding of the region, personalized POI suggestions are very relevant and valuable. The goal of this research is to provide new insights into POI-based recommender systems. This assists users in discovering new and intriguing attractions in their local city or in the area they want to visit. Almost all user recommendations will be tailored to the individual's tastes. The model will employ a cross technique, which combines many methodologies for developing a recommender system. Combining many methods into one is a useful technique since it helps you to minimize the flaws in each method that will be employed in the final algorithm. There are five parts to this paper. The first section describes what a recommender system is, the diverse types of recommender systems, and the methods that are utilized for developing recommender systems. The second section discusses the review of various research published on the topic and various domains of knowledge. The third section focuses on examining existing POI recommender systems and methods which you can use for building POI recommender systems. The fourth section discusses the results of the investigation, followed closely by the conclusion and future directions in this research.

Filtering strategies like content-based and collaborative require user information that isn't always available. It is difficult or impossible to have access to user interaction history or other user-related data. The traditional techniques would not be effective in such situations. In this situation, sessions can be used to handle interactions between a single, anonymous user and the system. Session-based recommendation techniques [2] are recommendation strategies that are totally dependent on the user's behaviors within a current session and adjust their recommendations accordingly. Each interaction that happens during a session is spread out over time. It may be for just 1 day, a few days, a week, or even months. A session, such as looking for a restaurant for tonight's dinner, listening to music in a certain style or mood, or deciding where to go on vacation,

Fig. 1 Effective website from the marketing effectiveness perspective

generally includes a time restriction. Context is a variable that earlier techniques overlooked. The circumstances in which a user communicates a want, preference, or requirement have a significant impact on their behavior and expectations. The built application should take this information into consideration and deliver accurate recommendations that are appropriate for the user's present situation. Therefore, context must be considered in recommender systems since it may have a subtle but significant impact on user behaviors and requirements. Although the user may have just one account, the behavior, and preferences of the logged-in user when surfing the site are dictated by the individual demands of the logged-in user. The user may find it useful to view suggestions of books that could be of interest to them when they are purchasing a product for others, but it would be more efficient to obtain ideas that are tailored to their present requirements based on recent purchases.

Users communicate and interact within a certain situation or scope, and their choices for various items may vary depending on the environment. A context-independent representation may lose its predictive power because potentially pertinent information from several contexts is frequently combined in application areas. Interactions between customers and things are multifaceted in a more formal sense. User preferences might change based on the situation and are often unpredictable. Almost anything may be regarded as context, including the time of year, the day of the week, the type of electronic gadget the user is using, and their mood. In terms of marketing efficacy perspective, the figure illustrates how successful website recommendations are using the Fishbone diagram (see Fig. 1).

2 Literature Review

Discussion Several sequence-aware recommendation issues [3] have been effectively addressed using sequence modelling methodologies. Our analysis, however, demonstrates that due to issues with data sparsity and computational complexity, for sequence-aware recommendation issues, Markov models often cannot be applied intuitively and directly. As a result, researchers typically use certain model variations or integrate heuristics into the study process. It is not always obvious, however, whether these model modifications apply to issues in the actual world.

In contrast, deep learning-based methods have recently received more attention. The current revival of neural networks [4] has been influenced by several factors, including the availability of enormous data sets for training and the improved processing power of contemporary computer technology.

Some of the earliest deep learning-based methods used today are occasionally just marginally superior to methods that need less processing power [5]. More study in this field is crucial since naïve approaches may have drawbacks like a bias toward recommending well-liked things that are not considered by modern problem formulations and typical performance indicators such as recall and accuracy.

A knowledge-based recommendation system with ontologies and semantics has been developed by Abhinav Agarwal et al. [6] for the promotion of Massive Open Online Courses (MOOC) platforms for use in electronic learning in 2022.

To help those who struggle with obesity, Ontology-based Dietary Recommendations (ODR) have axioms to represent different dishes and ingredients, and several diets were proposed by Dexon Mckensy et al. [7].

A thorough analysis of travel and related factors was shown in the survey by Kinjal et al. [8].

The paper by Christoph Trattner et al. [9], depending on weather data, recommended POIs. To do this, they have increased the number of meteorological-related variables, like temperature, cloudiness, moisture, and rainfall severity, in the cutting-edge Rank-GeoFM POI recommender algorithm.

In the article, Zhe Fu et al. [10] suggested TRACE (Travel Reinforcement Recommendations Based on Location-Aware Context Extraction), a location-aware recommender system that incorporates users' history preferences and current context. Using a location-aware context learning model and reinforcement learning; it extracts user characteristics and generates dynamic suggestions.

3 The Recommendation Model

3.1 *User Profile and Context Information*

This approach uses the hybrid recommendation system to generate POI recommendations depending on the user profile and relevant contextual data. Context enhances the information and sets the user and the item domain as follows: A rating function R_c is incorporated in the construction of the algorithm and encompasses the users and the item (POI) as the domain:

$$R_c = \text{Users} \times \text{Items (POI)} \times \text{Context} \tag{1}$$

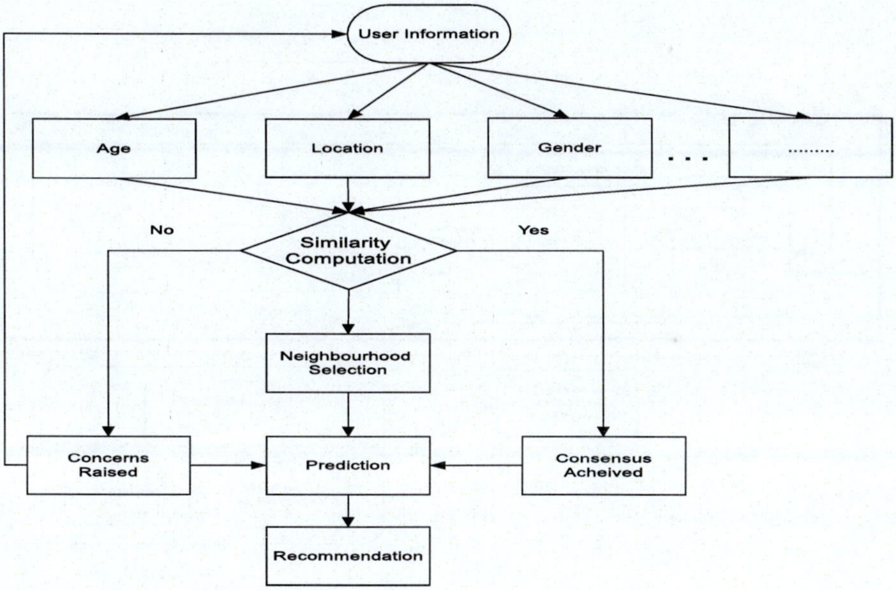

Fig. 2 Decision-making in recommendations

Recommender systems take advantage of the enormous quantity of user data available to provide efficient suggestions to the user. The profile data is analyzed, the similarity measure is calculated, and an appropriate decision and recommendation are generated as shown in the picture (see Fig. 2).

3.2 Situation or Context Awareness

Over the past 20 years, the psychology and HF communities have paid close attention to the SA construct. Although the military aviation field provided the first motivation for studying the construct, it has subsequently expanded to every field where actions performed by people are part of complex, dynamic systems. Military establishment operations [11, 12], aeronautics [13, 14], air traffic management (ATC) [15, 16], automotive [17], and the environments are just a few of the many fields where SA research is prevalent and [18, 19] asserts that there are three levels of operator SA: level 1 SA (perception of environmental factors), level 2 SA (understanding of those elements' meanings), and level 3 SA (the projection of their future status). Figure 1 depicts the SA's three-tier model. SA is portrayed in the model as being a crucial aspect of human decision-making. Individual and job characteristics, including knowledge, education, workload, and the interface, have an impact on the development and maintenance of SA. The three hierarchical levels are especially helpful for measuring the construct, and the model is simple

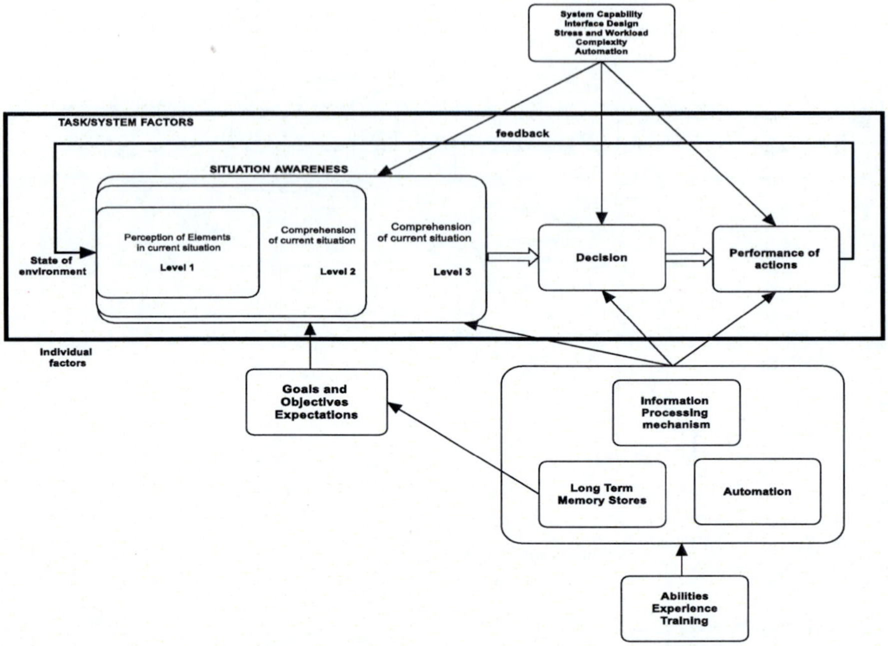

Fig. 3 Three-level measurement model

and logical. As a result, the three-level model serves as the foundation for many current SA measuring approaches (see Fig. 3).

3.3 *Use of Ontology*

The use of ontologies in the Semantic Web appears to be a clear response from which we may draw organizational and relational capabilities. Ontology-based recommender systems are a developing trend in the tourism industry [20, 21]. Some systems [22, 23] rely solely on ontologies to describe tourist POIs. In addition to tourist ontologies [24, 25], other publications also take user profile ontologies into consideration, such as [26]. An ontology to express user preferences and context elements is put out in [27].

Semantic Web Tim Berners-Lee, the creator of the World Wide Web, initially proposed a vision of the modern Web (see Fig. 4a), which served as the basis for the Semantic Web (see Fig. 4b). As a component of the present network architecture growth, it provides metadata to online contents and precisely characterizes them so that a computer may interpret their actual meaning through metadata.

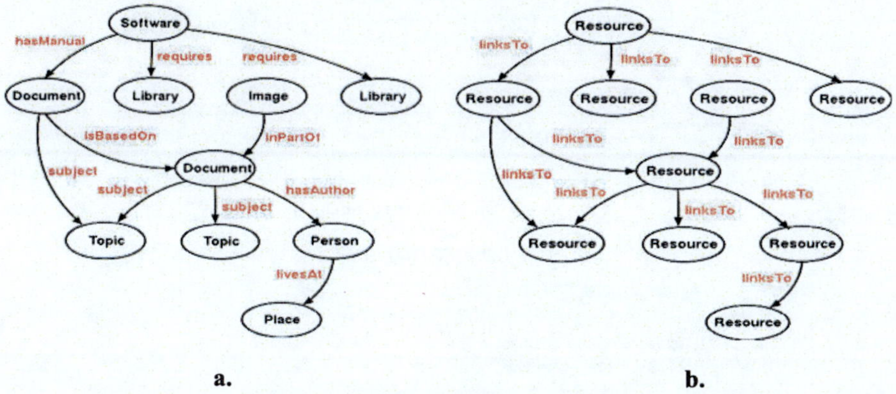

Fig. 4 (**a**) Current web architecture. (**b**) Semantic Web

The compound word ontology ("study of being") combines onto- "being" or "that which is" and logia- "logical discourse." For recommendations that are informed by knowledge, this approach employs a semantic knowledge base. With the aid of concept identification and relation extraction, the domain ontology is created from unstructured data. User's interaction with this system is gathered into the "User Interest Ontology" using log files. The fully represented knowledge base for recommender systems is created by mapping the domain and user ontologies together. The suggested process is made up of phases that enable connecting an ontological representation of the data with its representation using the travel dataset (see Fig. 5). From the integrated ontology, a knowledge graph is created. To find ideas that are contextually related and reflect the knowledge base in a vector space model, the similarity of nodes and a similarity measure are computed after that. Related items are grouped together using clustering and learning, and classification is employed using the MapReduce function to provide accurate recommendations using the CNN and AMFO algorithms.

- Compiling information from various sources and developing a knowledge base for users, including information on travel, context, and item ratings based on ontologies.

  The system either explicitly collects user preferences through direct participation or passively collects user preferences (by data mining, social network analysis, and user image analysis, for instance). These preferences are linked to specific POI categories in accordance with the POI ontology. From higher classes in the POI ontology to lower classes, user preferences are communicated down the hierarchy. The system receives the user's context either explicitly or indirectly through the context module (obtained from an API, mobile information, etc.). Using its recommender system model, the system provides the user with a list of suggestions.

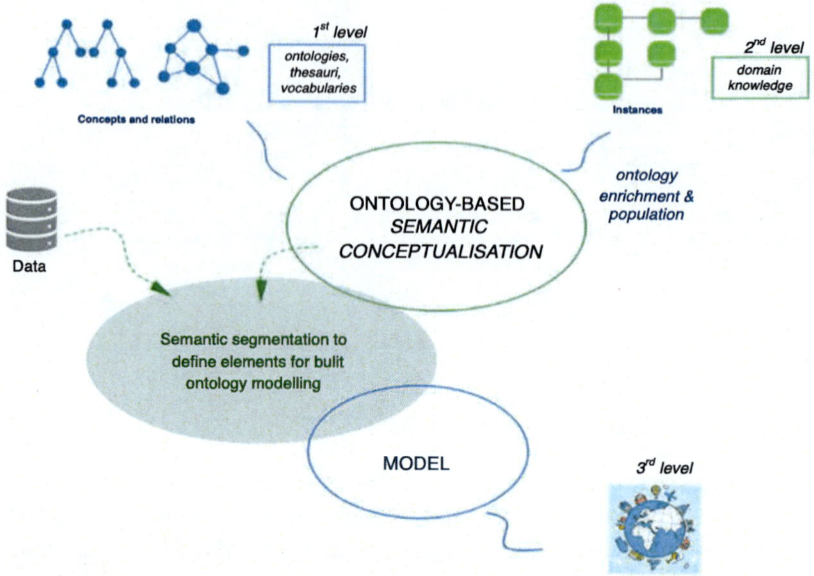

Fig. 5 Proposed methodological workflow and its levels

- To link descriptions of the user profile, context, and item profile based on the data already available, several ontologies are organized and mapped. The relational database is mapped to RDF triples for the user profile, location descriptions, contextual data, and interaction data of users. Analyzing user queried history and calculating similarities between the user profile and item descriptions by vectorization of user profiles and similarity calculation algorithms.

  By vectorizing user profiles and using similarity calculation algorithms, we can analyze user query histories and determine how closely item descriptions match user profiles.
- To find the top N users who are comparable to the present user, one can cluster ontology-based user profiles.
- Ranking and filtering a list of relevant recommendations for a user (see Fig. 6).

3.4 Experimental Setup

Dataset used The Gowalla dataset was gathered between February 2009 and October 2010. In accordance with the study, we preprocessed the Gowalla dataset by eliminating cold users, or those who visited places for fewer than 10 check-ins. The dataset is accessible from the website, https://snap.stanford.edu/data/loc-gowalla.html.

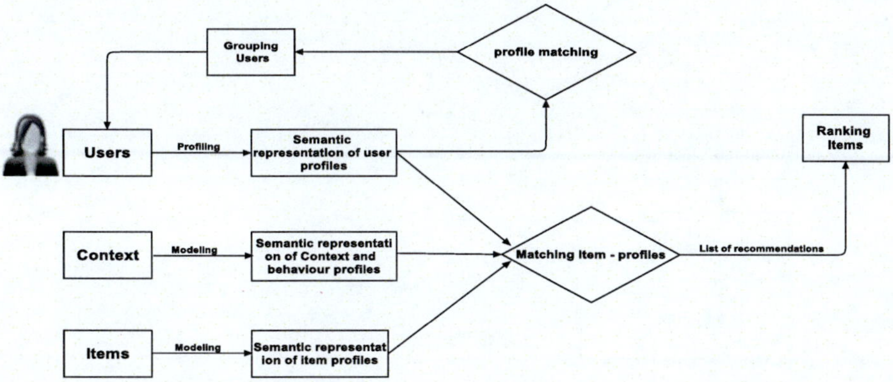

Fig. 6 The information filtering process

The outcomes of the simulation were used to document the model's effectiveness and uniqueness. Here, numerous performance metrics including recall, accuracy, and precision were used to implement the evaluation. Finally, the new model's performance is evaluated in comparison to several established models, including CNN, ontology-based CNN, CNN with MapReduce (CNN-MR), and CNN-MR with SA (CNN-MR-SA) (see Fig. 7 and Table 1).

4 Conclusion

In this paper, a model MR-AMFO-CNN method for intelligent recommendation systems is introduced. In addition, the Gowalla dataset is used to analyze the recommender system. In essence, it has several characteristics and has been condensed to reduce processing time. MapReduce is initially used for data preprocessing, which includes feature extraction and feature selection according to predetermined rules. The relevant features in the dataset's various features are examined and extracted for use as CNN inputs. To predict the recommendations for each user, CNN is used. Additionally, the MFO method helps improve CNN's performance by optimizing CNN's weight. The accuracy, specificity, and recall metrics are used to compare the predicted and actual ratings of the items to evaluate the efficacy of the recommended model and the suggested recommendation system. Sentiment analysis will be used after a trip to learn what visitors thought of the sights and about tourism facilities like vehicle parking, stores, and health services. Forecasting can be used before a trip to learn when tourists will arrive. Tourism Recommendation Systems can be used while on a trip to recommend more focused itineraries, reduce traffic conditions and pollution, and achieve better results in the short period of time possible for leisure travelers. Using fewer complicated algorithms would make the representation easier. Ontology creation might be made more automated in a variety of ways. To glean more insightful information, implicit connection extraction might be further researched.

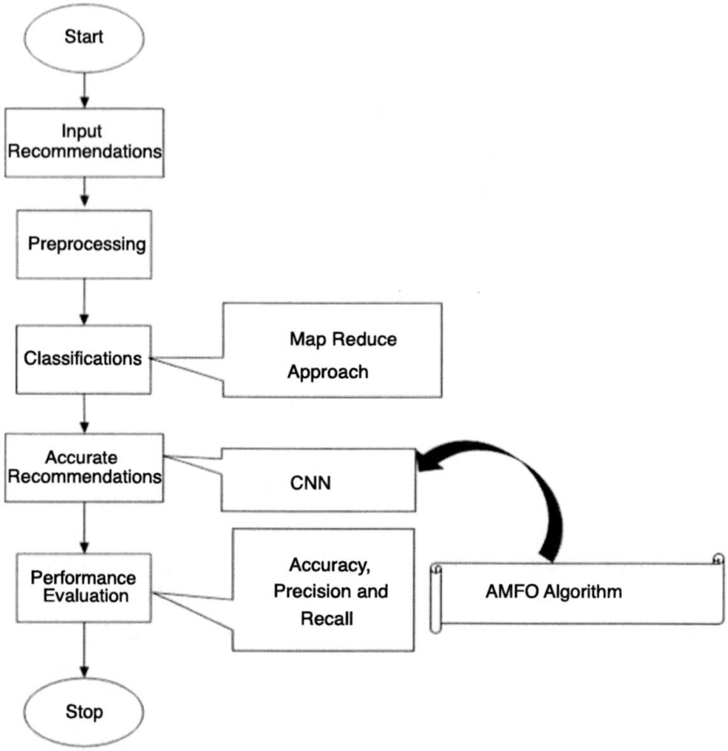

Fig. 7 The proposed recommender model

Table 1 Overall performance of the proposed MR-AMFO-CNN in terms of accuracy, precision, recall, and F1 score

Measures	CNN	OB-CNN	MR-CNN	MR-SA-CNN	MR-AMFO-CNN
Accuracy	0.926	0.930	0.929	0.933	0.959
Recall	0.857	0.857	0.833	0.833	1.000
F1 score	0.682	0.718	0.895	0.894	0.862
Specificity	0.983	0.996	0.996	0.996	1.000
NPV	0.927	0.924	0.926	0.923	0.908
Error rate	1.093	1.066	0.838	0.927	0.417
Iterations	3	3	3	3	3
Time	00:36:42	00:34:03	00:29:04	00:07:49	00:03:37

References

1. Goo, J., Huang, C. D., Yoo, C. W. & Koo, C. Smart Tourism Technologies' Ambidexterity: Balancing Tourist's Worries and Novelty Seeking for Travel Satisfaction. *Inf. Syst. Front.* 1–20 (2022).
2. Alom, M. Z., Hasan, M., Yakopcic, C., Taha, T. M. & Asari, V. K. Recurrent Residual Convolutional Neural Network based on U-Net (R2U-Net) for Medical Image Segmentation. *arXiv [cs.CV]* (2018).
3. Jannach, D., Ludewig, M. & Lerche, L. Session-based item recommendation in e-commerce: on short-term intents, reminders, trends, and discounts. *User Model. User-adapt Interact.***27**, 351–392 (2017).
4. Bobadilla, J., Ortega, F., Gutiérrez, A. & Alonso, S. Classification-based deep neural network architecture for collaborative filtering recommender systems. *Int. j. interact. multimed. artif. intell.***6**, 68 (2020).
5. Kumar, G., Jerbi, H. & O'Mahony, M. P. A sequence-based and context modelling framework for recommendation. *Expert Syst. Appl.***175**, 114665 (2021).
6. Agarwal, A., Mishra, D. S. & Kolekar, S. V. Knowledge-based recommendation system using semantic web rules based on Learning styles for MOOCs. *Cogent Engineering***9**, 2022568 (2022).
7. Mckensy-Sambola, D., Rodríguez-García, M. Á., García-Sánchez, F. & Valencia-García, R. Ontology-based nutritional recommender system. *Appl. Sci.***12**, 143 (2021).
8. Chaudhari, K. & Thakkar, A. A Comprehensive Survey on Travel Recommender Systems. *Arch. Comput. Methods Eng.***27**, 1545–1571 (2020).
9. Trattner, C., Oberegger, A., Marinho, L. & Parra, D. Investigating the utility of the weather context for point of interest recommendations. *Information Technology & Tourism***19**, 117–150 (2018).
10. Fu, Z., Yu, L. & Niu, X. TRACE: Travel Reinforcement Recommendation Based on Location-Aware Context Extraction. *ACM Trans. Knowl. Discov. Data***16**, 1–22 (2022).
11. Endsley, M. R. & Garland, D. J. *Situation Awareness Analysis and Measurement.* (CRC Press, 2000).
12. Matthews, K. R., Homer, D. B., Thies, F. & Calder, P. C. Effect of whole linseed (Linum usitatissimum) in the diet of finishing pigs on growth performance and on the quality and fatty acid composition of various tissues. *Br. J. Nutr.***83**, 637–643 (2000).
13. Ma, R. & Kaber, D. B. Situation awareness and driving performance in a simulated navigation task. *Ergonomics***50**, 1351–1364 (2007).
14. Keller, H. *et al.* Distal and proximal parenting as alternative parenting strategies during infants' early months of life: A cross-cultural study. *Int. J. Behav. Dev.***33**, 412–420 (2009).
15. Eyferth, K., Niessen, C. & Spaeth, O. A model of air traffic controllers' conflict detection and conflict resolution. *Aerosp. Sci. Technol.***7**, 409–416 (2003).
16. Blandford, A. & William Wong, B. L. Situation awareness in emergency medical dispatch. *Int. J. Hum. Comput. Stud.***61**, 421–452 (2004).
17. Zheng, X. S., McConkie, G. W. & Simons, D. J. Effects of Verbal and Spatial-Imagery Tasks on Drivers' Situation Awareness of Traffic Information. *Proc. Hum. Fact. Ergon. Soc. Annu. Meet.***49**, 1999–2003 (2005).
18. Walker, B., Holling, C. S., Carpenter, S. & Kinzig, A. Resilience, Adaptability and Transformability in Social-ecological Systems. *Ecol. Soc.***9**, (2004).
19. Endsley, M. R. Measurement of Situation Awareness in Dynamic Systems. Hum. Factors 37, 65–84 (1995).
20. Borràs, J., Moreno, A. & Valls, A. Intelligent tourism recommender systems: A survey. Expert Syst. Appl. 41, 7370–7389 (2014).
21. Yochum, P., Chang, L., Gu, T. & Zhu, M. Linked Open Data in Location-Based Recommendation System on Tourism Domain: A Survey. *IEEE Access***8**, 16409–16439 (2020).

22. García-Crespo, Á., Ruiz-Mezcua, B., López-Cuadrado, J. L. & González-Carrasco, I. A review of conventional and knowledge-based systems for machining price quotation. *J. Intell. Manuf.***22**, 823–841 (2011).
23. Bahramian, Z., Ali Abbaspour, R. & Claramunt, C. A CONTEXT-AWARE TOURISM RECOMMENDER SYSTEM BASED ON A SPREADING ACTIVATION METHOD. *The International Archives of the Photogrammetry, Remote Sensing and Spatial Information Sciences* vols XLII-4/W4 333–339 (2017).
24. Rizaldy Hafid Arigi, L., Abdurahman Baizal, Z. K. & Herdiani, A. Context-aware recommender system based on ontology for recommending tourist destinations at Bandung. *J. Phys. Conf. Ser.***971**, 012024 (2018).
25. Trung, L. H. https://euroasia-science.ru/pdf-arxiv/the-controllability-function-of-polynomial-for-descriptor-systems-23-31/. *EurasianUnionScientists* vol. 4 (2019).
26. Ruotsalo, T. *et al.* SciNet: Interactive Intent Modeling for Information Discovery. in *Proceedings of the 38th International ACM SIGIR Conference on Research and Development in Information Retrieval* 1043–1044 (Association for Computing Machinery, 2015).
27. Bernardo, M., Marimon, F. & Alonso-Almeida, M. del M. Functional quality, and hedonic quality: A study of the dimensions of e-service quality in online travel agencies. *Information & Management***49**, 342–347 (2012).

Simulation of GA-Based Harmonics Elimination in CHMLI Using DTC for Dynamic Load

Akhilesh Sharma and Sarsing Gao

1 Introduction

The limited availability of fossil fuels has forced engineers and researchers across the globe to identify and fruitfully harness electrical energy as most of the generating stations utilize conventional sources of energy. To meet the electrical energy demand, the consumption of fossil fuels has increased at an alarming rate. These sources are likely to vanish in half a century or so. Moreover, fuel consumption is creating an environmental threat. So, there should be a balance between energy generation and fuel consumption. To some extent, this is possible through the utilization of nonconventional sources of energy.

Nonconventional energy resources like wind, tide, geothermal, and solar power exist throughout the length and breadth of the earth. Their accessibility is restricted due to the geographical features that exist on earth. In some cases, like coastal areas, the availability of wind and tide is greater. Even though the sun's radiation is greater, its distribution is not the same throughout the globe. For hilly regions, the absence of tides and winds of variable speed has restricted the use of these sources. The sun is the only source available throughout the globe. At the same time, the solar intensity is high at one location while it may be low at another part of the earth. The solar radiation is intense at the tops of hills and mountains. It's less effective at the base. Thus, the energy associated with this radiation is nonuniform and even affected by weather conditions. The sun's radiation may be directly converted and stored as chemical energy, as in a cell or battery. Even so, it is directly utilized in the form of direct current. Most industrial drives require an AC source. So, the conversion of

A. Sharma (✉) · S. Gao
North Eastern Regional Institute of Science and Technology, Electrical Engineering, Naharlagun, Arunachal Pradesh, India

M. Mishra et al. (eds.), *Applications of Computational Intelligence in Management & Mathematics*, Springer Proceedings in Mathematics & Statistics 417,
https://doi.org/10.1007/978-3-031-25194-8_11

DC into AC is the need of the day. Therefore, the harnessed electrical energy from the sun needs to change into an AC for AC drive applications employing an inverter.

The application of inverters from domestic levels to grid systems is increasing day by day, from single phase to three phase and from two level to multilevel. A single-phase two-level inverter has many disadvantages, like low output power and low efficiency. It also suffers from power quality issues. Therefore, the two-level inverter is being replaced by multilevel inverters. The multilevel inverter has many advantages over the two-level inverter, like low total harmonic distortion, a better quality of power, high efficiency, etc.

There are numerous topologies of multilevel inverters, starting from conventional inverters like cascaded H-bridge multilevel inverters (CHMLI), diode-clamped inverters, and capacitor-clamped inverters, too symmetrical or asymmetrical 31st-level inverters, or even hybrid multilevel inverters [1–12]. Each of these inverter topologies has limitations in terms of the number of voltage levels, switching devices, voltage sources, etc. For example, a diode-clamped inverter or capacitor-clamped inverter may provide many voltage levels from a single DC source, but the number of components involved is higher, thus increasing the complexity of the circuit and losses. In the case of an asymmetrical inverter [6, 7], many variable sources are required. The beauty of this scheme is the ease of implementation.

The direct torque control method is widely used in the control of inverter-based industrial drives. In the early stages of drive control, two-level inverters were widely used. The two-level inverter has its limitations, like higher THD, less output power, etc. Therefore, a multilevel topology is preferred. Depending upon the number of components and number of voltages, one may decide which topology suits better. In this case, a CHMLI inverter has been chosen because it best suits asymmetrical voltage.

Modulation plays a significant role in the generation of gating signals for switching devices. Therefore, it is important to properly select the modulation index. It will not only help with proper triggering but also improve the output voltage waveform. It could be achieved by the selective harmonic elimination (SHE) method or other optimization techniques. There are various ways in which modulation could be done. Some notable modulation schemes are pulse width modulation, sinusoidal pulse width modulation, space vector approach, etc. Each of these modulation techniques has its own usefulness in terms of power quality [13–19].

In this paper, a CHMLI topology has been used, whose switching signals are based on direct torque control. The switching angles have been optimized by using Newton-Raphson to reduce the THD. A MATLAB simulation has been carried out with both symmetrical and asymmetrical voltages. A three-phase induction motor has been applied as a load to study the usefulness of the scheme. The results show that dominant orders are reduced at a frequency below 50 Hz while these harmonic orders are eliminated.

2 Cascaded H-Bridge Multilevel Inverter (CHMLI)

The cascaded H-bridge Multilevel Inverter (CHMLI) is the most basic and prevalent topology among multilevel inverters. This inverter topology may be implemented for both one-phase and three-phase devices. A multilevel inverter should have at least three voltage levels. The voltage levels could be easily achieved by considering a single H-bridge as shown in Fig. 1, where the stepped output voltage may be any one of the three voltages +Vs, 0 or –Vs. The switching is done in such a way that there is no dead short circuit. The switching pattern is as given in Table 1. If several such H-bridges are cascaded, they will form a cascaded H-bridge multilevel inverter (CHMLI). This inverter will provide multilevel stepped output whose number of steps will be equal to (2*n + 1), where "n" is the number of cascaded H-bridges. So, to obtain five levels of stepped output, two numbers of cascaded H bridges are needed. This arrangement will provide a stepped output voltage of +2Vs, Vs, 0, –Vs, and –2Vs for symmetrical and nine-level voltages for asymmetrical DC source. The switching pattern is as shown in Table 1. Although this type of inverter is easy to implement, isolation of individual power sources is one of the main problems associated with this type of inverter.

The number of stepped outputs determines the number of switches. For a "2*n + 1" level of stepped output voltage, the total number of switches is given by "4*n," where "n" is the number of voltages in CHB. The possible switching states for the cascaded H-bridge inverter are shown in Table 1. Based on the switching table, triggering pulses for the controlled semiconductor devices may be generated. Accordingly, the output voltage may be obtained.

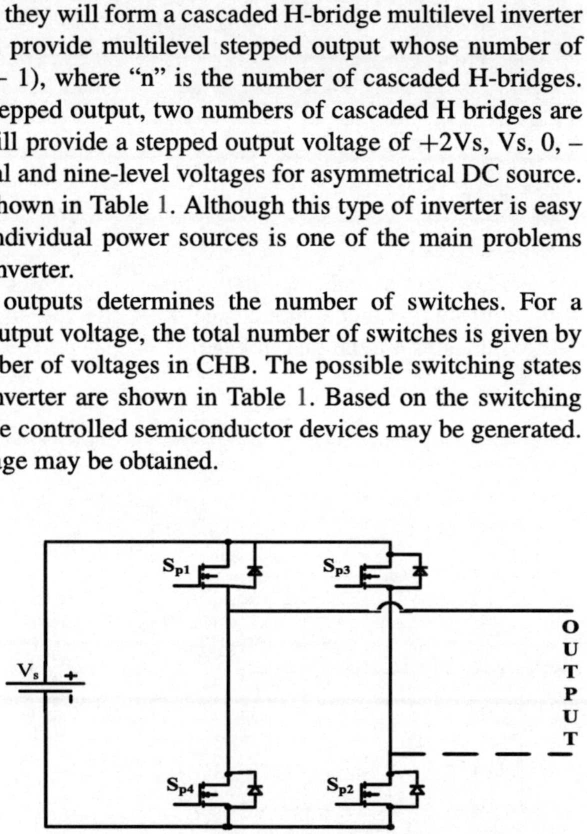

Fig. 1 Single H-bridge topology

Table 1 H-bridge inverter

Sl. No.	S_{p1}	S_{p2}	S_{p3}	S_{p4}	V_{out}
1	0	0	0	0	0
2	1	1	0	0	$+V_s$
3	0	0	1	1	$-V_s$
4	1/0	0/1	1/0	0/1	0

3 Dynamic Modelling of Induction Motor

To study the speed characteristic of an induction motor, the dynamic equations of the motor are necessary. These equations could be derived from the equivalent d-q model [20–29] of an induction motor as shown in Fig. 2. They are expressed as follows:

$$\begin{bmatrix} \frac{d\psi_{qs}}{dt} \\ \frac{d\psi_{ds}}{dt} \\ \frac{d\psi_{qr}}{dt} \\ \frac{d\psi_{dr}}{dt} \end{bmatrix} = \begin{bmatrix} -R_s & -(\omega_e L_{ls} + \omega_e L_m) & 0 & -\omega_e L_m \\ (\omega_e L_{ls} + \omega_e L_m) & -R_s & \omega_e L_m & 0 \\ 0 & -(\omega_e - \omega_r) L_m & -R_r & -(\omega_e - \omega_r)(L_m + L_{lr}) \\ (\omega_e - \omega_r) L_m & 0 & (\omega_e - \omega_r)(L_m + L_{lr}) & -R_r \end{bmatrix} \begin{bmatrix} i_{qs} \\ i_{ds} \\ i_{qr} \\ i_{dr} \end{bmatrix} + \begin{bmatrix} V_{qs} \\ V_{ds} \\ V_{qr} \\ V_{dr} \end{bmatrix} \tag{1}$$

Here, suffix "s" indicates stator, and "r" indicates rotor. "R," "L," "ψ," "i," and "V" indicate resistance, inductance, flux, current, and voltages, respectively, while "L_m" indicates mutual flux linkage

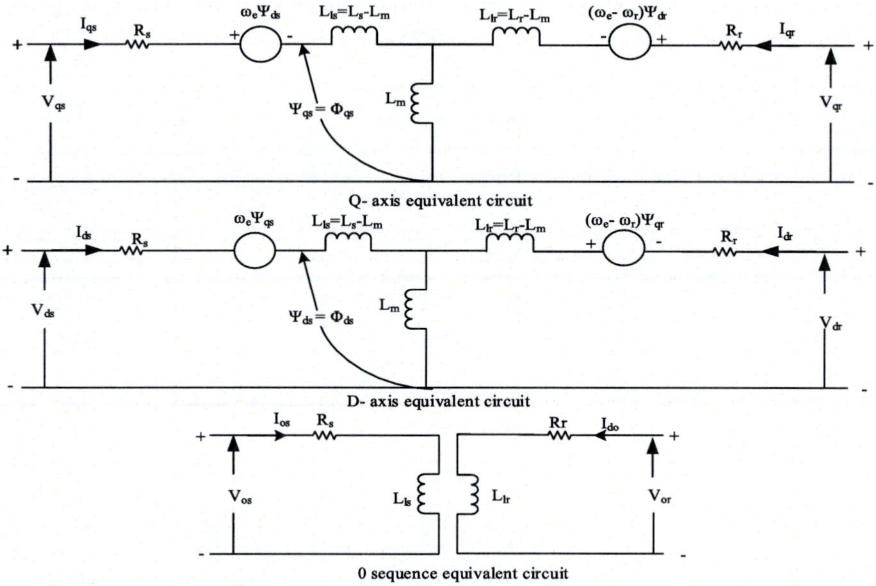

Fig. 2 Equivalent dq0 model of induction motor

$$\begin{bmatrix} \frac{d\lambda_{os}}{dt} \\ \frac{d\lambda_{or}}{dt} \end{bmatrix} = \begin{bmatrix} V_{os} & i_{os} R_s \\ V_{or} & i_{or} R_r \end{bmatrix} \begin{bmatrix} 1 \\ -1 \end{bmatrix} \tag{2}$$

Where,

$$\psi_{qs} = L_{ls} i_{qs} + L_m \left(i_{qs} + i_{qr} \right) \tag{3}$$

$$\psi_{qr} = L_{lr} i_{qr} + L_m \left(i_{qs} + i_{qr} \right) \tag{4}$$

$$\psi_{ds} = L_{ls} i_{ds} + L_m \left(i_{ds} + i_{dr} \right) \tag{5}$$

$$\psi_{dr} = L_{lr} i_{dr} + L_m \left(i_{ds} + i_{dr} \right) \tag{6}$$

$$\lambda_{os} = L_{ls} i_{os} \tag{7}$$

$$\lambda_{or} = L_{lr} i_{or} \tag{8}$$

λ_{os} and λ_{or} are the zero-sequence flux linkages.

where ω_e and ω_r are arbitrary angular and stator angular frequencies, respectively.

The expression of electromagnetic torque is:

$$T_e = \frac{3}{2}\frac{P}{2} L_m \left(i_{qs} i_{dr} - i_{ds} i_{qr} \right) \tag{9}$$

Also,

$$T_e = T_l + \frac{2}{P} J \frac{d\omega_r}{dt} \tag{10}$$

T_l and J represent load torque and moment of inertia, respectively.

3.1 Direct Torque Control (DTC)

In the case of direct torque control method, the controlling element is the torque of the motor. It also requires a flux component. These two components are related in Eqs. (11) to (13), allowing the stator current to be represented in d-q reference. These equations help in converting into two-phase stator current, I_{ds} and I_{qs}, and corresponding speeds. These currents are converted into a three-phase stator current.

$$I_{ds} = \frac{\varphi_{ds}}{L_m} \tag{11}$$

$$I_{qs} = \left(\frac{2}{3}\right)\left(\frac{L_r}{PL_m}\right)\left(\frac{T_e}{\varphi_{ds}}\right) \tag{12}$$

$$\omega_m = \left(\frac{2}{3}\right)\left(\frac{L_m}{Pt_r}\right)\left(\frac{T_e}{\varphi_{ds}^2}\right) \tag{13}$$

4 Research Methodology

The scheme of single-phase CHMLI has been extended to develop a three-phase CHMLI. This inverter is connected to a three-phase IM whose control is based on the direct torque control method. There are many methods of speed control for induction motors [19–27], a DTC scheme is as shown in Fig. 3. Based on the dynamic model equations of an induction motor and triggering an individual switch of CHMLI, a Simulink model in MATLAB 16a has been created, as shown in Fig. 4. By using the optimized switching angles: 6.016; 13.235; 24.523; 34.779; 46.467; 62.452; 87.491 (all in degrees), obtained by the GA method, flux and torque commands, sinusoidal carrier signals have been generated. These signals have been compared to zero for generating triggering pulses for three-phase CHMLI, i.e., to say that a sinusoidal pulse width modulation technique is implemented. The parameters considered for the motor and inverter are shown in Appendix I.

The simulation of the direct torque-based inverter has been modeled in MATLAB 16(a) Simulink at NERIST server-based asymmetrical voltages of magnitude 72 V, 60 V, 54 V, 48 V, 42 V, 36 V, and 24 V, respectively. As seen from Fig. 4, direct torque and flux commands have been considered for conversion into d-q currents. These d-q currents have been converted into three-phase signals in which optimum values, to reduce a particular order of harmonics, have been considered for obtaining the

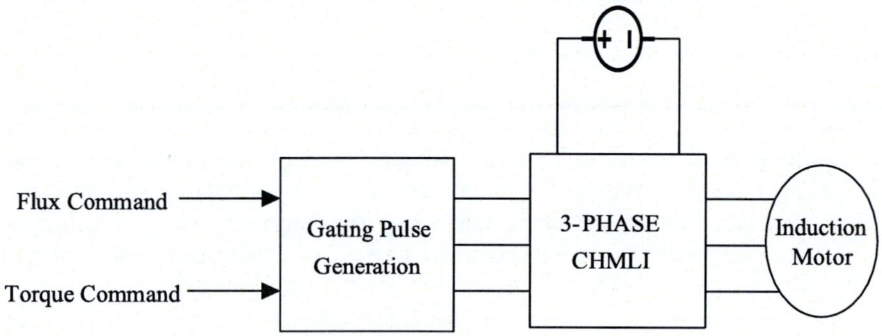

Fig. 3 General representation of the direct torque command

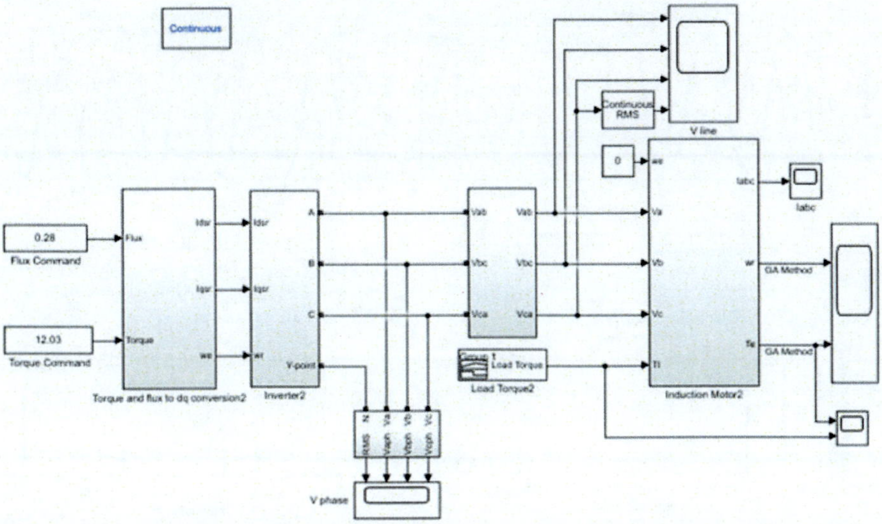

Fig. 4 Simulink model of three-phase CHMLI-based direct torque control of IM

switching pulses for an inverter. To achieve the optimized angles for the generation of pulses, the GA method solves the nonlinear equations at a fundamental frequency in such a way that all the optimized angles fall between 0 and 90 degrees.

5 Simulation and Result

The simulation in Fig. 4 provides the waveforms given in Figs. 5 and 6, respectively. Figure 5 indicates the phase and line voltages of a three-phase CHMLI. It is clear from Fig. 5a that the three-phase output voltage has 5.24% THD with an RMS value of 427.6 V. The THD increases to 9.83% with an RMS voltage of 247.7 V with almost 8% being due to the third order of harmonics. It is visible in Fig. 5b. It is also seen in Fig. 5a that individual orders of harmonics: 5th, 7th, 11th, 13th, 17th, and 19th) have been suppressed. This value of THD is acceptable as per the IEEE 514–2014 standard.

Figure 6 shows the torque and speed curves. It contains two different cases, as indicated in the figure. In Fig. 6a, a constant torque command has been applied. The load torque is varied in equal steps from 0 Nm to 10 Nm during the intervals 0–2 sec, 2–4 sec, and 4–6 sec, respectively. During this variation, the speed of the IM is almost constant. When the torque command is 8 Nm at time $t = 0$ sec, the corresponding motor speed is 1006 rpm. This speed changes to 1475 rpm when the torque command is 12 Nm at time $t = 2$ sec. When the torque command is 10 Nm at time $t = 4$ sec, it reduces to 1245 rpm. In all the above cases, the constant flux of 0.28 weber has been assumed.

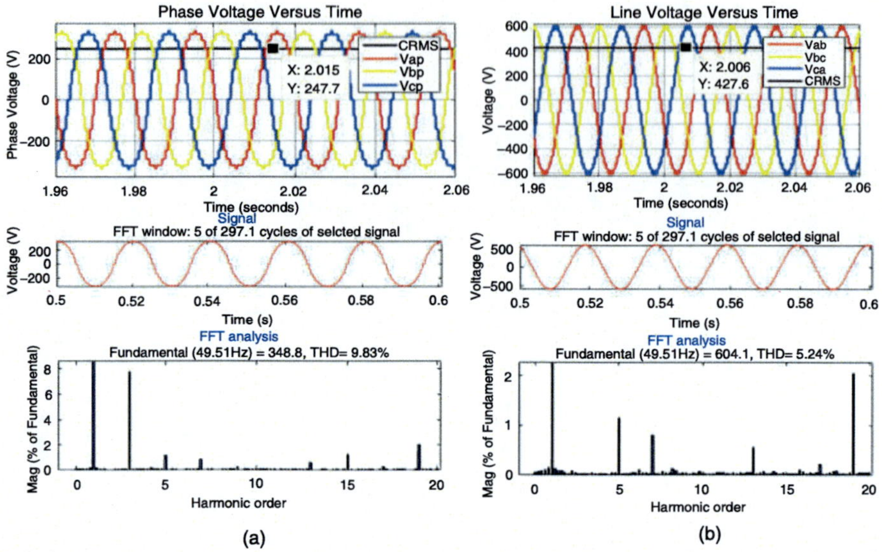

Fig. 5 Phase and line voltages with their THDs. (**a**) Phase voltage with THD. (**b**) Line voltage with THD

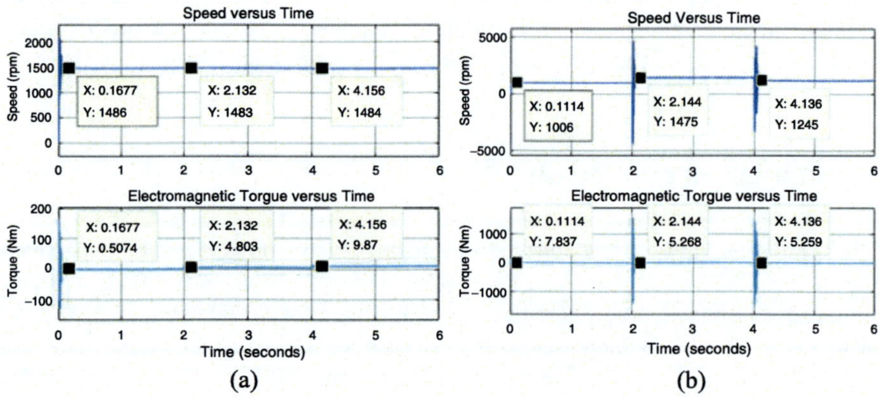

Fig. 6 Speed and torque characteristics. (**a**) Constant torque command. (**b**) Variable torque command

6 Conclusion

A direct torque control technique for the CHMLI is applied for a 15th-level CHMLI. The simulation result shows that the individual dominant orders of harmonics in the voltage waveform are suppressed, thereby smoothening the voltage waveform. This is possible due to the optimum angle obtained through the application of the genetic algorithm method. The THD in the waveform is also reduced, thereby improving the quality of the output voltage. It reduces the necessity of a filter, as needed at two levels of output voltage. At the same time, the speed of the motor is controlled only by torque commands. It is also seen that the speed of the induction motor becomes independent of the load torque for a particular load command.

A.1 Annexure I – Motor parameter

Sl. no.	Parameter	Value	Sl. no.	Parameter	Value
1	Stator resistance	6.03 Ω	5	Mutual inductance	489.3e-3 H
2	Stator inductance	489.3e-3 H	6	No. of poles	4
3	Rotor resistance	6.085 Ω	7	Moment of inertia	0.00488 kgm^2
4	Rotor inductance	450.3e-3 H	8	DC voltage source (sym.)	54 V

References

1. J. Rodriguez, J.-S. Lai, F.Z. Peng. Multilevel inverters: a survey of topologies, controls, and applications. IEEE Transactions on industrial electronics, 49(4), 724–738 (2002).
2. Gaddafi Sani Shehu, Abdullahi Bala Kunya, Ibrahim Haruna Shanono, and Tankut Yalcinoz. A Review of Multilevel Inverter Topology and Control Techniques. Journal of Automation and Control Engineering 4(3), 233-241 (2016).
3. Lipika Nanda, A. Dasgupta, and U.K. Rout. A Comparative Studies of Cascaded Multilevel Inverters Having Reduced Number of Switches with R and RL-Load. International Journal of Power Electronics and Drive System 8(1), 40-50 (2017).
4. Domingo A. Ruiz-Caballero, Reynaldo M. Ramos-Astudillo, and Samir Ahmad Mussa, "Symmetrical Hybrid Multilevel DC–AC Converters with Reduced Number of Insulated DC Supplies", IEEE Transactions on Industrial Electronics, Vol. 57, No. 7, pp. 2307 – 2314 (2010).
5. Kannan Ramani, Mohd. Ali Jagabar Sathik, and Selvam Sivakumar. A New Symmetric Multilevel Inverter Topology Using Single and Double Source Sub-Multilevel Inverters. Journal of power electronics 15(1), 96-105 (2015).
6. Kennedy Adinbo Aganah, Cristopher Luciano, Mandoye Ndoye, and Gregory Murphy. New Switched-Dual-Source Multilevel Inverter for Symmetrical and Asymmetrical Operation. Energies 11(8), 1-13 (2018).
7. Angga Muhamad Andreh, Subiyanto, and Said Sunardiyo. Simulation Model of Harmonics Reduction Technique using Shunt Active Filter by Cascade Multilevel Inverter Method. International Conference on Engineering, Science and Nanotechnology (ICESNANO 2016) AIP Conf. Proc. (2016) https://doi.org/10.1063/1.4968318.
8. Naresh K. Pilli, M. Raghuram, Avneet Kumar, Santosh K. Singh. Single dc-source-based seven-level boost inverter for electric vehicle application. IET Power Electronics 12(13), 3331-3339 (2019), https://doi.org/10.1049/IET-PEL.2019.0255.
9. C. Dhanamjayulu, G. Arun Kumar, B. Jaganatha Pandian, C. V. Ravi Kumar, M. Praveen Kumar, A. Rini Ann Jerin And P. Venugopal. Real-time implementation of a 31-level asymmetrical cascaded multilevel inverter for dynamic loads. IEEE Access 7, 51254-51266 (2019), https://doi.org/10.1109/ACCESS.2019.2909831.

10. B. D. Reddy, M. P. Selvan and S. Moorthi. Design, Operation, and Control of S3 Inverter for Single-Phase Microgrid Applications. IEEE Transactions on Industrial Electronics 62(9) 5569-5577 (2015), https://doi.org/10.1109/TIE.2015.2414898.
11. M. Leon M. Tolbert, and Thomas G. Habetler. Novel Multilevel Inverter Carrier-Based PWM Method. IEEE Transactions On Industry Applications 35(5), 1098- 1107 (1999).
12. Ebrahim Babaei, and Seyed Hossein Hosseini. New cascaded multilevel inverter topology with a minimum number of switches. Energy Conversion and Management 50, 2761–2767 (2009).
13. Kaumil B. Shah and Hina Chandwani. Comparison of Different Discontinuous PWM Technique for Switching Losses Reduction in Modular Multilevel Converters. International Journal of Energy and Power Engineering 12(11), 852- 859 (2018).
14. B. G. Fernandes and S. K. Pillai. Programmed PWM inverter using PC for induction motor drive," [1992] Proceedings of the IEEE International Symposium on Industrial Electronics, 2, 823-824 (1992), https://doi.org/10.1109/ISIE.1992.279712.
15. D. Singh, A. Sharma, P. D. Singh, and S. Gao. Selective Harmonic Elimination for Cascaded Three-Phase Multilevel Inverter. 2nd IEEE International Conference on Power Electronics, Intelligent Control and Energy Systems, pp. 437–442, (2018), https://doi.org/10.1109/ICPEICES.2018.8897407.
16. Tapan Trivedi, Pramod Agarwal, Rajendrasinh Jadeja, Pragnesh Bhatt. FPGA Based Implementation of Simplified Space Vector PWM Algorithm for Multilevel Inverter Fed Induction Motor Drives. World Academy of Science, Engineering and Technology 9(8), 1144-1149 (2015).
17. A. Sharma, D. Singh, P. Devachandra Singh, and S. Gao. Analysis of Sinusoidal PWM and Space Vector PWM based diode clamped multilevel inverter. 5th IEEE Uttar Pradesh Section International Conference on Electrical, Electronics and Computer Engineering (UPCON), pp. 1-6, (2018), https://doi.org/10.1109/UPCON.2018.8596899.
18. Lazhar Manai, Faouzi Armi, Mongi Besbes. Optimization-based selective harmonic elimination for capacitor voltages balancing in multilevel inverters considering load power factor. Electrical Engineering, (2020), https://doi.org/10.1007/s00202-020-00960-5.
19. Muhammad Salman, Inzamam Ul Haq, Tanvir Ahmad, Haider Ali, Affaq Qamar, Abdul Basit, Murad Khan, and Javed Iqbal. Minimization of total harmonic distortions of cascaded H-bridge multilevel inverter by utilizing bio-inspired AI algorithm. EURASIP Journal on Wireless Communications and Networking, (2020), https://doi.org/10.1186/s13638-020-016865.
20. J.N. Nash. Direct torque control, induction motor vector control without an encoder. IEEE Transactions on Industry Applications 33(2), 333–341 (1997), https://doi.org/10.1109/28.567792.
21. B. K. Bose. Neural Network Applications in Power Electronics and Motor Drives—An Introduction and Perspective. IEEE Transactions on Industrial Electronics 54(1), 14-33, (2007), https://doi.org/10.1109/TIE.2006.888683.
22. M. Mengoni, A. Amerise, L. Zarri, et al. Control scheme for open-ended induction motor drives with a floating capacitor bridge over a wide speed range. IEEE Trans. Ind., 53(5), 4504–4514 (2017).
23. K.K. Venkata Praveen, K.M. Ravi Eswar, T. Vinay Kumar. Improvised predictive torque control strategy for an open-end winding induction motor drive fed with four-level inversion using normalized weighted sum model. IET Power Electron. 11(5), 808–816 (2018).
24. K.M. Ravi Eswar, K.V. Praveen Kumar, T. Vinay Kumar. Modified predictive torque and flux control for open-end winding induction motor drive based on ranking method. IET Electronics Power 12(4), 463– 473 (2018).
25. Manish Verma, Neeraj Bhatia, Scott D. Holdridge & Tood O'Neal. Isolation Techniques for medium voltage adjustable speed drives. IEEE Industry Application Magazine, 25(6), 92-100, (2019), https://doi.org/10.1109/MIAS.2019.2923114.
26. Najib El Ouanjli, Aziz Derouich, Abdelaziz El Ghzizal, Saad Motahhir, Ali Chebabhi, Youness El Mourabit, and Mohammed Taoussi. Modern improvement techniques of direct torque control for induction motor drives - a review. Protection and Control of Modern Power Systems 4, 1-12 (2019)

27. David B. Durocher & Christopher Thompson. Medium Voltage adjustable speed drives upgrade. IEEE Industry Application Magazine 25(6), 35-43, (2019) https://doi.org/10.1109/MIAS.2018.2875183.
28. Isao Takahashi, Toshihiko Noguchi. A New Quick-Response and High-Efficiency Control Strategy of an Induction Motor. IEEE Transactions on Industry Applications 22(5), 820–827 (1986) https://doi.org/10.1109/tia.1986.4504799.
29. Yukai Wang, Yuying Shi, Yang Xu, and Robert D. Lorenz. A Comparative Overview of Indirect Field Oriented Control (IFOC) and Deadbeat-Direct Torque and Flux Control (DB-DTFC) for AC Motor Drives. Chinese Journal of Electrical Engineering 1(1), 9-20 (2015).

Discussing the Future Perspective of Machine Learning and Artificial Intelligence in COVID-19 Vaccination: A Review

Rita Roy, Kavitha Chekuri, Jammana Lalu Prasad, and Subhodeep Mukherjee

1 Introduction

Machine learning (ML) is a vital part of the rapidly growing field of data science. Algorithms are instructed to categorize or forecast using statistical methods, uncovering essential insights in data mining projects [1]. These perspectives drive decision-making within implementations and businesses, ideally influencing key growth performance measures [2]. ML is used in various healthcare applications, from crisis intervention of prevalent infectious conditions to utilization of patient health information in line with external factors such as air pollutants and weather [3]. Artificial intelligence (AI) will affect physicians and hospitals because it will be essential in decision support, enabling earlier disease detection and tailor-made treatment plans to ensure optimum performance [4].

Traditional vaccine discovery has been a complex challenge that attempts to develop innovative molecules with a wide range of desirable properties. As a result, the importance of forming pharmaceutical drugs is highly costly and complex,

R. Roy (✉)
Department of Computer Science and Engineering, GITAM Institute of Technology, GITAM (Deemed to be University), Visakhapatnam, Andhra Pradesh, India

K. Chekuri
Department of Computer Science and Engineering, Raghu Engineering College, Visakhapatnam, Andhra Pradesh, India

J. L. Prasad
Department of Computer Science and Engineering, Centurion University of technology and management, Vizianagaram, Andhra Pradesh, India

S. Mukherjee
Department of Operation management, GITAM School of Business, GITAM (Deemed to be University), Visakhapatnam, Andhra Pradesh, India

M. Mishra et al. (eds.), *Applications of Computational Intelligence in Management & Mathematics*, Springer Proceedings in Mathematics & Statistics 417,
https://doi.org/10.1007/978-3-031-25194-8_12

with a low chance of success [5]. To help mitigate this difficulty, AI-/ML-based features have been paradigm-shifted to explore many plausible, varied, and novel candidate particles in vast molecular space. The AI/ML employs computation structures that interpret and learn these features from the feedback information, making impartial decisions to achieve its specific goals. Furthermore, the AI and ML identify potentially toxic drugs, providing faster drug target verification and drug framework design and optimization [6].

Over the last century, AI-based models have revolutionized drug development [7]. AI has also resulted in countless reverse vaccinology (RV) with virtual paradigms, widely categorized as rule-based refinement models. ML enables the development of modelling techniques to gain knowledge, generalize patterns in available data, and draw conclusions from unseen data. With the advent of deep learning (DL), the form involved can include feature extraction from raw data. Data science and ML can be used to assist scientists in understanding disease. Even though our expertise had been constrained, such tools became even more essential for COVID-19. At the moment, there's no theory. Advanced AI/ML algorithms to create unbiased knowledge about the disease based on facts.

2 Basic Concept and Terminology

2.1 Machine Learning/Artificial Intelligence

ML is a data analysis method that automatically performs the formation of analytical models [8]. The main objective is for computer systems to learn instantly without human involvement or guidance and adapt their behavior correspondingly [9].

Supervised ML algorithms can forecast future occurrences by applying what they have learned in the past to new data using classification methods [10]. Unsupervised learning investigates how processes can infer a function from unmarked data to reveal a hidden structure. Semi-supervised ML algorithms are between unsupervised and supervised, including that they train both with labelled and unlabeled – in fact, a single quantity of labelled data of unsupervised learning. Reinforcement ML algorithms are a form of ML that interacts with its surroundings by conducting concrete steps and detecting errors or benefits [11].

2.2 Machine Learning and Artificial Intelligence in COVID-19

AI/ML models can help doctors streamline diagnostic tests and decrease manual work by ensuring accurate and fast diagnoses [12]. Using training data, AI models could identify patients at increased risk, classify the prevalence and incidence of COVID-19, and prototype the disease for its transmission. AI-/ML-based methodologies, such as reusing existing drugs, using health checks as targets for vaccines based on various potential scenarios of the mutation model in the SARS-CoV-2, and screening substances as potential adjuvants, could aid in identifying novel drugs and vaccines. AI-powered chatbots were used effectively in clinical situations. They can also advise more people, which can be operated through call centres, reducing the burden on medical hotlines [13].

In COVID-19, AI/ML has been used in the diagnosis, clinical decision support, social control, public health, therapeutics, monitoring, operation of other core healthcare, vaccine development, big data, and management of COVID-19 patients [14]. AI techniques are being used to detect features extracted from the chest during X-rays to aid in the prognosis of COVID-19. COVID-19 has extensively used AI in various areas, including diagnosis, public health, clinical decision-making, social control, therapeutics, vaccine development, surveillance, combination with big data, operation of other critical clinical services, and management of COVID-19 patients. AI and ML will have both immediate and long-term effects on COVID-19. As ML is used to diagnose the virus, the impact will initially be modest, especially on the healthcare front [15].

Nevertheless, it will significantly advance medical technology in the long run and aid in the fight against pandemics [16]. There was a lot of enthusiasm in the early stages of the pandemic for using AI for diagnosis, prognosis, and the identification and development of a COVID-19 drug. Sharing data, developing standardized formats for data collection, and developing transparent AI models are all necessary for researchers and healthcare professionals to achieve that. AI algorithms could have been trained and used to screen drug treatments for COVID-19 treatment effectiveness.

3 Materials and Methodology

3.1 Research Questions

A total of two research questions were formulated for the present study:

RQ1: How can the AI/ML techniques be used in the COVID-19 vaccine?
RQ2: Future research trends in AI/ML techniques for the COVID-19 vaccine?

3.2 Search Strategy

A search for peer-reviewed journal articles on IoT-based UAVs to improve agriculture which were published in well-reputed journals was carried out in January 2022. The investigation was conducted in three different databases: the Scopus database, the web of science database, and the IEEE database, as shown in Fig. 1. The search for the articles was done using the keywords: TITLE-ABS-

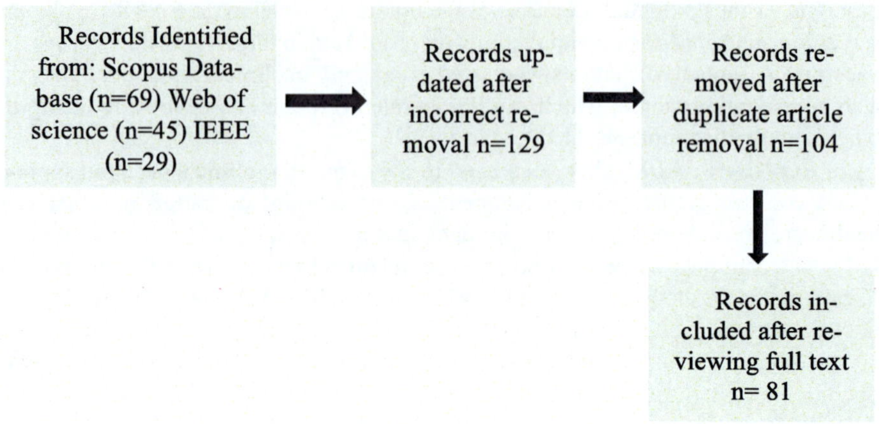

Fig. 1 Literature review search and selection process

KEY ((“Machine Learning” OR “ML”) OR (“Artificial Intelligence” OR “AI” OR “Artificial Intelligent”) OR (“Deep Learning”) AND (“COVID-19 Vaccine” OR “COVID-19 Vaccination”)).

The criteria used for selecting the articles were that they should be in the English language, they should be published on or before January 2022, they should be from a well-reputed journal or good publishers, and they should appear in the area of AI/ML in the area of COVID-19 vaccination.

4 Results

4.1 Year-Wise Publication Process

AI/ML in COVID-19 vaccination is an emerging field in computer science. COVID-19 vaccination requires technological advancement. So, in this chapter, we have highlighted the AI/ML in the area of COVID-19 vaccination. In Fig. 2, we can see that most of the publications were done in 2021 and 2020.

4.2 Highly Cited Paper (Global Citations)

Table 1 shows the global citations of the most cited by the authors globally. The concept of Global Citation Impact (GCI) is proposed to treat paper citations similarly.

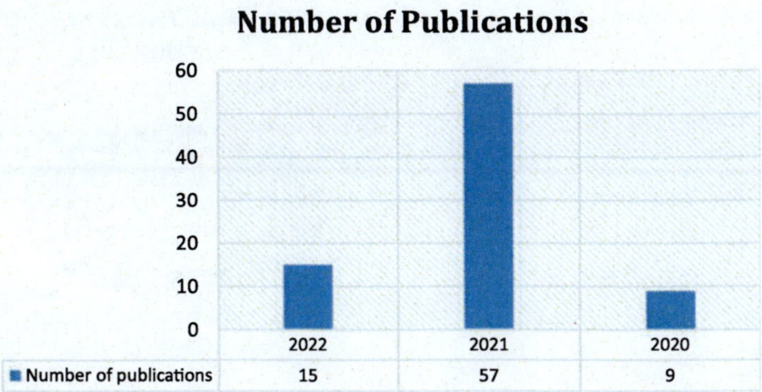

Fig. 2 Year-wise publication trends for the IoT-based UAVs for Indian agriculture

Table 1 List of articles with the most citations

Author	Title	Global citations
[17]	COVID-19 Coronavirus Vaccine Design Using Reverse Vaccinology and Machine Learning	228
[18]	Artificial intelligence-enabled analysis of public attitudes on Facebook and Twitter toward COVID-19 vaccines in the United Kingdom and the United States	33
[19]	Side effects and perceptions following COVID-19 vaccination in Jordan: A randomized, cross-sectional study implementing machine learning for predicting the severity of side effects	18
[20]	The role of artificial intelligence and machine learning techniques: Race for COVID-19 vaccine	17
[21]	Artificial intelligence model of drive-through vaccination simulation	11
[22]	Deep Learning Techniques and COVID-19 Drug Discovery: Fundamentals, State-of-the-Art and Future Directions	9
[23]	Application of machine intelligence technology in the detection of vaccines and medicines for SARS-CoV-2	7
[24]	Vaxign2: The second generation of the first Web-based vaccine design program using reverse vaccinology and machine learning	4
[25]	Application of artificial intelligence and machine learning for COVID-19 drug discovery and vaccine design	2

4.3 Authors Keyword Occurrence

These are keywords chosen by the author(s) that perfectly represent the contents of their manuscript from their viewpoint. Figure 3 shows the keywords that occurred mainly in the revived manuscripts.

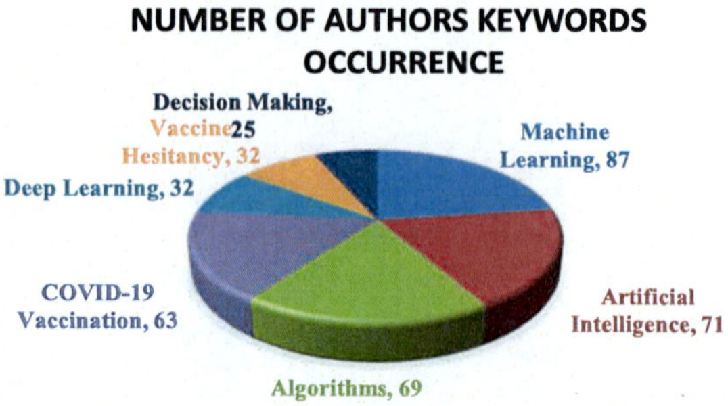

Fig. 3 Authors' keywords occurrence

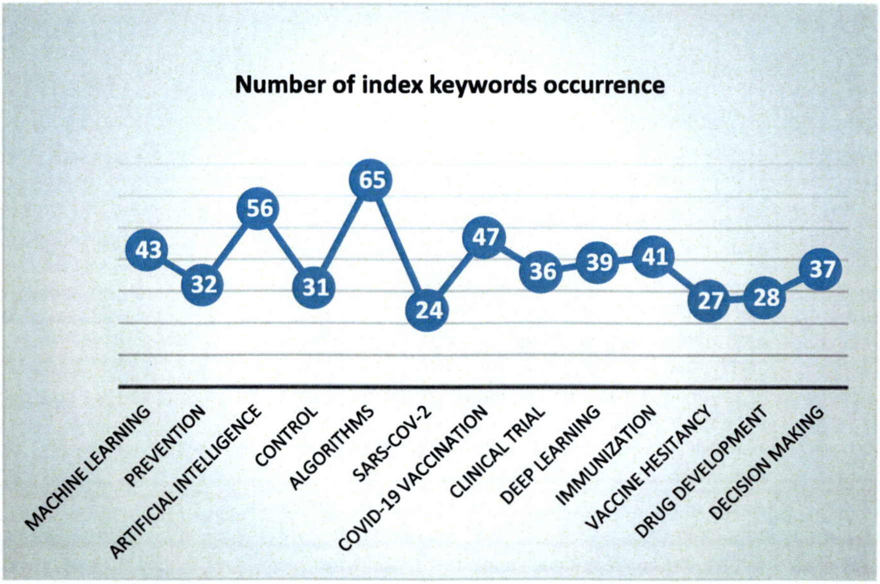

Fig. 4 Index keywords occurrence

4.4 Index Keyword Occurrence

These are keywords chosen by content providers and normalized using available public vocabulary. Figure 4 shows the number of index keywords in most of the journals.

5 Discussion and Future Trends

The AI/ML functionality for the vaccine can be used in a variety of ways, including determining which populations to target to halt the pandemic sooner, adjusting supply chain and distribution logistics to ensure the most significant number of people are vaccinated in the shortest amount of time, and tracking adverse reactions and side effects [26]. The AI system will indeed aid in creating actionable data sets that enable physicians to investigate root causes or issues that investigators do not have time to explore. AI has been used to advance components, among other things. It has been central to the process and will be critical to any vaccine adjustments, which appear to be essential at some point [7].

5.1 AI-/ML-Powered Vaccine and Antibody Development for COVID-19 Therapies

AI/ML has a meaningful impact on vaccine discovery due to data volume and the need for automatic conceptual extracting features. The AI/ML models for COVID-19 vaccine development use various methods to anticipate possible biomarkers, including artificial neural networks, gradient boosted trees, and convolutional neural networks [27]. The authors analyzed the current peptide-HLA forecasting tools to detect SARS-CoV-2 epitopes.

Because of the global COVID-19 pandemic, there's been significant worry on how to translate research results into powerful new medications and developments. As a result, new drug supplies and short growth times have climbed to the forefront of a study project's priority list. The rise of AI/ML has advanced the often long and complicated drug approval. With the increasing number of computational methods to treat COVID-19, their significant prediction power is crucial for early diagnosis to fight the rise in cases [28].

5.2 COVID-19 Vaccine Discovery

It is critical to determine the best potential targets for vaccine development to fight a virus's high infection rate. The human immune reaction reacts to a diseased cell by producing antibody levels by B cells or attempting to attack directly by T cells. To recognize immune cells from protein sequences, ML algorithms such as Support Vector Machine (SVM), Recursive Feature Selection (RFE), and Random Forest (RF) have been broadly used [29]. Deep convolutional neural networks (DCNN) are becoming an extra viable alternative for MHC and peptide conditional forecasting due to their low responsiveness in predicting locally grouped conversations in some cases. Natural language processing structures, especially language modelling methods, also have influenced the COVID-19 vaccine revelation.

5.3 ML Being Used to Counter COVID-19 Vaccine Hesitancy

Combining ML with behavioral psychology and configuration can aid in developing an adequate reaction to impact a person's attitudes. Researchers and practitioners can use ML to flag and recognize misleading information and assist in crafting the correct message and initiatives to help those who are hesitant to vaccinate. ML/AI model predicts which tests people have to take, whether they should get vaccinations, and when and where they should wear a mask [30].

5.4 AI/ML in Delivery

ML-enabled technology enables the automated assessment of defects in manufacturing machinery. The computerized method can identify flaws in vaccine production and vaccines using facial recognition. The benefit of these transferable skill quality checks is that the opportunities of delivering defective vaccines and equipment to end consumers are reduced. ML can help reduce the complex nature of production plans using current production data and mentoring practical algorithms [31]. ML models and methodologies can be used to achieve the optimal solution in which the entire country is vaccinated to recognize and minimize ineffectiveness and waste [32].

6 Conclusion

This study aims to perform research on AI/ML in COVID-19 vaccination. The COVID-19 vaccination drive is the world's largest vaccination drive required to vaccinate the world's entire population. There is a requirement for technological innovation in the COVID-19 vaccination drive. The latest technologies like AI/ML can help speed up the vaccination drive worldwide. The study identified 143 articles from the different databases and used 81 articles for further analysis. The study highlights the importance of AI/ML in the COVID-19 vaccination. This study will help the vaccine manufacturer know what technological innovations they can use in the vaccination process and its implementation. The policymakers can also use the AI/ML in COVID-19 vaccination to reach out to remote places for vaccination.

References

1. Piccialli, F., Di Cola, VS., Giampaolo, F., Cuomo, S.: The role of artificial intelligence in fighting the COVID-19 pandemic. Information Systems Frontiers. (2021)

2. Busulwa, R., Pickering, M., Mao, I.: Digital transformation and hospitality management competencies: Toward an integrative framework. International Journal of Hospitality Management. (2022)
3. Nayyar, A., Gadhavi, L., Zaman, N.: Machine learning in healthcare: review, opportunities and challenges. Machine Learning and the Internet of Medical Things in Healthcare. (2021)
4. Asif, M., Xu, Y., Xiao, F., Sun, Y.: Diagnosis of COVID-19, vitality of emerging technologies and preventive measures. Chemical Engineering Journal. 2021 Nov 1;423:130189.
5. Mustafa A, Rahimi Azghadi M. Automated machine learning for healthcare and clinical notes analysis. Computers. (2021).
6. Mukherjee, S., Venkataiah, C., Baral, MM., Pal, SK.: Analyzing the factors that will impact the supply chain of the COVID-19 vaccine: A structural equation modeling approach. Journal of Statistics and Management Systems. (2022).
7. Mukherjee, S., Chittipaka, V.: Analysing the adoption of intelligent agent technology in food supply chain management: an empirical evidence. FIIB Business Review. (2021)
8. Carleo, G., Cirac, I., Cranmer, K., Daudet, L., Schuld, M., Tishby, N., Vogt-Maranto, L., Zdeborová, L.: Machine learning and the physical sciences. Reviews of Modern Physics. (2019).
9. Mukherjee, S., Chittipaka, V., Baral, MM.: Developing a Model to Highlight the Relation of Digital Trust With Privacy and Security for the Blockchain Technology. InBlockchain Technology and Applications for Digital Marketing (2021).
10. Roy, R., Giduturi, A.: Survey on pre-processing web log files in web usage mining. Int. J. Adv. Sci. Technol. (2019).
11. Mukherjee, S., Baral, MM., Venkataiah, C., Pal, SK., Nagariya, R.: Service robots are an option for contactless services due to the COVID-19 pandemic in the hotels. Decision. (2021)
12. Mukherjee, S., Baral, MM., Chittipaka, V., Pal, SK., Nagariya, R.: Investigating sustainable development for the COVID-19 vaccine supply chain: a structural equation modelling approach. Journal of Humanitarian Logistics and Supply Chain Management. (2022).
13. Mukherjee, S., Chittipaka, V., Baral, MM., Srivastava, SC.: Can the Supply Chain of Indian SMEs Adopt the Technologies of Industry 4.0? InAdvances in Mechanical and Industrial Engineering (2022)
14. Mohanty, E., Mohanty, A.: Role of artificial intelligence in peptide vaccine design against RNA viruses. Informatics in Medicine Unlocked. (2021).
15. Mukherjee, S., Venkataiah, C., Baral, MM., Pal, SK.: Measuring the Organizational Performance of Various Retail Formats in the Adoption of Business Intelligent. InBusiness Intelligence and Human Resource Management (2022).
16. Mukherjee, S., Baral, MM., Chittipaka, V., Pal, SK.: A Structural Equation Modelling Approach to Develop a Resilient Supply Chain Strategy for the COVID-19 Disruptions. InHandbook of Research on Supply Chain Resiliency, Efficiency, and Visibility in the Post-Pandemic Era (2022).
17. Ong, E., Wong, MU., Huffman, A., He, Y.: COVID-19 coronavirus vaccine design using reverse vaccinology and machine learning. Front Immunol. (2021).
18. Hussain, A., Tahir, A., Hussain, Z., Sheikh, Z., Gogate, M., Dashtipour, K., Ali, A., Sheikh, A.: Artificial intelligence–enabled analysis of public attitudes on facebook and twitter toward covid-19 vaccines in the united kingdom and the united states: Observational study. Journal of medical Internet research. (2021).
19. Hatmal, MM., Al-Hatamleh, MA., Olaimat, AN., Hatmal, M., Alhaj-Qasem, DM., Olaimat, TM., Mohamud, R.: Side effects and perceptions following COVID-19 vaccination in Jordan: a randomized, cross-sectional study implementing machine learning for predicting severity of side effects. Vaccines. (2021).
20. Kannan, S., Subbaram, K., Ali, S., Kannan, H.: The role of artificial intelligence and machine learning techniques: Race for covid-19 vaccine. Archives of Clinical Infectious Diseases. (2020).

21. Asgary, A., Valtchev, SZ., Chen, M., Najafabadi, MM., Wu, J.: Artificial intelligence model of drive-through vaccination simulation. International journal of environmental research and public health. (2021).
22. Jamshidi, MB., Lalbakhsh, A., Talla, J., Peroutka, Z., Roshani, S., Matousek, V., Roshani, S., Mirmozafari, M., Malek, Z., Spada, LL., Sabet, A.: Deep learning techniques and COVID-19 drug discovery: Fundamentals, state-of-the-art and future directions. InEmerging Technologies During the Era of COVID-19 Pandemic (2021).
23. Alsharif, MH., Alsharif, YH., Albreem, MA., Jahid, A., Solyman, AA., Yahya, K., Alomari, OA., Hossain, MS.: Application of machine intelligence technology in the detection of vaccines and medicines for SARS-CoV-2. Eur. Rev. Med. Pharmacol. Sci. (2020).
24. Ong, E., Cooke, MF., Huffman, A., Xiang, Z., Wong, MU., Wang, H., Seetharaman, M., Valdez, N., He, Y.: Vaxign2: The second generation of the first Web-based vaccine design program using reverse vaccinology and machine learning. Nucleic acids research. (2021).
25. Arora, G., Joshi, J., Mandal, RS., Shrivastava, N., Virmani, R., Sethi, T.: Artificial intelligence in surveillance, diagnosis, drug discovery and vaccine development against COVID-19. Pathogens. (2021).
26. Pal, SK., Baral, MM., Mukherjee, S., Venkataiah, C., Jana, B.; Analyzing the impact of supply chain innovation as a mediator for healthcare firms' performance. Materials Today: Proceedings. (2022).
27. Baral, MM., Mukherjee, S., Nagariya, R., Patel, BS., Pathak, A., Chittipaka, V.: Analysis of factors impacting firm performance of MSMEs: lessons learnt from COVID-19. Benchmarking: An International Journal. (2022).
28. Mukherjee, S., Baral, MM., Venkataiah, C.: Supply Chain Strategies for Achieving Resilience in the MSMEs: An Empirical Study. InExternal Events and Crises That Impact Firms and Other Entities (2022).
29. Barbieri, D., Giuliani, E., Del Prete, A., Losi, A., Villani, M., Barbieri, A.: How artificial intelligence and new technologies can help the management of the COVID-19 pandemic. International Journal of Environmental Research and Public Health. (2021).
30. Aronskyy, I., Masoudi-Sobhanzadeh, Y., Cappuccio, A., Zaslavsky, E.: Advances in the computational landscape for repurposed drugs against COVID-19. Drug Discovery Today. (2021).
31. Pal, SK., Mukherjee, S., Baral, MM., Aggarwal, S.: Problems of big data adoption in the healthcare industries. Asia Pacific Journal of Health Management. (2021).
32. Dutta, S., Kumar, A., Dutta, M., Walsh, C.: Tracking COVID-19 vaccine hesitancy and logistical challenges: A machine learning approach. PloS one. (2021).

Embedded Platform-Based Heart-Lung Sound Separation Using Variational Mode Decomposition

Venkatesh Vakamullu, Aswini Kumar Patra, and Madhusudhan Mishra

1 Introduction

In recent years, the auditory information preserved in biological signals is paving the way for development of intelligent medical devices. For example, acoustic signals emanated from human heart and lungs can infer the information that facilitates the diagnosis of cardiovascular diseases and respiratory problems [1–5]. Quite few studies have already been carried out on the tangible physical models pertaining to heart and lung sound signal analysis. For example, some popular signal processing techniques, viz., time domain signal analysis techniques [3, 6], frequency-based and time-frequency resolution-based methods [7], and higher-order statistical and machine learning-based approaches, have been extensively used for heart and lung sound analysis and abnormality detection. In the clinical practice, the heart sounds recorded from each subject predominantly consist of fundamental heart sound S1 and S2 and rarely consist of secondary heart sounds S3 and S4 [8]. On the other hand, the lung sounds consist of different audio frequency components characterized by breathing patterns, viz., inspiration sound, expiration sound, and abnormal sounds such as crackles, wheezes, and whistles [9]. In clinical practice, both these sounds are conventionally heard by medical doctors using a widely

V. Vakamullu
Department of Electrical Engineering, Indian Institute of Technology Kharagpur, Kharagpur, West Bengal, India

A. K. Patra
Department of Computer Science & Engineering, North Eastern Regional Institute of Science and Technology, Nirjuli, Arunachal Pradesh, India

M. Mishra (✉)
Electronics & Communication Engineering, North Eastern Regional Institute of Science and Technology (NERIST), Itanagar, Nirjuli, Arunachal Pradesh, India

M. Mishra et al. (eds.), *Applications of Computational Intelligence in Management & Mathematics*, Springer Proceedings in Mathematics & Statistics 417,
https://doi.org/10.1007/978-3-031-25194-8_13

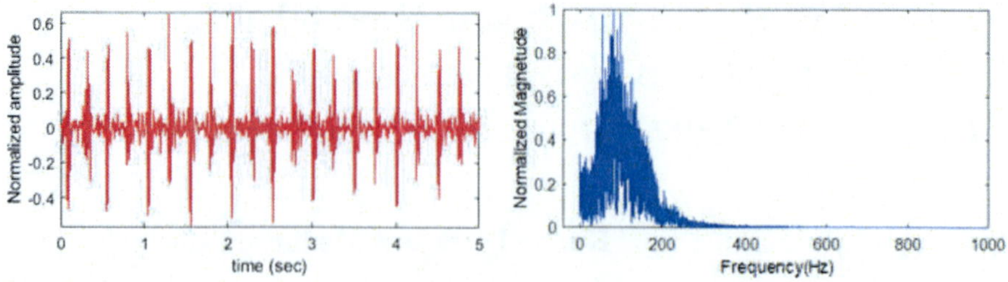

Fig. 1 Fundamental heart sounds of healthy subject and its frequency spectrum

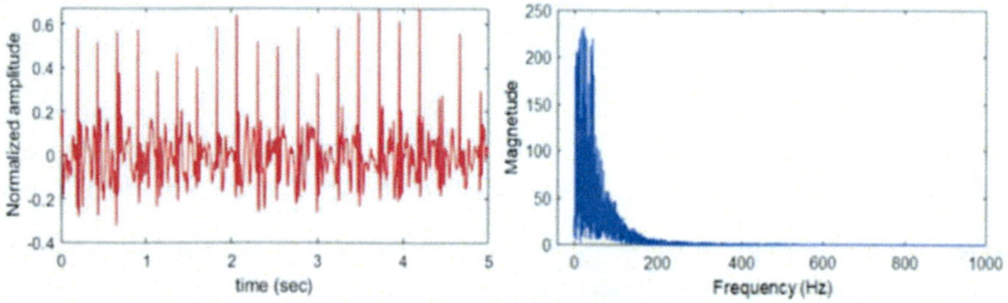

Fig. 2 Lung sounds of healthy subject and its frequency spectrum

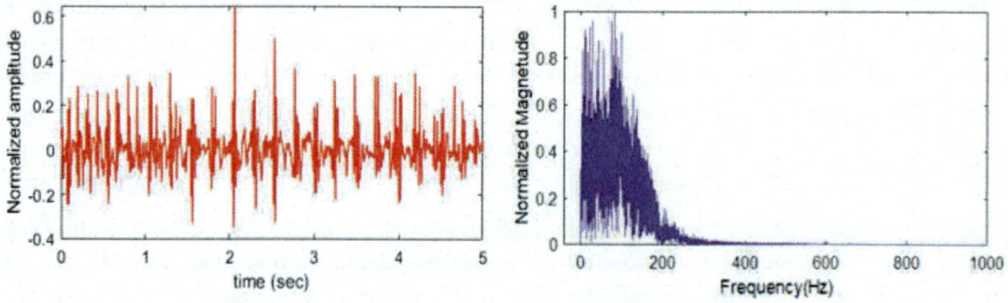

Fig. 3 Mixed version of heart and lung sounds and its frequency spectrum

accepted instrument, viz., stethoscope. However, the discrimination and analysis of these sounds are ambiguous even for experienced medical specialists. Quite a few times, this task becomes even harder as these sounds often mix with each other [10]. On the other hand, electronic stethoscope can facilitate the recording of heart and lung sound signals by means of audio waves on computers. Further, these recorded waves can be processed and visualized using sophisticated signal processing tools on standalone computer [11]. Nevertheless, it's a tedious task for the signal processing tools to accurately distinguish and separate the heart and lung sound signals from mixed heart-lung sounds. Figures 1 and 2 represent the individual heart and lung sounds, while Fig. 3 represents the heart-lung mixed sound signal.

In heart-lung sound signal analysis, separation of heart and lung sounds is the predominant task. Since the recorded signal is usually a mixture of individual heart and lung sounds, it is seldom to have the accessibility of pure heart-lung sounds, which in turn makes the separation of heart and lung sounds a challenging task. Primarily, the fundamental heart sounds of a healthy subject (S1 and S2) lie in the frequency range 20–150 Hz. Further, some high-frequency heart sounds such as murmurs lie in the range 100–600 Hz, or at times the frequencies are up to 1000 Hz [12]. On the other hand, the normal lung sounds of a healthy subsect spread in the frequencies 100–1000 Hz (tracheal sounds occupy 850–1000 Hz frequencies), while abnormal lung sounds such as wheeze spread over the frequencies between 400 and 1600 Hz, and crackle and rales occupy the range 100–500 Hz [13]. Hence, the high degree of overlap is perceived in heart and lung sounds in real clinical recordings. As a result of this, the heart-lung sound analysis and abnormality detection may end up with unreliable results and even at times may lead to false diagnosis. Therefore, it is extremely important to have sophisticated signal processing and decomposition tools to obtain the satisfactory level of heart and lung sound separation.

2 Related Work

Various decomposition techniques and sound separation algorithms facilitated for heart-lung sound decomposition have been studied so far. Objectives mentioned in [14, 15] delineate the adaptive filter-based techniques, while Mondal et al. [15] exploited the empirical mode decomposition (EMD) for heart-lung sound bifurcation. Hossain et al. [16, 17] discussed the use of discrete wavelet transform to filter out the lung sounds from mixed heart-lung signals. Pourazad et al. [18] presented a method signal to time-frequency domain (STFT) and continuous wavelet transform (CWT) ensemble method to discord the lung sounds from mixed signals. However, all the classical filtering techniques mentioned above offer poor performance due to inherently existing overlapped frequency components. The works accomplished in [19] discussed typical blind source extraction techniques, which majorly include independent component analysis (ICA) and its derivations. The crux of these methods relies on the fact that prior information about the source is not needed. However, ICA-based approaches require a minimum of two channels of recordings and hence not suitable for the recordings grabbed from a single channel [20]. Further, the optimum level of independence is assumed in between the channels. In recent times, the supervised nonnegative matrix factorization (NMF) method aiding the single channel was implemented for separating two individual sources [21]. It manifested its capacity in handling overlapped frequency components [22].

All the methods discussed in the literature end up with some disadvantages in one way or the other, and hence no idle method was tailored for complete and reliable separation of heart and lung sounds. On the other hand, these methods were exclusively implemented on standalone computers which primarily have some limitations to explore the heart-lung sound analysis. The standalone computers are bulky in nature; they require high power and are not portable and not suitable for handheld devices. Hence, it is not possible to handheld the device with the subject to monitor the recordings whenever and wherever it is needed. Hence, standalone computers cannot serve better the remotely located patients. Therefore, there is need for a portable, low-power, and handheld embedded platform-based implementation for heart-lung sound separation. In light of the above investigation, in this work, we proposed a novel framework assisted with variational mode decomposition algorithm for heart-lung sound separation. We implemented the same on Raspberry PI-based embedded platform to emulate the real-time analysis of heart and lung sounds. The following sections of this article delineates the methodology, data acquisition, experimental work, results and discussion, and conclusion.

3 Methodology

3.1 Variational Mode Decomposition

Variational mode decomposition [23] is a signal decomposition algorithm which predominantly applies on the non-linear and non-stationary signals to decompose into group of individual narrowband subcomponents called modes. The primary objective of the VMD algorithm is to decompose any non-linear and non-stationary multifrequency toned signal f into a finite number of narrowband sub-signals (modes), u_k; each mode k is distributed around a center frequency ω_k, which is to be computed along with the decomposition. The resulting constrained variational problem is the following:

$$min_{\{u_k\},\{\omega_k\}} \left\{ \sum_k \left\| \partial_t \left[\left(\delta(t) + \frac{j}{\pi t} \right) * u_k(t) \right] e^{-j\omega_k t} \right\|_2^2 \right\}$$
$$such\ that \quad \sum_k u_k = f \tag{1}$$

where $\{u_k\} = \{u_1, \ldots, u_K\}$ and $\{\omega_k\} = \{\omega_1, \ldots, \omega_K\}$ are shorthand notations for the set of all modes and their center frequencies, respectively. Equally, $\Sigma_k = \sum_{k=1}^{K}$is understood as the summation over all modes.

The individual narrow band sub-signal components/modes can be derived as

$$u_k^{n+1} = argmin1_{u_k \in X} \left\{ \alpha \left\| \partial_t \left[(\delta(t) + j/\pi t) * u_k(t) \right] e^{-j\omega_k t} \right\|_2^2 + \left\| f(t) - \sum_i u_i(t) + \lambda(t)/2 \right\|_2^2 \right\} \tag{2}$$

The center frequencies ω_k do not appear in the reconstruction fidelity term, but only in the bandwidth prior. The relevant problem thus reads:

$$\omega_k^{n+1} = argmin_{\omega_k} \left\{ \left\| \partial_t \left[\left(\delta(t) + \frac{j}{\pi t} \right) * u_k(t) \right] e^{-j\omega_k t} \right\|_2^2 \right\} \quad (3)$$

Hence the VMD algorithm described in this section is directly applied to the mixed version of heart-lung sound signals to decompose the individual modes and separate the heart and lung sounds.

3.2 Data Acquisition

In this work, the proposed framework is subjected to the mixed heart-lung sound signals obtained from two standard and publicly available online databases, viz. Michigan heart and sound library [24] and Littmann lung sound library [25]. Michigan library constitutes 23 heart sound signals obtained from various healthy and abnormal subjects by placing the electronic stethoscope at the apex area, supine and left decubitus; aortic area, sitting; and pulmonary area, supine. The recordings elapse for 5–20 seconds and all the recordings have a resolution of 44.1 kHz sampling rate. Similarly, the Littmann library constitutes six different lung sounds obtained from normal and abnormal subjects. In addition, these sounds are annotated with labels for clear identification of breathing cycles and respective class of sounds. All the lung recordings have a resolution of 44.1 kHz sampling frequency. Further, heart sounds and lung sounds obtained from the above-mentioned databases are clubbed up to form the mixed version of heart and lung sounds.

3.3 Experimental Setup

Figure 4 depicts the flow diagram representation of proposed framework. The mixed signals prepared from the pure heart and lung sounds obtained from Michigan and Littmann databases are fed to the proposed framework shown in Fig. 5. At first, since the mixed signals have an inherent sampling rate of 44.1 kHz, they are fed to the decimator where the mixed signals are downsampled to 4 kHz. Next, it passes through a fourth-order Butterworth lowpass filter with 2 kHz cut-off frequency to alleviate the high-frequency noises and other artifacts. Next, it passes through variational mode decomposition algorithm (VMD) where the mixed heart-lung sound signals are decomposed into narrowband sub-signals called modes. At last, these individual modes are fed to mode selection and combiner block to combine the modes which exclusively forms the heart sounds and lung sounds separately. In this way, the proposed framework separated the heart sounds and lung sounds from the mixed version of heart and lung sounds.

Figure 5 represents the experimental setup depicting the implementation of proposed framework on Raspberry PI. Raspberry PI is a single-board computer which facilitates the signal processing operation to compete with the standalone

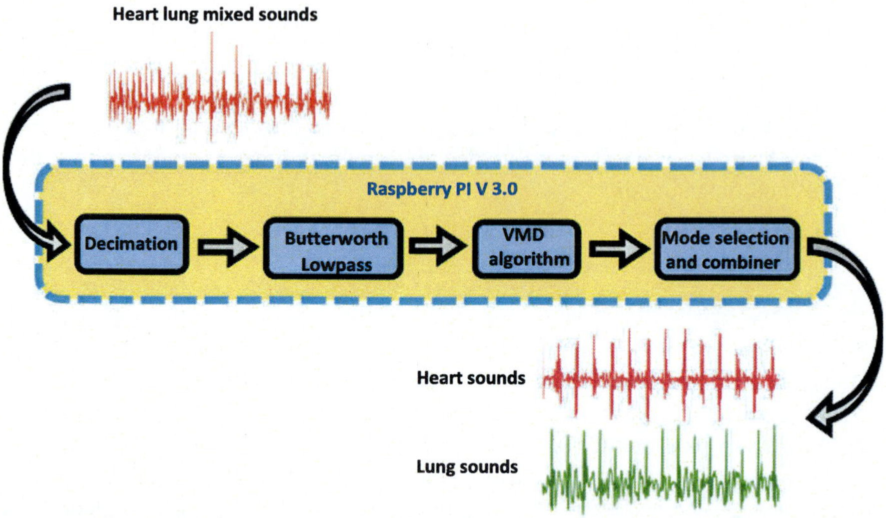

Fig. 4 Flow diagram representation of embedded platform-based heart-lung sound separation framework

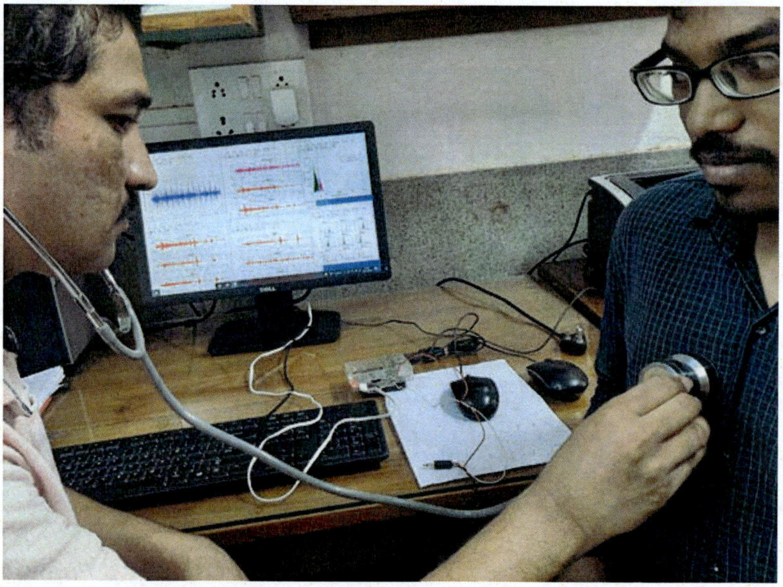

Fig. 5 Experimental step to facilitate the implementation of proposed framework on Raspberry PI

PC. Though the Raspberry PI module is a tiny and low-power device, it offers sufficiently good processing speed to emulate the real-time operation of the heart-lung sound separation. Hence, Raspberry PI-based VMD algorithm can perfectly decompose the individual modes from mixed heart signals.

4 Results and Discussion

In this work, Raspberry PI is used as an embedded module to implement the proposed framework. The framework uses two publicly available online databases to manifest the performance of the proposed algorithm. The entire work is implemented by using Python programming language on Raspberry PI module. Figure 6 shows the output of the VMD algorithm. As shown in Fig. 6, the mixed heart sound signal is decomposed in 8 modes, in which the first 4 modes represent the high-frequency sub-signals, while the last 4 modes represent the low-frequency sub-signals. Hence, the mode selection and combiner block intuitively select the modes 1, 2, 3, and 5 to reconstruct the high-frequency heart sounds, while the modes 4, 6, 7, and 8 are combined to form low-frequency lung sounds.

Figure 7 depicts the frequency spectrum of each individual mode. Figure 8 depicts the reconstructed heart and lung signals from mixed heart and lung sound signals. Table 1 displays the performance metrics of the proposed algorithm on Windows PC as well as Raspberry PI module. From the table, it is clearly observed that the proposed methods exhibit the same performance on both Windows and Raspberry PI platforms. However, there is significantly small difference in mean absolute error and mean square error values observed on Raspberry PI and its counterpart. The reason behind this difference is the 64bit double datatype is used for data representation on Windows, while the 32-bit floating-point representation is used on Raspberry PI.

Since Raspberry PI is a resource-constrained and low-power computing platform, it is inevitable to maintain the trade-off between accuracy and resource utilization. Hence, 32-bit floating-point representation is comfortably allowed for highly

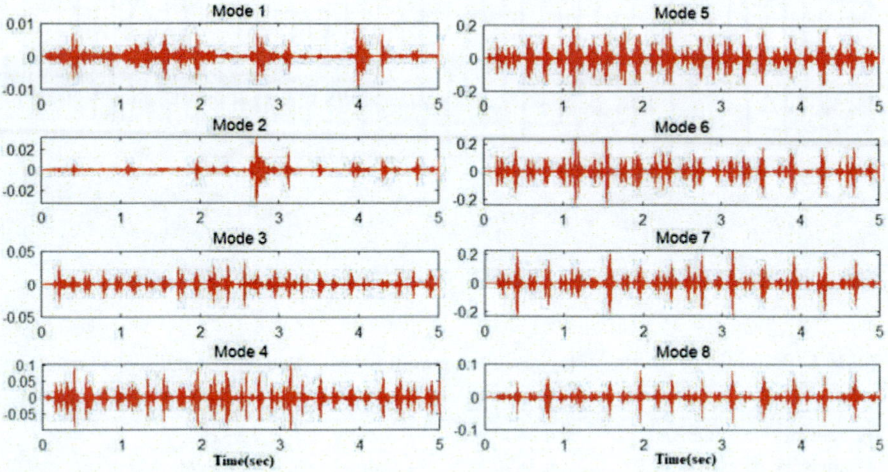

Fig. 6 Decomposed modes representing the individual sub-band signals obtained from VMD algorithm

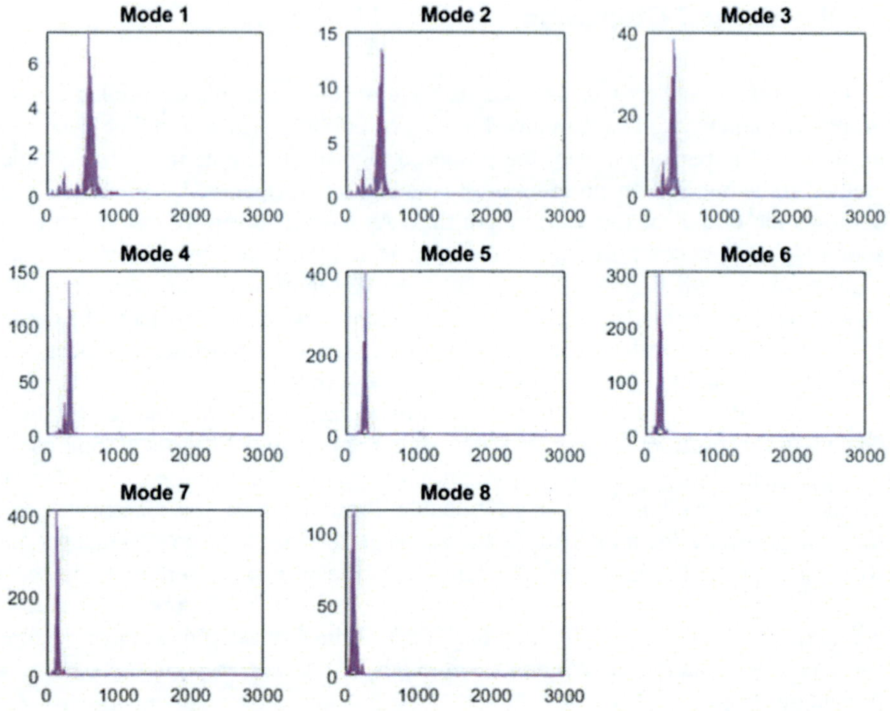

Fig. 7 Spectrum of decomposed modes shown in Fig. 6. Obtained from VMD algorithm

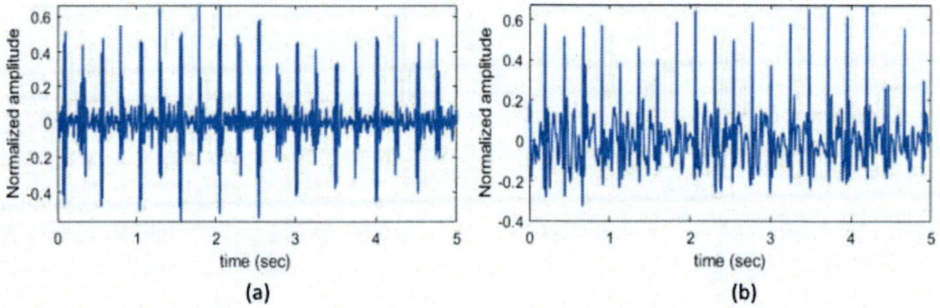

Fig. 8 Reconstructed heart and lung sound from decomposed modes of VMD algorithm

Table 1 Performance metrics of proposed framework for heart and lung sound separation

	MAE (%)	MSE	Computational time (s)
Framework on windows	4.3849	3.4669	19.3292
Framework on raspberry PI	4.3896	3.4674	179.5521

MAE mean absolute error, *MSE* mean squared error

intensive computations. In addition, it is clearly observed that the computational time for signal decomposition and recognition on Raspberry PI is relatively high as compared to that of Windows PC. The reasons for the huge difference in computation time are Raspberry PI operates at 1.4 G Hz clock speed, while the Windows PC operates at 3.5 G Hz clock speed, and Windows PC is boomed with 8 core processor and 8 GB RAM, while Raspberry PI constitutes quad core processor with 512 MB RAM. Finally, it is observed that Raspberry PI has competence in heart-lung sound signal separation.

5 Conclusion

In this work, VMD algorithm implementation on Raspberry PI for heart-lung sound separation imparts some fascinating conclusions: firstly, a novel framework was proposed to decompose and reconstruct the heart and lung sounds from heart-lung sound signals. Further, Michigan heart sound library and Littmann lung sound library were probed to the proposed algorithm to test the efficacy. Further, the same work was implemented on Raspberry PI-based embedded platform. The embedded implementation has almost imparted equal performance with standalone PC. Since Raspberry PI is a low-power handheld device, the precise implementation of the proposed framework can assist the remotely located patients to verify their heart-lung sounds on their own and can share the details of the recordings to medical doctors for further diagnosis. Finally, the detailed implementation of this work can find numerous applications in hospitals, clinics, and healthcare centers to assist the doctors and validate the manual diagnosis.

References

1. A. B. Bohadana, R. Peslin, H. Uffholtz and G. Pauli, "Potential for lung sound monitoring during bronchial provocation testing", *Thorax*, vol. 50, no. 9, pp. 955–961, 1995.
2. J. Hardin and J. Patterson Jr, "Monitoring the state of the human airways by analysis of respiratory sound", *Acta Astronautica*, vol. 6, no. 9, pp. 1137–1151, 1979.
3. C. Ahlstrom et al., "Feature extraction for systolic heart murmur classification", *Ann. Biomed. Eng.*, vol. 34, no. 11, pp. 1666–1677, 2006.
4. A. Jones, "A brief overview of the analysis of lung sounds", *Physiotherapy*, vol. 81, no. 1, pp. 37–42, 1995.
5. J. Cummiskey, T. C. Williams, P. E. Krumpe and C. Guilleminault, "The detection and quantification of sleep apnea by tracheal sound recordings", *Amer. Rev. Respir. Dis.*, vol. 126, no. 2, pp. 221–224, 1982.
6. H. Liang, S. Lukkarinen and I. Hartimo, "Heart sound segmentation algorithm based on heart sound envelogram", *Proc. Comput. Cardiol.*, pp. 105–108, 1997.
7. D. Kumar et al., "Detection of S1 and S2 heart sounds by high frequency signatures", *Proc. Int. Conf. IEEE Eng. Med. Biol. Soc.*, pp. 1410–1416, 2006.
8. Reddy, P. S., Salerni, R., & Shaver, J. A. (1985). Normal and abnormal heart sounds in cardiac diagnosis: Part II. Diastolic sounds. *Current problems in Cardiology*, *10*(4), 1–55.

9. Sarkar, M., Madabhavi, I., Niranjan, N., & Dogra, M. (2015). Auscultation of the respiratory system. *Annals of thoracic medicine*, *10*(3), 158.
10. Tavel, M. E. (1996). Cardiac auscultation: a glorious past—but does it have a future? *Circulation*, *93*(6), 1250–1253.
11. Hoyte, H., Jensen, T., & Gjesdal, K. (2005). Cardiac auscultation training of medical students: a comparison of electronic sensor-based and acoustic stethoscopes. *BMC medical education*, *5*(1), 1–6.
12. H. Ren, H. Jin, C. Chen, H. Ghayvat and W. Chen, "A novel cardiac auscultation monitoring system based on wireless sensing for healthcare", *IEEE J. Translational Eng. Health Med.*, vol. 6, Jun. 2018.
13. D. Emmanouilidou, E. D. McCollum, D. E. Park and M. Elhilali, "Computerized lung sound screening for pediatric auscultation in noisy field environments", *IEEE Trans. Biomed. Eng.*, vol. 65, no. 7, pp. 1564–1574, Jul. 2017.
14. V. K. Iyer, P. Ramamoorthy, H. Fan and Y. Ploysongsang, "Reduction of heart sounds from lung sounds by adaptive filtering", *IEEE Trans. Biomed. Eng.*, vol. 12, pp. 1141–1148, Dec. 1986.
15. Z. Wang, J. N. da Cruz and F. Wan, "Adaptive Fourier decomposition approach for lung-heart sound separation", *Proc. IEEE Int. Conf. Comput. Intell. Virtual Environ. Measure. Syst. Appl.*, pp. 1–5, 2015.
16. C. Lin, W. A. Tanumihardja and H. Shih, "Lung-heart sound separation using noise assisted multivariate empirical mode decomposition", *Proc. Int. Symp. Intell. Signal Process. Commun. Syst.*, pp. 726–730, 2013.
17. I. Hossain and Z. Moussavi, "An overview of heart-noise reduction of lung sound using wavelet transform based filter", *Proc. 25th Annu. Int. Conf. IEEE Eng. Med. Biol. Soc. (IEEE Cat. No. 03CH37439)*, pp. 458–461, 2003.
18. L. J. Hadjileontiadis and S. M. Panas, "A wavelet-based reduction of heart sound noise from lung sounds", *Int. J. Med. Inform.*, vol. 52, no. 1–3, pp. 183–190, 1998.
19. M. Pourazad, Z. Moussavi and G. Thomas, "Heart sound cancellation from lung sound recordings using time-frequency filtering", *Med. Biol. Eng. Comput.*, vol. 44, no. 3, pp. 216–225, 2006.
20. F. Ayari, M. Ksouri and A. T. Alouani, "Lung sound extraction from mixed lung and heart sounds FASTICA algorithm", *Proc. IEEE Mediterranean Electrotechnical Conf.*, pp. 339–342, 2012.
21. C. Lin and E. Hasting, "Blind source separation of heart and lung sounds based on nonnegative matrix factorization", *Proc. Int. Symp. Intell. Signal Process. Commun. Syst.*, pp. 731–736, 2013.
22. F. Weninger, J. L. Roux, J. R. Hershey and S. Watanabe, "Discriminative NMF and its application to single-channel source separation", *Proc. 15th Annu. Conf. Int. Speech Commun. Assoc.*, pp. 3749–3753, 2014.
23. Dragomiretskiy, K., & Zosso, D. (2013). Variational mode decomposition. *IEEE transactions on signal processing*, *62*(3), 531–544.
24. https://www.med.umich.edu/lrc/psb_open/html/repo/primer_heartsound/primer_heartsound.html
25. https://www.thinklabs.com/copy-of-lung-sounds.

MIMO: Modulation Schemes for Visible Light Communication in Indoor Scenarios

Chiranjeevulu Divvala, Venkatesh Vakamullu, and Madhusudhan Mishra

1 Introduction

In recent times, data traffic and bandwidth requirements are rapidly increasing due to significant growth in the number of smartphones and IoT devices. According to formal statistics, it is claimed that around 70% of traffic in the wireless networks is perceived in the indoor environment (home/office, etc.) [1]. Therefore, there is a need for highly reliable and cost-effective technologies to facilitate a persistent data transmission in indoor wireless systems. Predominantly, three approaches are explained to increase the capacity of wireless radio systems, viz., release of a new spectrum, ample number of nodes, and abstracting away the interference [1, 2].

Since RF spectrum is limited and expensive, optical wireless communication (OWC) can be emerged as an alternative approach to RF. OWC offers ultra-high bandwidth, less spurious to electromagnetic contamination, full spatial confinement, frequency reuse, and greater security in indoor environments. OWC systems that operate in the range of 380–750 nm are defined as visible light communication (VLC). In principle, the preliminary design of PHY (physical layer) of VLC systems is quite similar to RF. However, the intensity modulation (IM) and direct detection (DD) is used in between source (LED) and the receiver (photo diode).

C. Divvala
Department of Electronics and Communication Engineering, North Eastern Regional Institute of Science and Technology (NERIST), Nirjuli, Arunachal Pradesh, India

V. Vakamullu (✉)
Department of Electrical Engineering, Indian Institute of Technology Kharagpur, Kharagpur, West Bengal, India

M. Mishra
Electronics & Communication Engineering, North Eastern Regional Institute of Science and Technology (NERIST), Itanagar, Nirjuli, Arunachal Pradesh, India

M. Mishra et al. (eds.), *Applications of Computational Intelligence in Management & Mathematics*, Springer Proceedings in Mathematics & Statistics 417,
https://doi.org/10.1007/978-3-031-25194-8_14

This fascinating technique makes visible light communication technology attractive and takes the advantage of illumination and communication inside the room. This unique functionality of the VLC system finds numerous applications in offices, car-to-car communication, aeroplanes, etc., where the indoor light and high-speed data transmission are essential requirements [2].

Designing the channel model of optical signal propagating through wireless medium is a hassle task. Particularly, the reflection properties of light waveform from discontinued surface greatly influence energy distribution of the light wave. Changes occur in the position, azimuthal angle, and elevation angle of the source and/or receiver which alter the channel parameters. On the other hand, user motion, blockage of signals, and shadow also alter the channel characteristics. Performance analysis of various channels is delineated in Sect. 2.1. In VLC bandwidth of the system limits the transmission rate. However, the key solution to enhance the transmission rate without expanding the power and bandwidth is to employ an orthogonal frequency division multiplexing (OFDM) scheme. The generic OFDM symbols are complex and bipolar quantities. Hence, the conventional RF-OFDM scheme needs to be revised to make it apposite for intensity modulation/direction detection systems. Usually, multiple LEDs are overlaid in the system to provide ample light, and these multiple LEDs function as multiple transmitters to manifest light MIMO communication. Realization of MIMO systems in VLC is quite difficult as compared to that of RF systems. In RF-MIMO systems, spatial diversity attributes the throughput (multiple spatial paths propagate in multiple directions with significant diversity). On the other hand, VLC MIMO light wave propagation experiences less diversity due to the similarity in paths across source and destination, particularly in indoor scenarios, Hence, spatial diversity is not aggressively seen in VLC MIMO systems. The implementation schemes of various MIMO-VLC systems are illustrated in Sect. 2.2.

The objective of this work is to study and extend the transmission competences of the multi-channel modulation schemes for indoor VLC systems and address the challenges associated with physical layer design assisted for VLC system and visible light access point deployment in the indoor scenarios to cater the secure, high data rate, and energy efficient communication for home and business environments.

2 Literature Survey

2.1 Channel Models for VLC

Communication channel characterization is assessed by the impulse response, which derives the effect of signal propagated through channel. Nemours works have been reported pertaining to channel characterization such as infrared (IR) modeling based on recursive method [3], iterative site technique [4], Monte Carlo ray tracing [5], modified Monte Carlo ray tracing [6], and combined deterministic and modified Monte Carlo (CDMMC) method [7]. Based on the works done in [3, 6, 7], VLC channel models were proposed in [8–12]. In [8], Monte Carlo ray tracing was used to study the channel impulse response (CIR) for VLC system applied in an empty room. In this work, the fixed reflectance values are assumed for surface. In [9], adopted recursive method for CIR was discussed. In [10], wavelength dependency is considered in recursive scheme to compute the CIR.

In recent VLC communication, novel systems molded with realistic approaches are being developed by ray tracing mechanism [11]. proposed techniques based on ZEMAX, which is a commercial illumination design software. The simulation ambiance developed by ZEMAX specifies the geometric layout of the ambiance. However, this approach offers greater flexibility and less computational complexity as compared to popular recursive methods. This extensive study of VLC channel characteristics for different indoor scenarios accomplishes and quantifies the primary channel attributes, viz., RMS delay spread, excess mean delay, DC gain, and coherence bandwidth [12].

Although the VLC system exhibits multipath transmission in the indoor environment, data transmission may be encountered by the dispersion, and sudden blockages occur due to multipath reflection. This blockage further leads to intersymbol interference (ISI) at high data rates. Usually, ISI occurs due to multipath transverse light waves through the existing medium. However, OFDM caters salient advantages in the existence of ISI caused due to the frequency selective channel. Since the transmitted signal in OFDM system is bipolar in nature, in VLC system, it is merely impossible to pass them without modification [13, 14].

2.2 *MIMO Techniques for VLC*

The existence of MIMO systems in RF communications predominantly (in IEEE 802.11n, long-term evolution – LTE) imparts high data rates. In a similar passion, a group of LEDs can be incorporated in VLC system to achieve higher spectral efficiency. MIMO scheme accompanied with optical sources proposed in [15, 16] improves the spectral efficiency by 2 times and discards the effects caused due to DC bias as compared to existing SISO-OFDM-based optical system.

In [17], multiple LEDs arranged in a 4×4 grid with different spacings and different locations are considered for studying the effect of various MIMO techniques, viz., repetition coding (RC), spatial multiplexing (SMP), and spatial modulation (SM). Particularly, authors proposed a framework for approximating the bit error ratios (BERs) and validating the theoretical boundaries by appropriate simulations. The obtained results proved that diversity in gains provokes to exhibit robustness in RC against transmitter-receiver alignments. Since spatial multiplexing gains are not imparted by RC, it needs huge signal constellation areas to cater high spectral efficiencies. In contrast, SMP provides high data rates by utilizing multiplexed gains. However, to achieve these gains, SMP requires to exhibit significantly less channel correlation. SM is refined as combined technique crafted from MIMO and digital modulation schemes. In [17] it was presented that SM has greater robustness to high channel correlation than that of SMP and enables wider spectral efficiency than that of RC. It was observed that substantial improvement in the performance of

SMP and SM can be seen due to power imbalance. In this scenario, it is observed that mere blocking of some links is not an obviously followed method to achieve the low channel correlation. Though the received energy is remarkably reduced by blocking, it makes the excess degradation by improving the channel conditions facilitated for SMP and SM. In [18], a generalized spatial modulation was proposed; this scheme is the extension of SM, in which multiple transmitters are active for observed symbol duration. However, in this scheme, higher spectral efficiency is achieved due to activation of multiple LEDs at moderate to high SNRs. However, this design puts the burden of additional complexity [19] delineates about an active space collaborative constellation-based GSM. The obtained results in this work yield the better power gain without losing the multiplexing gain.

Advanced MIMO schemes for multi-user MIMO (MU-MIMO) are nuanced and 4000 are yet to be developed at full pace for VLC. Some previous works reported in [12, 13] dealt with the multi-user MISO (multiple-input single-output) schemes for VLC, where multiple LEDs (coupled through power lines in indoor scenarios) transmit the data to multiple receivers while alleviating the inter-receiver interference by the aid of zero-forcing precoding [14] proposed a precoding technique to facilitate the communication through MU-MIMO-VLC system. In this work diagonalization algorithm was used to abstract away the multi-receiver interference. It manifested that the proposed precoding scheme reduces the computational burden and power requirement at the receiver end. The block diagonalization algorithm needs precoding that imparts inconsistent performance, which in turn needs more receiving antennas (photodiodes) than the transmitting antennas [15, 16]. The study in [17] has examined the receiver's FOV (field of view) and further studied the SNR vs BER statistics.

However, almost all the MIMO techniques are enforced in line of sight (LOS) communication. Therefore, it is essential to study the MIMO for non-LOS paths (transmitters and receivers are precisely aligned) and persistence user motion. Therefore, this arrangement resembles practical scenarios of channels and further deals with the problems associated with reliable data transmission. Hence, it is extremely important to design proper transmission constellation in coordination with channel condition while targeting for significantly lesser computational complexity.

3 Work Done

3.1 Impulse Response for Indoor Channel

In the present work, we computed the impulse response of channel for empty rectangular room. The first-order reflections with constant transmitter and receiver positions were considered for analysis.

Figure 1 depicts the arrangement of transmitters and receivers in empty rectangular room to facilitate VLC.

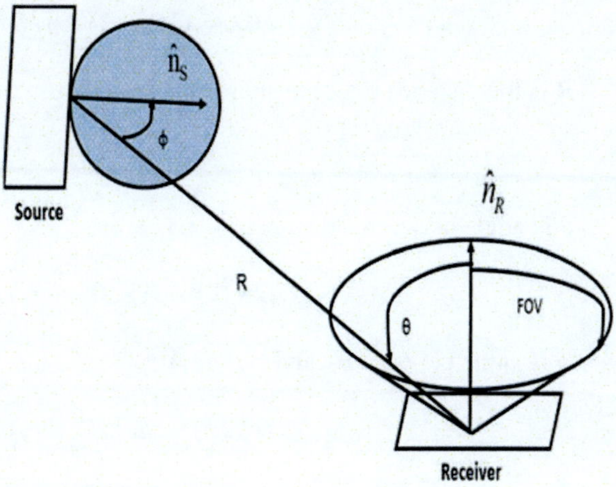

Fig. 1 Geometry of source and detectors without reflectors

Consider an optical source S represented by a position vector r_s and a unit length orientation vector $\hat{n}_s$ as depicted in Fig. 1. The radiation pattern of source having uniaxial symmetry is given by [5]

$$R\,(\Phi) = \frac{n+1}{2\pi} P_S COS^n\,(\phi) \tag{1}$$

where n is mode number, which is given by

$$n = -\frac{\ln(2)}{\ln COS\phi^1/_2} \tag{2}$$

$\Phi 1/2$ is half-power semi-angle

A unit impulse of optical intensity emitted by the point source is denoted by a third-order tuple as

$$S = \{r_s, \hat{n}_s, s\}$$

Similarly, a receiving element R with position r_r, orientation $n^\wedge_R$, area A_r, and field of view FOV will be denoted by a four-tuple

$$R = \{r_{r,}\hat{n}_r, A_r, FOV\}$$

Line of sight response is given by

$$h\,(t; S, R) \approx \frac{n+1}{2\pi} COS^n\,(\Phi)\, d\Omega rect\,(\theta/FOV)\,\delta\,(t - R/C) \tag{3}$$

where dΩ represents a solid angle formed by receiver differential area

$$d\Omega \approx COS\,(\theta)\,A_R/R^2 \tag{4}$$

R is the distance between source and receiver

$$R = \|r_s - r_r\| \tag{5}$$

θ is angle between $\hat{n}_R$ and $(r_s - r_r)$

$$COS\,(\theta) = \hat{n}_R.\,(r_s - r_r)\,/R \tag{6}$$

ϕ is angle between $\hat{n}_S$ and $(r_r - r_s)$

$$COS\,(\phi) = \hat{n}_S.\,(r_r - r_s)\,/R \tag{7}$$

Multiple bounce is response given by

$$h\,(t;\,S,\,R) = \sum_{k=0}^{\infty} h^{(k)}\,(t;\,S,\,R) \tag{8}$$

where $h^{(k)}(t)$ is the response of light subjected to k reflections. The line of sight response $h^{(0)}(t)$ is given by (3), while higher-order terms ($k > 0$) are computed in a repetitive manner:

$$h^{(k)}\,(t;\,S,\,R) = \int_s h^{(0)}\left(t;\,S,\left\{r,\hat{n},\,\pi/2,\,d\,r^2\right\}\right) \otimes h^{(k-1)}\,(t;\,\{r,n,\,1\}\,,\,R) \tag{9}$$

where symbol $\otimes$ denotes convolution in our case $k = 1$, we calculated first-order reflection by segmenting the reflection surface into tiny reflecting elements each with area ΔA

$$h^{(k)}\,(t;\,S,\,R) \approx \sum_{i=1}^{n} h^0\,(t;\,S,\,\in_i) \otimes h^{(k-1)}\,(t;\,\in_i,\,R) \tag{10}$$

where $\in_i$ confines ith element and N is the total number of elements. Determination of higher-order reflections by using Eq. (10) is not efficient, as identical computations have to be performed multiple times. For $k > 1$, we can use modified Monte Carlo method [7].

3.2 Results and Discussions

The system parameters that are used in this work are listed in Table 1. Figure 2 shows line of sight response scaled by 1.23×10^{-6} W. The reflected response shown in Fig. 3 indicates four peaks obtained from four walls of the room, and it also specifies that reflected power from four walls is 0.505×10^{-6} W.

It can be observed from Eq. (9) that power decreases for each of higher-order reflections. Further, this power is received much later than all other lower-order reflections (Fig. 4).

Table 1 System parameters

	Length (X)	5 m
Room	Width (Y)	5 m 3.5 m
	Height (Z)	
	Mode number	1
Source	X Y Z	2.5 m 2.5 m 3 m
	Elevation	−90°
	Azimuth	0
Receiver	Area	1 cm^2
	X, Y, Z	0.5 m, 1 m, 0
	FOV	85°
	Elevation	90°
	Azimuth	0
Reflectivity		0.8
	N_x	500
	N_y	500
First bounce	N_z	300

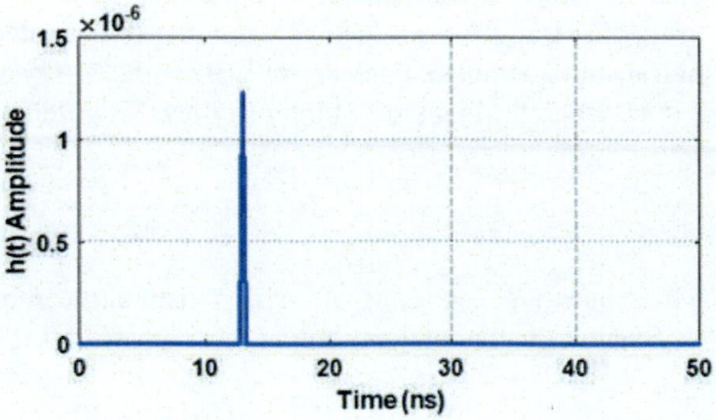

Fig. 2 LOS response

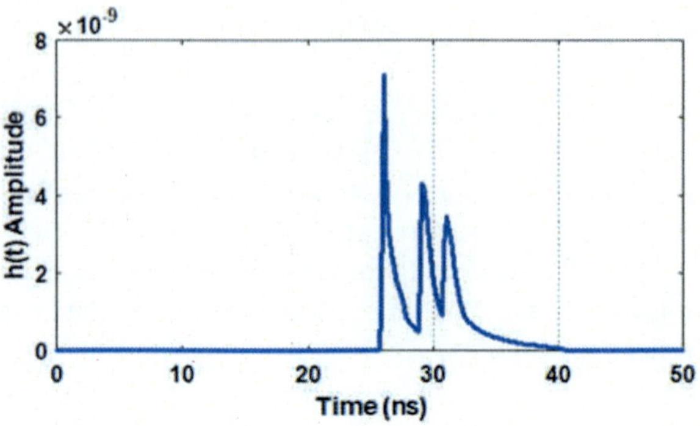

Fig. 3 NLOS (first-order reflections) response

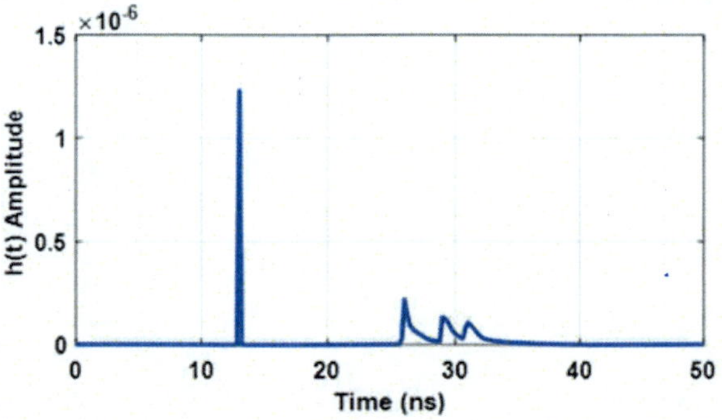

Fig. 4 Channel response (LOS and NLOS)

3.3 Generalized Spatial Modulation (GSM)

In GSM scheme, information is not only merely conveyed as modulation symbols sent on active LEDs, but also through indices of the active LEDs. In each channel usage, the transmitter selects N_a (number of activated antennas) out of N_t (number of transmit antennas). Each active LED emits M-ary an intensity modulation symbol $I \in M$, where M is the set of intensity level is given by

$$I_m = \frac{2I_p m}{M+1} \tag{11}$$

where m = 1,2 . . . .M and I_p is statistical mean of optical power.

Eventually, the total volume of bits conveyed through channel using GSM is:

$$\eta_{gsm} = \log_2 \left\lfloor \left(\frac{N_t!}{N_a!\,(N_t - N_a)!} \right\rfloor + N_a \lfloor \log_2 M \rfloor bpcu \tag{12}$$

3.4 System Model

Let $N_t = 4$, $N_a = 2$ and $M = 2$. In this setup, four bits can be transmitted per channel use. Out of all transmitted bits, the first couple of bits are related to the LED activation pattern, and the last couple of bits are related to the intensity levels of active LEDs as shown in Fig. 5. The LEDs and the photo detectors are located in a room of dimensions 5 × 5 × 3.5 m as displayed in Table 2. The LEDs are located at a height of 0.5 m under the ceiling, and the photo detectors are positioned on a bench of height 0.8 m.

Assuming perfect synchronization, the received signal vector at the receiver is given by

$$Y = rHX + n$$

where x is an N_t- dimensional vector with exactly N_a non-zero elements.

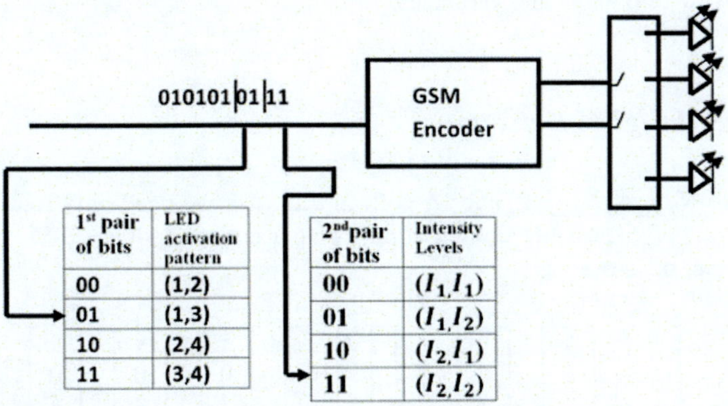

Fig. 5 GSM transmitter for VLC system with $N_t = 4$, $N_a = 2$, and $M = 2$

Table 2 System parameters

	Length (X)	5 m
Room	Width (Y)	5 m
	Height (Z)	3.5 m
	Height from floor	3
Transmitter	Mode number	1
	Elevation azimuth	−90°
		0°
	Height from floor	0.8 m
	Area	1
Receiver	FOV	85°
	Responsivity, r	0.75 A/W
	Elevation	90°
	Azimuth	0

H denotes the $N_r X N_t$ optical MIMO channel matrix

$$H = \begin{bmatrix} h_{11} & h_{12} & h_{13} & .. & h_{1N_t} \\ h_{21} & h_{22} & h_{23} & .. & h_2 N_t \\ h_{31} & h_{32} & h_{33} & .. & h_3 N_t \\ . & . & . & .. & . \\ h_{N_r 1} & h_{N_r 2} & h_{N_r 3} & .. & h_{N_r N_t} \end{bmatrix}$$

$$h_{ij} = \frac{n+1}{2\pi} COS^n \phi_{ij} COS\theta_{ij} \frac{A}{R_{ij}^2} rect\left(\frac{\theta_{ij}}{FOV}\right) \tag{13}$$

where h_{ij} is the LOS (line of sight) channel gain between jth LED and ith photo detector, j = 1,2,.....N_t, and i = 1,2...N_r; further higher-order reflections h_{ij} can be calculated by using Eq. (3). r is the responsivity of the detector and n is the noise vector of dimension $N_r \times 1$. Each element in the noise vector n is the sum of received thermal noise and ambient shot noise, which can be modeled as i.i.d. real AWGN with zero mean and variance σ^2.

3.5 GSM Signal Detection

If we take $N_t = 4$, $N_a = 2$, and M = 2, in this case we need only four activation patterns out of 6 possible combinations, and a corresponding GSM set for this example can be chosen as

$$S_{N_t,M}^{N_a} = \begin{bmatrix} \frac{2}{3} \\ \frac{2}{3} \\ 0 \\ 0 \end{bmatrix}, \begin{bmatrix} \frac{2}{3} \\ \frac{4}{3} \\ 0 \\ 0 \end{bmatrix}, \begin{bmatrix} \frac{4}{3} \\ \frac{2}{3} \\ 0 \\ 0 \end{bmatrix}, \begin{bmatrix} \frac{4}{3} \\ \frac{4}{3} \\ 0 \\ 0 \end{bmatrix}, \begin{bmatrix} \frac{2}{3} \\ 0 \\ \frac{2}{3} \\ 0 \end{bmatrix}, \begin{bmatrix} \frac{2}{3} \\ 0 \\ \frac{4}{3} \\ 0 \end{bmatrix}, \begin{bmatrix} \frac{4}{3} \\ 0 \\ \frac{2}{3} \\ 0 \end{bmatrix}, \begin{bmatrix} \frac{4}{3} \\ 0 \\ \frac{4}{3} \\ 0 \end{bmatrix}, \begin{bmatrix} 0 \\ \frac{2}{3} \\ 0 \\ \frac{2}{3} \end{bmatrix},$$

$$\times \begin{bmatrix} 0 \\ \frac{2}{3} \\ 0 \\ \frac{4}{3} \end{bmatrix}, \begin{bmatrix} 0 \\ \frac{4}{3} \\ 0 \\ \frac{2}{3} \end{bmatrix}, \begin{bmatrix} 0 \\ \frac{4}{3} \\ 0 \\ \frac{4}{3} \end{bmatrix}, \begin{bmatrix} 0 \\ 0 \\ \frac{2}{3} \\ \frac{2}{3} \end{bmatrix}, \begin{bmatrix} 0 \\ 0 \\ \frac{2}{3} \\ \frac{4}{3} \end{bmatrix}, \begin{bmatrix} 0 \\ 0 \\ \frac{2}{3} \\ \frac{2}{3} \end{bmatrix}, \begin{bmatrix} 0 \\ 0 \\ \frac{2}{3} \\ \frac{4}{3} \end{bmatrix}$$

where $I_1 = \frac{2}{3}$, $I_2 = \frac{4}{3}$

The choice of activation pattern will determine the performance of GSM system, since choosing a particular activation pattern can alter minimum Euclidean distance between any two GSM signal vectors X_1 and X_2 for a given H, which is given by

$$d_{\min,H} \triangleq \underset{X_1, X_2 \in S_{N_t,M}^{N_a}}{\arg\min} \left\| H(X_2 - X_1) \right\|^2 \tag{14}$$

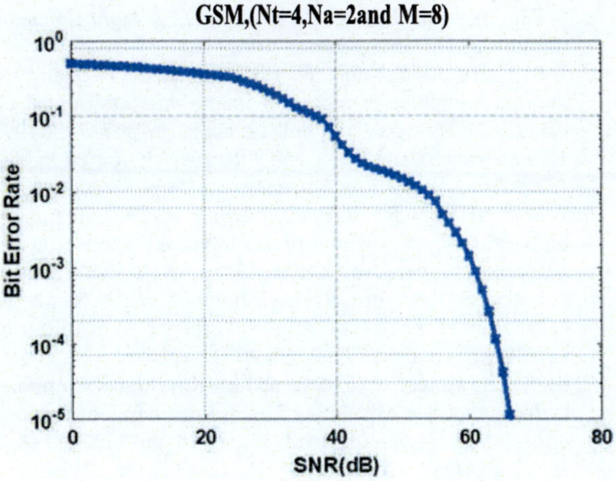

Fig. 6 BER plot for $N_t = 4$, $N_a = 2$, $N_r = 4$, and $M = 8$. (spectral efficiency: 8 bpcu)

The activation pattern set and mapping between elements of set and antenna selection bits are known at both transmitter and receiver. The ML decision rule for GSM signal detection is given by

$$\hat{X} = \underset{X \in S^{N_a}_{N_t,M}}{\arg\min} \left\| Y - rH\,X \right\|^2 \tag{15}$$

For small N_t and N_a, the GSM signal set may be fully enumerated, and ML detection can be done. But for large N_t and N_a, brute force computation of $X^{\wedge}$ becomes computationally infeasible. Therefore it is required to design a low complexity detection scheme.

3.6 Results and Discussions

In our work, we have considered the GSM system with $N_t = 4$, $N_a = 2$, $N_r = 4$, and $M = 8$. The system parameters that are used in this work are listed in Table II. The simulated results in Figure 6 show that BER reaches 10^{-6} at SNR 74 dB which verifies with original work. In this work we find optimum placement of LEDs by using Eq. (14).

References

1. N. Chi, H. Haas, M. Kavehrad, T. D. Little, and X.-L. Huang, "Visible light communications: demand factors, benefits and opportunities [guest editorial]," *IEEE Wireless Communications*, vol. 22, no. 2, pp. 5–7, 2015.

2. S. Dimitrov and H. Haas, *Principles of LED Light Communications: Towards Networked Li-Fi*. Cambridge University Press, 2015.
3. J. R. Barry, J. M. Kahn, W. J. Krause, E. A. Lee, and D. G. Messerschmitt, "Simulation of multipath impulse response for indoor wireless optical channels," *IEEE journal on selected areas in communications*, vol. 11, no. 3, pp. 367–379, 1993.
4. J. B. Carruthers and P. Kannan, "Iterative site-based modeling for wireless infrared channels," *IEEE Trans- actions on Antennas and Propagation*, vol. 50, no. 5, pp. 759–765, 2002.
5. F. Lopez-Hernandez, R. Perez-Jimenez, and A. Santamaria, "Modified monte carlo scheme for high-efficiency simulation of the impulse response on diffuse or wireless indoor channels," *Electronics Letters*, vol. 34, no. 19, pp. 1819–1820, 1998.
6. F. J. Lopez-Hernandez, R. Perez-Jimenez, and A. Santamaria, "Ray-tracing algorithms for fast calculation of the channel impulse response on diffuse ir wireless indoor channels," *Opt. Eng*, vol. 39, no. 10, pp. 2775–2780, 2000.
7. M. S. Chowdhury, W. Zhang, and M. Kavehrad, "Combined deterministic and modified monte carlo method for calculating impulse responses of indoor optical wireless channels," *Journal of Lightwave Technology*, vol. 32, no. 18, pp. 3132–3148, 2014.
8. H. Chun, C.-J. Chiang, and D. C. O'Brien, "Visible light communication using oleds: Illumination and channel modeling," in *Optical Wireless Communications (IWOW)*. IEEE, 2012, pp. 1–3.
9. H. Nguyen, J.-H. Choi, M. Kang, Z. Ghassemlooy, D. Kim, S.-K. Lim, T.-G. Kang, and C. G. Lee, "A matlab-based simulation program for indoor visible light communication system," in *Communication Systems Networks and Digital Signal Processing (CSNDSP), 7th International Symposium on*. IEEE, 2010, pp. 537–541.
10. K. Lee, H. Park, and J. R. Barry, "Indoor channel characteristics for visible light communications," *IEEE Communications Letters*, vol. 15, no. 2, pp. 217–219, 2011.
11. F. Miramirkhani and M. Uysal, "Channel modeling and characterization for visible light communications," *IEEE Photonics Journal*, vol. 7, no. 6, pp. 1–16, 2015.
12. A. Yesilkaya, E. Basar, F. Miramirkhani, E. Panayirci, M. Uysal, and H. Haas, "Optica MIMO-OFDM with generalized led index modulation," *IEEE Transactions on Communications*, 2017.
13. T. Fath and H. Haas, "Performance comparison of MIMO techniques for optical wireless communications in indoor environments," *IEEE Transactions on Communications*, vol. 61, no. 2, pp. 733–742, 2013.
14. S. Alaka, T. L. Narasimhan, and A. Chockalingam, "Generalized Spatial Modulation in indoor wireless visible light communication," in *Global Communications Conference (GLOBECOM)*. IEEE, 2015, pp. 1–7.
15. C. R. Kumar and R. Jeyachitra, "Power efficient generalized spatial modulation MIMO for indoor visible light communications," *IEEE Photonics Technology Letters*, vol. 29, no. 11, pp. 921–924, 2016.
16. J. Chen, N. Ma, Y. Hong, and C. Yu, "On the performance of MU-MIMO indoor visible light communication system based on thp algorithm," in *Communications in China (ICCC)*. IEEE, 2014, pp. 136–140.
17. Y. Hong, J. Chen, Z. Wang, and C. Yu, "Performance of a precoding MIMO system for decentralized multiuser indoor visible light communications," *IEEE Photonics Journal*, vol. 5, no. 4, pp. 7 800 211–7 800 211, 2013.
18. Z. Yu, R. J. Baxley, and G. T. Zhou, "Multi-user MISO broadcasting for indoor visible light communication," in *Acoustics, Speech and Signal Processing (ICASSP), 2013 IEEE International Conference on*. IEEE, 2013, pp. 4849–4853.
19. R. Tejaswi, T. L. Narasimhan, and A. Chockalingam, "Quad-led complex modulation (QCM) for visible light wireless communication," in *Wireless Communications and Networking Conference Workshops (WCNCW)*. IEEE, 2016, pp. 18–23.

Image Resampling Forensics: A Review on Techniques for Image Authentication

Vijayakumar Kadha, Venkatesh Vakamullu, Santos Kumar Das, Madhusudhan Mishra, and Joyatri Bora

1 Introduction

In present digital era, images and videos are manifested as predominant source of information that is often carried through communication channels. Image, unlike text, represents an effective and natural communication media for humans because of their instantaneous and ease to understand the contents. However, some people intentionally create tampered images from the original images and put up them to social networking websites and applications to spread over the wrong information. Enormous modern tampering techniques and image editing software such as Photoshop intelligently change the information laid in images, which in turn makes the detection of manipulations hard and hassle task even for experts by the perception of their vision. So, the authenticity of multimedia data (audio/image/video) is a vital aspect of the current world. Therefore, in order to examine the authenticity

V. Kadha (✉) · S. K. Das
Department of Electronics and Communication Engineering, National Institute of Technology Rourkela, Odisha, India
e-mail: dassk@nitrkl.ac.in

V. Vakamullu
Department of Electrical Engineering, Indian Institute of Technology Kharagpur, Kharagpur, West Bengal, India

M. Mishra
Electronics & Communication Engineering, North Eastern Regional Institute of Science and Technology (NERIST), Itanagar, Nirjuli, Arunachal Pradesh, India

J. Bora
Department of Electronics and Communication Engineering, North Eastern Regional Institute of Science and Technology, Nirjuli, India
e-mail: jb@nerist.ac.in

M. Mishra et al. (eds.), *Applications of Computational Intelligence in Management & Mathematics*, Springer Proceedings in Mathematics & Statistics 417,
https://doi.org/10.1007/978-3-031-25194-8_15

of digital multimedia content such as identifying forgeries, the processing tracing history, identifying the origin, the field of information forensics was born [1–4]. For example, as shown in Fig. 1, content was copied and moved to form tampered image.

Image forensics is a subset of the general information forensics domain, and it deals with investigating likely causes and detection of forgeries w.r.t. images. Image forensics has several sub-sections depending on what aspect the forger is trying to manipulate. As shown in Fig. 2, image manipulation includes various sub-categories viz. image steganography, image forgery, image tampering, and image generation. All image editing and painting activities are referred to as image manipulation. Image steganography [7] is not the same thing as image forgery. In fact, in this technique, the information is kept hidden by changing the pixels in an image in some way. On the other hand, image tampering belongs to the category of image forgery as it creates fake information in an image in order to fool past events. While image generation techniques can be used for forgeries, they are not always used for that purpose.

Fig. 1 Copy–move forgery (original image left, tampered image right). One of the soldiers is hidden by coping with other content in the same image [6]

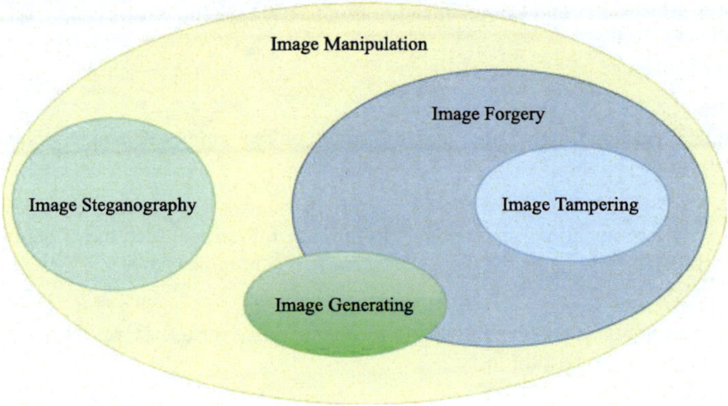

Fig. 2 The overview of image forensics [5]

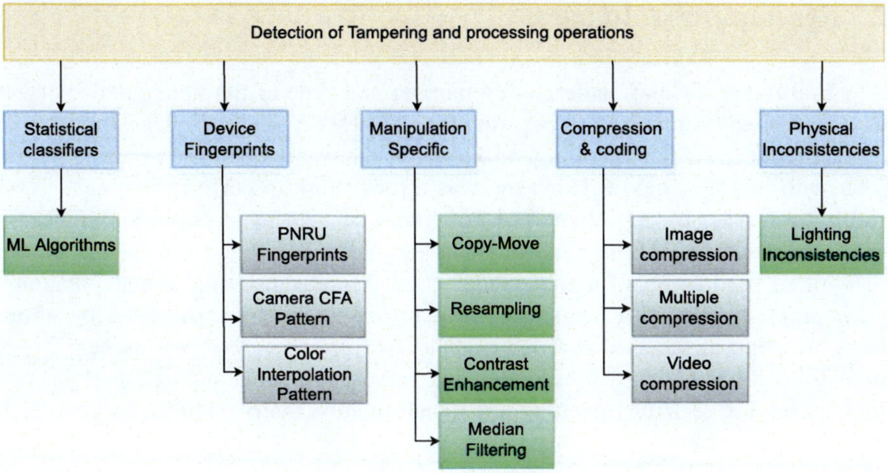

Fig. 3 The review on detection techniques for identifying image tampering and processing operations

Forensic approaches for tracing the processing history, detecting the modification, and identifying forgeries of multimedia files are divided into five groups based on how they work. Some predominant works are pertaining to statistical based in which statistical classifiers in association with efficient machine learning algorithms are used on a wide list of image statistics. Some other techniques facilitate the searching of inconsistencies found in fingerprints laid on capturing devices, manipulation-specific fingerprints, compression and coding fingerprints, and some techniques dedicated for searching of physical inconsistencies in multimedia content. However, all these manipulations cannot be detected single-handed by any particular forensic tool. In fact, the combination of two or more tools can work together to achieve the reliable detection of multiple varieties of forgeries and tampering. In this chapter, we elaborately present the overview of forensic procedures pertaining to the above-mentioned five areas. Figure 3 depicts the flow diagram representation of overview of image tampering detection schemes.

Common manipulations include copy–paste, splicing, filtering, resampling, etc., in both the presence and absence of compressions. Among the various kinds of fingerprints, the common ones which the signal analyst tries to catch are compression and coding fingerprints, resampling fingerprints, device fingerprints, etc. In this chapter, our discussion is emphasized on the review of various manipulation detection techniques of image authentication using resampling clues.

2 Resampling of Images

A digital image I_1 may undergo any number and type of the manipulations such as non-aligned cropping, resizing, rotation, successive geometric transformations. Perfect reconstruction, according to sampling theory, is obtained by using a sinc filter with infinite support. However, it is a tough task to perform in reality. As a result, various finite-support interpolation filters, such as linear, cubic, and truncated sinc, are commonly utilized as alternatives. Table 1 imparts the details of the most often used kernels for forgery detection. Moreover, by these manipulations, new sampling grid with new pixel intensity values will be generated by affine transformation of the axes $(n_1, n_2) \in Z^2$ to $A(n_1, n_2)^T + (\phi_1, \phi_2)^T$, where 2×2 transformation matrix (A) is used to resize, rotation, skewing and $(\phi_1, \phi_2)^T$ is used to translate. For resizing operation, the transformation matrix is in the form

$$A_\xi = \begin{bmatrix} \xi & 0 \\ 0 & \xi \end{bmatrix} \tag{1}$$

where ξ is the resizing factor. For image rotation, the affine transformation matrix with θ rotation angle is given as

$$A_\theta = \begin{bmatrix} \cos\theta & \sin\theta \\ -\sin\theta & \cos\theta \end{bmatrix} \tag{2}$$

While resampling takes place, the new resampling grid of the image (m_1, m_2) and intensity values of (m_1, m_2) can be obtained by performing convolution with interpolation kernel $h(n_1, n_2)$.

$$I_2 = h(n_1, n_2) * I_1(n_1, n_2) \tag{3}$$

where I_2 is the doctored image, i.e., resampled image, and I_1 is the original image. The forensics of resampling is an essential task in image forgery, image tampering detection, steganography, and steganalysis. In reality, even the entire image is resampled; it does not imply the meaning of image forgery; in fact, it gives the indication of processing the part of or the whole image. Moreover, when some part of the image is copied into the investigated image, it is essential to carry out some profound geometric transformations such as resizing and rotating the images to hide traces of forgery [1, 8–10]. In some scenarios, forger alters an image by substituting a portion of the image with content copied from elsewhere within the same image. Hence, to create forged image spatial transformations (i.e. resampling operations) are utilized which inherently leaves characteristic fingerprints that are typically not present in the genuine images. Due to the improper use of resampling in image tampering, resampling forensics attracts research attention.

Table 1 Different interpolation kernel expressions and its Fourier transforms

Name of kernel	Expression of the kernel $h(x)$	Fourier transform $H(f)$
Ideal sinc function	$\text{sinc}(t) = \frac{\sin(\pi t)}{\pi t}$	$\text{Rect}(f) = \begin{cases} 1 & \|f\| \leq 0.5 \\ 0 & \|f\| > 0.5 \end{cases}$
Linear interpolation	$\begin{cases} 1 - \|x\| & \|x\| \leq 1 \\ 0 & \|x\| > 1 \end{cases}$	$\text{sinc}^2(f)$
Cubic interpolation	$\begin{cases} \frac{3}{2}\|x\|^3 - \frac{5}{2}\|x\|^2 + 1 & \|x\| \leq 1 \\ -\frac{1}{2}\|x\|^3 + \frac{5}{2}\|x\|^2 - 4\|x\| + 2 & 1 < \|x\| \leq 2 \\ 0 & \|x\| > 2 \end{cases}$	$(3\,\text{sinc}(f) - 2\cos(f\pi))\,\text{sinc}^3(f)$
Lanczos3 windowed sinc	$\begin{cases} \text{sinc}(x)\,\text{sinc}\left(\frac{x}{3}\right) & \|x\| \leq 3 \\ 0 & \|x\| > 3 \end{cases}$	$18\,\text{Rect}(f) * \text{Rect}(3f) * \text{sinc}(6f)$

3 Resampling Detection Using Conventional Detection Methods

Several authors proposed resampling detection algorithms to detect forgeries in images. In [1], authors discussed that every pixel in a resampling image is correlated with its neighboring pixels. They used an expectation/maximization (EM) algorithm to compute the periodic correlations among the image pixels. The algorithm is based on a series of initialization parameters, and it takes a long time to get over it. Kirchner [8] presented an automatic detector based on the maximum spectral slope probability map *(p-map)* to use a linear predictor and reduce computational complexity. However, this detector got better results in upscaling but could not detect downscaled images. Gallagher [20], Mahdian and Saic [11] have proved that second-order derivatives in the interpolated signal exhibit periodicity. This feature is computed by estimating the DFT of a statistical averaged signal. However, the limitation of this feature is that the performance is decreased in a compressed scenario.

Moving away from the spectrum-based techniques, Padin et al. [21, 22] adopted other linear and 1-D signal methods. They presented a detector that can detect the difference between upscaled and genuine images. However, this resampling detector is analyzed to the upsampled case, and downscaled case was left untouched for future development. Kirchner et al. [9] presented a method to detect the resampling of compressed images using JPEG pre-compression artifacts. The techniques used for detecting and estimating the scaling factor of JPEG were refined by combining the energy-based [12] and predictor-based methods [23]. However, resampling detection is challenging to handle in recompressed images [24].

Later, Bianchi and Piva [13] delineated a technique that estimates the quantization matrix as well as the scaling factor in recompressed images. However, the spectrum contains a combination of resampling and shifted JPEG peaks [9]. So, it is not possible to get a resampling peak simply by selecting a more prominent peak (i.e., the resampling peak is considered as JPEG peak, and an incorrect estimate is made). On the other hand, shifted JPEG peak size is related to the quality factor of the previously compressed JPEG image. If the quality factor is high, it is pathetic to see the JPEG peaks. Further, conventional techniques fail to detect resampling processes in recompressed images, and even their performance is also not remarkable. Therefore, it is inevitable to adopt the real scenario where manipulations are directly employed on JPEG compressed image.

4 Resampling Detection Using Deep-Learning-Based Approaches

Convolutional neural networks are predominantly applied in media forensics due to their tremendous developments in the field of image processing and computer vision [25–27]. Unlike conventional vision problems, most image forensics research has focused on low-level patterns that are more nuisance than anything else. This philosophy is commonly used in deep learning setups. Many current deep learning approaches start by feeding residual images into a pool of networks, which constrains the learning process to learn a residual layer. However, this can be accomplished by an ensemble of initial layer with fixed parameters [28], trainable [29], or high-pass filters [30]. A two-stream network developed by [31] computes both low-level and high-level features by connecting the pre-eminent network to a general-purpose deep CNN. The summary of related work on resampling detection and estimation using both classical and deep learning approaches is shown in Table 2.

5 Datasets

The original and altered images are included in the tampering dataset. In general, the images laid in the tampered image data are labeled as "0" for the legitimate image and "1" for the tampered image. To benchmark the detection, the image alteration dataset includes ground truth. However, in the earlier days of research, the availability and scope of the image manipulation dataset were limited [32, 33]. During the initial days of research, people used to consider significantly small database with a limited number of images, and only a single type of tampering to employ on machine learning and deep learning architectures. Previously, the images were examined to identify whether they are original or tampered by means of simple and easy set of features extracted from the distinctive traces left behind by the specific sort of tampering, and then the obtained results were compared with the original mask. Hence, the old studies mentioned above have the limited applicability and generality. Further, the early reported methods work well for specific datasets and fail for most of the other datasets which are new to the above methods. Further, creating an ideal dataset that includes a diverse range of tampering activities, several image formats, and photographs taken under a variety of different circumstances is always a challenge.

However, as deep-learning-based methods have seamlessly evolved over the past decade, they demand significantly large datasets for efficient computation and classification [34], only few large datasets consisting of thousands of images are publicly available online databases, and even their repository is not adequate to manifest the effectiveness of deep learning algorithms. As a result, most researchers create synthetic images from the publicly available online datasets [30, 35–37]. Using these datasets, real-world image alteration detection may be detected by detecting the multiple tampering operations conducted and by identifying and pin-pointing the tampered location in the image. We looked at several publicly available datasets on image manipulation in this part (see Table 2). Image modification datasets are described in detail in Table 2, including information about image size,

Table 2 Summary of the dataset and interpolation techniques utilized on image resampling detection in the literature

Literature	# Images	Interpolation technique	Feature extraction technique	Classifier
Popescu and Farid [1], 2005	200	Bicubic	Probability maps (*p-map*) spectrum	Expectation/ Maximization (EM) Algorithm
Mahdian and Saic [11], 2008	40	Bicubic	Variance of *n*th order derivative of image using Radon transform	Threshold-based peak detector
Kirchner [8], 2008	200	Bilinear	Probability maps (*p-map*) spectrum	Cumulative periodograms
Feng et al. [12], 2012	7500	Bicubic	Normalized energy density (DFT)	SVM classifier
Bianchi and Piva [13], 2012	500	Bilinear	Integer periodicity property (IPM)	Threshold-based peak detector
Padin et al. [14], 2017	1317	Bilinear, Bicubic Lanczos	Asymptotic eigenvalue distribution of sample autocorrelation matrix	Threshold-based peak detector
Bayar and Stamm [15], 2018	3334	Bilinear	Constrained CNN	ET classifier, fully connected neural network
Qiao et al. [16], 2019	1338	Bilinear and nearest neighbor	Pixel-level texture map, block-level texture map, region-level texture map	LRT test
Cao et al. [17], 2019	50,000	Bilinear	Horizontal, vertical, and interleaved stream	Fully connected neural network
Liang et al. [18], 2019	1700	Bicubic	Residual neural network	Fully connected neural network
Liu et al. [19], 2019	1445	Bilinear	VGG Net with 25 convolutional layers	Fully connected neural network

count of images laid in the dataset, and a comprehensive description about the alterations made on the existing images of the dataset.

6 Conclusions and Future Directions

The study presented in this chapter examines several image forgeries techniques, publicly available image datasets, and deep learning approaches for image manipulation detection. Detecting and locating tampering in an image has gotten increasingly difficult due to the development of image alteration technologies. Handcrafted and transform domain features are used before the adoption of deep learning for manipulation detection, but in the recent times, deep learning approaches have conquered the top position in the studies pertaining to detection and assessment of image manipulation. The convolutional neural network (CNN) learns features from images' content and categorizes them using discriminative qualities. Deep learning architectures are predominantly used in image alteration detection and assessment. On the other hand, the CNN efficiently suppresses the image content to locate the traces left behind by the tampering process in the case of image manipulation detection methods by utilizing pre-processing layers.

Many tampering procedures are carried out nowadays, and post-processing is carried out in order to remove the traces left behind by each operation, guaranteeing that no evidence of the manipulation is left behind. Developing a most generic image modification detection scheme that can discriminate between authentic and modified images is extremely difficult. However, various deep learning methods have been developed to facilitate the paradigm of general-purpose image modification detection. These models have been demonstrated to be quite accurate in categorizing images as authentic or tampered. When it comes to classification, deep learning architectures require huge datasets that contain a sufficiently large number of images to manifest its performance to be effective. Though various datasets pertaining to image manipulation detection are reported in the literature, the disclosed datasets have a significantly less number of images. Hence, it is extremely difficult to develop, train, and test the deep learning classifiers with the existing datasets. Therefore, creating a synthesized image dataset can fix the problem and help to accomplish the successful training of the classifier.

A realistic manipulation detection system is difficult, but an efficient and generic image manipulation detection system can be constructed by combining numerous models of various sizes and utilizing the transfer learning approach. Furthermore, forgers are seeking to develop new methods of making tampered photographs in such a way that existing detection technology will be unable to detect tampering with the images. GAN (Generative Adversarial Network) [38] and DeepFakes [39] are two examples of sophisticated networks built by researchers to manufacture the fake content, such as the fake images generated by the GAN and the fake photos generated by the DeepFakes looking like real-world photos to aid the multimedia forensics. As a result, researchers should continue to investigate anti-forensics, anti-tampering countermeasures, and manipulation detection systems.

References

1. A. C. Popescu and H. Farid, "Exposing digital forgeries by detecting traces of resampling," IEEE Transactions on Signal Processing, vol. 53, no. 2, pp. 758–767, 2005.
2. H. Farid, "Image forgery detection," IEEE Signal Processing Magazine, vol. 26, no. 2, pp. 16–25, 2009.

3. A. C. Popescu and H. Farid, "Statistical tools for digital forensics," in International Workshop on Information Hiding. Springer, 2004, pp. 128–147.
4. M. C. Stamm, M. Wu, and K. R. Liu, "Information forensics: An overview of the first decade," IEEE Access, vol. 1, pp. 167–200, 2013.
5. L. Zheng, Y. Zhang, and V. L. Thing, "A survey on image tampering and its detection in real-world photos," Journal of Visual Communication and Image Representation, vol. 58, pp. 380–399, 2019.
6. V. Christlein, C. Riess, and E. Angelopoulou, "A study on features for the detection of copy-move forgeries," in Sicherheit 2010. Sicherheit, Schutz und Zuverlssigkeit, F. C. Freiling, Ed. Bonn: Gesellschaft fr Informatik e.V., 2010, pp. 105–116.
7. S. Katzenbeisser and F. Petitcolas, "Digital watermarking," Artech House, London, vol. 2, 2000.
8. M. Kirchner, "Fast and reliable resampling detection by spectral analysis of fixed linear predictor residue," in ACM Workshop on Multimedia and Security, 2008, pp. 11–20.
9. M. Kirchner and T. Gloe, "On resampling detection in re-compressed images," in IEEE International Workshop on Information Forensics and Security (WIFS), 2009, pp. 21–25.
10. C. Chen, J. Ni, Z. Shen, and Y. Q. Shi, "Blind forensics of successive geometric transformations in digital images using spectral method: Theory and applications," IEEE Transactions on Image Processing, vol. 26, no. 6, pp. 2811–2824, 2017.
11. B. Mahdian and S. Saic, "Blind authentication using periodic properties of interpolation," IEEE Transactions on Information Forensics and Security, vol. 3, no. 3, pp. 529–538, 2008.
12. X. Feng, I. J. Cox, and G. Doërr, "An energy-based method for the forensic detection of re-sampled images," in IEEE International Conference on Multimedia and Expo, 2011, pp. 1–6.
13. T. Bianchi and A. Piva, "Reverse engineering of double JPEG compression in the presence of image resizing," in IEEE International Workshop on Information Forensics and Security (WIFS), 2012, pp. 127–132.
14. D. Vazquez-Padin, F. Pérez-González, and P. Comesana- Alfaro, "A random matrix approach to the forensic analysis of upscaled images," IEEE Transactions on Information Forensics and Security, vol. 12, no. 9, pp. 2115– 2130, 2017.
15. B. Bayar and M. C. Stamm, "Constrained convolutional neural networks: A new approach towards general purpose image manipulation detection," IEEE Transactions on Information Forensics and Security, vol. 13, no. 11, pp. 2691–2706, 2018.
16. T. Qiao, R. Shi, X. Luo, M. Xu, N. Zheng, and Y. Wu, "Statistical model-based detector via texture weight map: Application in re-sampling authentication," IEEE Transactions on Multimedia, vol. 21, no. 5, pp. 1077–1092, 2019.
17. Gang Cao, Antao Zhou, Xianglin Huang, Gege Song, Lifang Yang, Yonggui Zhu. Resampling detection of recompressed images via dual-stream convolutional neural network[J]. Mathematical Biosciences and Engineering, 2019, 16(5): 5022–5040. https://doi.org/10.3934/mbe.2019253
18. Y. Liang, Y. Fang, S. Luo and B. Chen, "Image Resampling Detection Based on Convolutional Neural Network," 2019 15th International Conference on Computational Intelligence and Security (CIS), 2019, pp. 257–261, https://doi.org/10.1109/CIS.2019.00061.
19. Chang Liu and Matthias Kirchner. 2019. CNN-based Rescaling Factor Estimation. In *Proceedings of the ACM Workshop on Information Hiding and Multimedia Security* (*IH&MMSec'19*). Association for Computing Machinery, New York, NY, USA, 119–124. https://doi.org/10.1145/3335203.3335725
20. A. C. Gallagher, "Detection of linear and cubic interpolation in JPEG compressed images." in Canadian Conference on Computer and Robot Vision (CRV'05), vol. 5. IEEE, 2005, pp. 65–72.
21. D. Vazquez-Padin and P. Comesana, "ML estimation of the resampling factor," in IEEE International Workshop on Information Forensics and Security (WIFS), 2012, pp. 205–210.
22. D. Vázquez-Padón, P. Comesana, and F. Pérez-González, "An SVD approach to forensic image resampling detection," in European Signal Processing Conference (EUSIPCO). IEEE, 2015, pp. 2067–2071.

23. S. Pfennig and M. Kirchner, "Spectral methods to determine the exact scaling factor of resampled digital images," in International Symposium on Communications, Control and Signal Processing. IEEE, 2012, pp. 1–6.
24. H. C. Nguyen and S. Katzenbeisser, "Detecting resized double JPEG compressed images–using support vector machine," in IFIP International Conference on Communications and Multimedia Security. Springer, 2013, pp. 113–122.
25. K. He, X. Zhang, S. Ren, and J. Sun, "Deep residual learning for image recognition," in IEEE Conference on Computer Vision and Pattern Recognition (CVPR), 2016, pp. 770–778.
26. A. Krizhevsky, I. Sutskever, and G. E. Hinton, "ImageNet classification with deep convolutional neural networks," in International Conference on Neural Information Processing Systems - Volume 1, ser. NIPS'12. Red Hook, NY, USA: Curran Associates Inc., 2012, p. 10971105.
27. K. Simonyan and A. Zisserman, "Very deep convolutional networks for large-scale image recognition," in International Conference on Learning Representations, Y. Bengio and Y. LeCun, Eds., 2015. [Online]. Available: http://arxiv.org/abs/1409.1556
28. Y. Rao and J. Ni, "A deep learning approach to detection of splicing and copy-move forgeries in images," in IEEE International Workshop on Information Forensics and Security (WIFS), 2016, pp. 1–6.
29. Y. Liu, Q. Guan, X. Zhao, and Y. Cao, "Image forgery localization based on multi-scale convolutional neural networks," in ACM Workshop on Information Hiding and Multimedia Security, 2018, pp. 85–90.
30. B. Bayar and M. C. Stamm, "A deep learning approach to universal image manipulation detection using a new convolutional layer," in ACM Workshop on Information Hiding and Multimedia Security, 2016, pp. 5–10.
31. P. Zhou, X. Han, V. I. Morariu, and L. S. Davis, "Learning rich features for image manipulation detection," in Proceedings of the IEEE Conference on Computer Vision and Pattern Recognition, 2018, pp. 1053–1061.
32. Y.-F. Hsu and S.-F. Chang, "Detecting image splicing using geometry invariants and camera characteristics consistency," in 2006 IEEE International Conference on Multimedia and Expo. IEEE, 2006, pp. 549–552.
33. T.-T. Ng, S.-F. Chang, and Q. Sun, "A data set of authentic and spliced image blocks," Columbia University, ADVENT Technical Report, pp. 203–2004, 2004.
34. H. Guan, M. Kozak, E. Robertson, Y. Lee, A. N. Yates, A. Delgado, D. Zhou, T. Kheyrkhah, J. Smith, and J. Fiscus, "MFC datasets: Large-scale benchmark datasets for media forensic challenge evaluation," in 2019 IEEE Winter Applications of Computer Vision Workshops (WACVW). IEEE, 2019, pp. 63–72.
35. D. Cozzolino, G. Poggi, and L. Verdoliva, "Recasting residual-based local descriptors as convolutional neural networks: an application to image forgery detection," in ACM Workshop on Information Hiding and Multimedia Security, 2017, pp. 159–164.
36. Y. Yan,W. Ren, and X. Cao, "Recolored image detection via a deep discriminative model," IEEE Transactions on Information Forensics and Security, vol. 14, no. 1, pp. 5–17, 2018.
37. J. H. Bappy, C. Simons, L. Nataraj, B. Manjunath, and A. K. Roy-Chowdhury, "Hybrid LSTM and encoder– decoder architecture for detection of image forgeries," IEEE Transactions on Image Processing, vol. 28, no. 7, pp. 3286–3300, 2019.
38. F. Marra, D. Gragnaniello, D. Cozzolino, and L. Verdoliva, "Detection of GAN-generated fake images over social networks," in 2018 IEEE Conference on Multimedia Information Processing and Retrieval (MIPR). IEEE, 2018, pp. 384–389.
39. P. Korshunov and S. Marcel, "DeepFakes: a new threat to face recognition? assessment and detection," arXiv preprint arXiv:1812.08685, 2018.

The Impact of Perceived Justice and Customer Trust on Customer Loyalty

Akuthota Sankar Rao, Venkatesh Vakamullu, Madhusudhan Mishra, and Damodar Suar

1 Introduction

Banks, as financial institutions, are essential to trade and commerce. The presence of a phenomenal administrative process streamlines activities and makes being distinct and unique. Despite this, there remains some gap in meeting the customer's expectations. Even the leading banks are not always sometimes unable to meet the customers' desired requirements or provide error-free services [32]. Some of the service problems faced by banks are server down, wrong transactions, ATM out of order, few ATMs, long queues, slow approval of loans, and ignorant employees. These failures can harm banks such as customer dissatisfaction, spreading negative word of mouth, customers switching over, complaints to the bank, or continuation with the bank despite dissatisfaction [14]. Therefore, addressing customer complaints is crucial for banks along with converting dissatisfied customers into satisfied ones. This will build strong relationships with banks and enhance customer loyalty [3, 24].

Banks cannot serve without failures. But they can effectually reply to such failures with a perceived justice [26]. The framework of justice theory is the most popular tool to measure the providers' recovery ([9, 11, 18, 23, 17, 25, 26]). The perceived justice has been widely used to resolve customer complaints about service recovery in the service sector. First, distributive justice concentrates on the compensation concerning the service failure to the complainers [25]. Procedural

A. S. Rao (✉) · V. Vakamullu · D. Suar
Indian Institute of Technology Kharagpur, Department of Humanities and Social Sciences, Kharagpur, West Bengal, India

M. Mishra
Electronics & Communication Engineering, North Eastern Regional Institute of Science and Technology (NERIST), Itanagar, Nirjuli, Arunachal Pradesh, India

M. Mishra et al. (eds.), *Applications of Computational Intelligence in Management & Mathematics*, Springer Proceedings in Mathematics & Statistics 417,
https://doi.org/10.1007/978-3-031-25194-8_16

justice estimates how the firm can justify the policies and procedures to solve the service complaints of the customers. Interactional justice includes the extent of the fair treatment of the complaints [28]. Extent literature concentrates on how perceived justice affects the customer reactions like customer repurchase intentions [15, 27], customer loyalty [4], WOM behavior, revisit intentions [10], and customer satisfaction ([12, 20, 27]). However, various types of service failures warrant different types of perceived justice and how these recoveries address customers' complaints and make them loyal to the bank. Customers' complaints stem from perceived justice. For example, distributive justice is pertinent when customers have done wrong transactions, charges to inactivated services, and money deducted without his/their acceptance. Failures like delay in checkbook issue and loans are addressed through procedural justice. Interactional justice is relevant to failures like miscommunication, employee unavailability during working hours, and inadequate number of employees.

2 Literature Review and Hypothesis

After a failure, service recovery is a tool of converting an aggrieved customer into a satisfied one. Banks need to carry on proper recovery steps for failure [13]. It is a thought-out and planned process of returning aggrieved or dissatisfied customers to a state of the satisfied customer with a service provider or service. The properties of the three justices are correlated and are not mutually exclusive [8]. It has been found that perceived justice relates to a timely action to retain loyal customers [29]. Effective service recovery measures can strengthen customer satisfaction with the quality of purchased products or services and thus increase customer loyalty [4]. Recovery efforts are very important to resolve customer complaints, establish long-term relationship [9], and affect customer's behavioral intentions [20].

Customer loyalty includes revisiting intentions and positive word of mouth. Evidence results that distributive, procedural, and interactional justice in service recovery influences the complaints' satisfaction which in turn leads to customer loyalty. Also, distributive and procedural justices are more influential in service recovery than interactional justice [21]. Strong evidence is also found from Blodgett et al. [3] that international justice influences customer behavioral intentions like revisit intention and positive word of mouth. Also, service recovery efforts are found to maintain the relationship with existing customers [9]. In the context of controversial findings of service recovery, we impose a multipurpose hypothesis:

H1. The (a) distributive, (b) procedural, and (c) interactional justice in service recovery will positively relate to customer loyalty.

Perceived justice additionally leads to customer trust and commitment [5, 26]. The trust exists between the service provider and the customer when the customer has the assurance of reliability and integrity of the service provider [19]. Of the three forms of justice, perceived justice is most influential to build customer trust, and it is useful for measuring customer loyalty [5]. The study of Wen and Chi [30] has used survey method in the airline industry to test the customer behavioral intentions on service recovery, and the study concluded that the three dimensions of justice are related to trust between customer and service provider. The findings and multiple hypotheses are proposed in the context of the banking sector.

H2. The (a) distributive, (b) procedural, and (c) interactional justice in service recovery will positively influence customer trust.

If customers have trust in a service provider, they have revisit intentions, spread favorable word of mouth, and build a long-term relationship with the service provider [14, 19]. It will positively influence customers' behavioral and attitudinal loyalty towards the service provider [5]. Customers have trust in banks and their services when banks provide error-free services and rectify customers' complaints with recovery strategies [26]. If the service provider continuously provides consistent and competent services, customers can trust the provider and carry on their relationship with the provider. This will further customer revisit intention and positive word of mouth about the service provider. Therefore, the service provider may use customer trust as a tool for creating customer loyalty. Accordingly, the following hypothesis is proposed:

H3. Customers' trust will positively influence customer loyalty.

Customer trust enhances or builds when the customer has the assurance of the provider's honesty, sincerity, and consistency of service [31]. Customers' trust is necessary for continuing the relationship with the service provider because customers take the decision based on their explained feelings before consumption [22]. Customers' trust was found to be a mediating variable between the relationship with the service provider and customer loyalty in an automobile retailer sector [19] and a hospitality sector [5]. Trust and commitment are the most influential variables of the customer repurchase intentions [7], positive word of mouth, and long-term relationships. Customers' trust focuses on positive expectations in the context of service recovery [6], and customers perceive the service provider as untrustworthy when customers have experienced a poor recovery (Fig. 1). Hence,

H4. Customer trust will mediate the relationship between the (1) distributive, (2) procedural, and (3) interactional justice in service recovery and customer loyalty.

3 Method

3.1 Participants

The convenience sampling technique was adopted to collect data. The bank customers who were available on the day of the visit were requested to respond to the questionnaire. The purpose of the study was explained to them, and it was mentioned that their participation was voluntary. The information they provide will be kept confidential in an aggregate manner without revealing their identity.

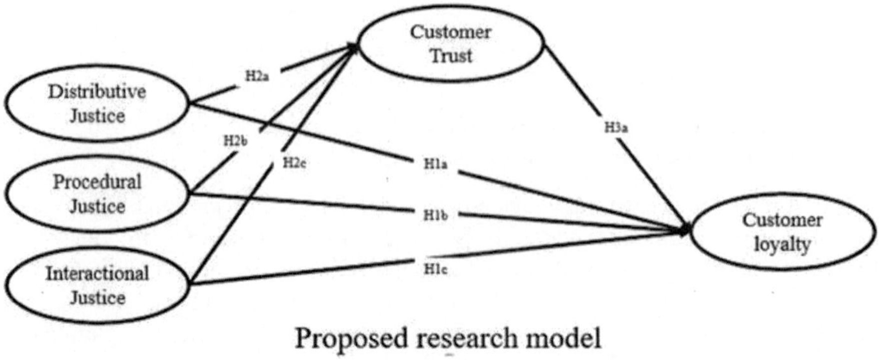

Fig. 1 The framework demonstrated for establishing the relationship between perceived justice, customer trust, and customer satisfaction

About 15 customers from a bank branch were given the questionnaire. Of the 350 participants, 290 were valid responses to the return. The effective return rate was 82.85%. The customers were from the metro cities of Hyderabad, Chennai, and Bangalore in southern India. A total of 68.96% of customers visited the bank 2–5 times in a month, and also 31% visited 10 times or more in a month. Of the participants, 63.1% were males, 49.32% were in the age group 21–30, 49.31% were in the age group 31–40, and 18.19% were in the age group 41–50. A total of 54.82% of the participants were technical graduates, and 23% were technical post-graduates. Overall, 22.08% of the total responses were nontechnical graduates. The service failures reported were long waiting time in ATM, bank queue, unsatisfactory Internet services, and solving the wrong transactions.

3.2 Measures

The data was collected using a questionnaire. The first part of the questionnaire contained socio-demographic information: gender, age, education, bank visits per month, and complaints about service recovery. The subsequent parts assessed distributive, procedural, and interactional justice, customer trust, and loyalty with 5, 5, 4, 5, and 5 items, respectively. The participants were given a questionnaire to fill up, and after a week, the questionnaire was collected. All returned the questionnaire. Along with descriptive statistics, the Pearson correlation and confirming factor analysis were used. Path analysis was used to examine the criteria and explain variance relationships and mediating effects.

4 Results

4.1 Scale Reliability and Validity

Confirmatory factor analysis (CFA) was carried out to estimate the factor loadings, scale reliability, and convergent validity in descriptive, procedural, and interactional justice, customer trust, and customer loyalty of items. The single factor modal was estimated for each construct to minimize common method bias (see Table 1).

The standard item loadings of constructs ranged from as low as 0.54 to as high as 0.98. Each construct loads the acceptance composite reliability greater than the 0.70 and convergent validity greater than 0.60. The model has acceptable fit indices ($\chi 2/df = 2.081$, CFI = 0.977, GFI = 0.872, NFI = 0.956, RMSEA = 0.062). The descriptive statistics and correlations among the variables are shown in Table 2. All the five variables positively related to one another.

Because correlations suggest bidirectional relations, a covariance-based SEM model was used to examine the path diagram. Mediation analysis of the proposed model was keenly justified by the Baron and Kenny's [2]. It is the threestep framework states that whether customer trust fully or partially mediates the perceived justice and customer loyalty or not. According to this framework, the existence of mediation relationship is possible when the following criteria are satisfied: First, a significant relationship between the perceived justice and customer trust exists; next, a significant association between the customer trust and customer loyalty exists; and last, a prominent relationship between perceived justice and customer loyalty exists in the absence of customer trust. However, the presence of a mediator diminishes the regression value of this relationship.

4.2 Path Relations

The results confirm the proposed hypothesis relations. High (low) levels of distributive, procedural, and interactional justice were associated with high (low) customer loyalty. The direct effects between them are significantly confirmed. For the second hypothesis, the three forms of perceived justice are again positively associated with customer trust. In the third hypothesis, the exploratory variables were the three forms of justice and trust, and the outcome variable was customer loyalty. When the full model was run, aligned with the third hypothesis, high (low) customer trust was associated with high (low) customer loyalty. The different forms of justice were positively associated with customer trust as in the second step. However, interactional justice which positively predicted customer loyalty in step one becomes insignificant, suggesting that the effect of interactional justice fully passed through customer trust to positively predict customer loyalty. Furthermore, the effect of distributive and procedural justice positively predicts the customer loyalty and decreased in the full model significantly that their effect partially passed

Table 1 Standardized weights, convergent validity, and composite reliability

Indicator	Loadings	Composite reliability	AVE
Distributive justice		0.89	0.65
Banks receive and review the customer's complaints properly	0.68***		
Banks maintain trustworthy in handling the complaints	0.66***		
Banks put efforts to respond and reply to the customers politely	0.98***		
Banks try to rectify the customer's complaints	0.98***		
Banks refund the amount if the service is not addressed properly	0.64***		
Procedural justice		0.93	0.781
Banks maintain fair policies to resolve the complaints	0.92***		
Banks implement the policies for timely follow-up and execution of complaints	0.92***		
Banks are accessible to convey the knowledge of process and maintain flexibility	0.87***		
Banks categorize the requests whether they are usual or not, mistake, or error and subsequently give the commitments towards the problem	0.83***		
Interactional justice		0.93	0.769
Banks are polite to the customers while listening and addressing the complaints	0.77***		
Banks are honest with customers when it comes to solving the complaints	0.95***		
Banks put maximum efforts and maintain a positive attitude towards the customers	0.87***		
Banks always respect the customer's opinions and suggestions in the recovery process	0.91**		
Banks always show interest to solve the customer's complaints	0.95***		
Customer trust		0.88	0.619
I believe that the bank is able to provide service recovery that customers need	0.90***		
I believe that the bank can provide service recovery of high quality for customers	0.98***		
I believe that the bank can effectively solve problems caused by service failures	0.78***		
I believe that the bank is very concerned with customers' interests	0.65***		
I believe that the bank can keep its promise for customers	0.54***		
Customer loyalty		0.90	0.656
I would recommend this bank to someone who seeks your advice	0.90***		
I would encourage friends and relatives to use this bank	0.97***		
I would say positive things about this bank to other people	0.72***		
I consider this bank my first choice to work with	0.72***		
I consider this bank to work with when I need banking services	0.71***		

$^{***}p < 0.001$

Table 2 Descriptive statistics and correlations among the study variables

	Mean	SD	1	2	3	4	5
1. Customer loyalty	3.65	1.09	1				
2. Customer trust	3.53	1.17	0.47***	1			
3. Distributive justice	3.80	1.02	0.39***	0.33***	1		
4. Interactional justice	4.26	0.80	0.24***	0.26***	0.15***	1	
5. Procedural justice	3.55	1.21	0.34***	0.42***	0.21***	0.20***	1

***$p < 0.001$

Table 3 Path coefficients

Path			B	SEB	CR	P	β	χ2/df	CFI	NFI	GFI	RMSEA
DJ	→	CL	0.337	0.057	5.962	***	0.325	2.395	0.978	0.963	0.881	0.07
PJ	→	CL	0.220	0.047	4.648	***	0.251					
IJ	→	CL	0.190	0.072	2.637	0.008	0.144					
DJ	→	CT	0.266	0.059	4.492	***	0.241	2.173	0.974	0.954	0.863	0.065
PJ	→	CT	0.333	0.050	6.691	***	0.357					
IJ	→	CT	0.228	0.076	3.009	0.003	0.162					
CT	→	CL	0.293	0.055	5.355	***	0.312					
DJ	→	CL	0.259	0.056	4.648	***	0.25					
PJ	→	CL	0.123	0.049	2.519	0.012	0.14					
IJ	→	CL	0.123	0.070	1.766	0.077	0.093					
DJ	→	CT	0.266	0.059	4.502	***	0.242	2.275	0.972	0.951	0.852	0.067
PJ	→	CT	0.334	0.050	6.721	***	0.358					
IJ	→	CT	0.228	0.076	3.017	0.003	0.162					
CT	→	CL	0.447	0.052	8.673	***	0.464					

Notes: B is an unstandardized beta; SEB is the standardized beta; CR is the critical ratio; P is the probability; β is the standardized beta
***$p < 0.001$

with customer trust to predict customer loyalty. In all three steps, the model had acceptable fit indices. Path coefficients and fit indices are shown in Table 3.

5 Discussion

The study found out the direct and indirect effects of the three forms of perceived justice on customer loyalty through customer trust. The results confirmed that there is a positive and significant relationship between three forms of justice and customer loyalty in the Indian banking condition. The results support the study of Wang et al. [29], who identified that justice has a positive effect on customer loyalty. The results demonstrate that ensuring the perceived justice is enhancing customer satisfaction and retention. Consequently, mitigating the failure effect and providing a good recovery seems to be the best action for complainers to retain. Especially, methods used to overcome the failures in interactional justice made a strong customer relationship. The results supported that the complainers who receive the recovery through procedural justice in a timely manner will show higher loyalty. Distributive justice has a direct and positive significance on loyalty behavior. This result may be most e-tailers response to failures by providing basic distributive recovery, such as free-of-charge commodity replacements, as opposed to advanced distributive recoveries, such as monetary compensation, coupons for future consumption, clear explanations and sincere apologies.

Moving to customer trust, the findings confirm the past results. It recommends that customer trust has played a mediating role. Protecting customer trust is the prime practice of relationship management, which improves loyalty [16]. It supports the explanation of why customers stay loyal to the bank, based on relevant literature as mentioned earlier in the field of marketing [1]. The customers build a loyal relationship with the bank by improving the quality services and a good recovery for failures. It makes a positive opinion and improves trust. Trust builds only when mitigating the failure severity with the support of frontline employees.

6 Managerial Implications for Practice

In the Indian banking context, a bank needs to retain the customers as the competition among the banks is getting increased. Since the Indian customers are adaptive in nature, they easily get connected to better services, which force the banks to take this advantage into account for developing good relationships with customers. Because of this, banks should make their customers satisfied and achieve the customer's attention for positive WOM. Moreover, Indian customers prefer to continue their relationships; hence, it makes them privileged when they receive good recovery. The conventional wisdom also implies that banks should endeavor in enhancing loyalty so that past experiences help to retain customers for the long term, particularly in failure cases.

Although failures are unavoidable, banks should try to mitigate the failure severity; to achieve this, regular assessment of recovery procedures, proper training to employees, and setting standard working conditions are essential. In addition, banks should encourage the employees to behave like pro-active communication, take the customer feedback, and make the necessary improvements. This helps in market expansion by grabbing new customers with the positive recommendation. A good recovery for a failure can perform a key role in maintaining loyalty. When a service failure happens, the immediate reactions can take advantage and lead to an increase in the trust and consequently loyalty, whereas poor reactions may push the customer to switch over to alternatives. Distributive justice has recorded the strongest recovery effect as compared to the other two. Banks must direct their employees to maintain trustworthiness in handling complaints, enhance their efforts to resolve them, and finally provide the appropriate compensation against the failed cases.

Procedural justice has the next strongest effect on post-recovery satisfaction. Banks should implement with fair policies, procedures, and criteria based on ethical and moral standards to recover the loss of the customer. Frontline employees need rigorous and efficient training on the policies and practices that help to respond to the complaints and convince the customer by explaining rules and regulations.

Interactional justice affects satisfaction but not by the fame of the organization. Frontline employees should implement the justice by executing tasks such as giving exact information, being honest, fulfilling the promises, and seeking suggestions from the customer for a good recovery. The employee plays a vital role in this phase, and he develops the qualities like patience, courtesy, and spontaneity towards the situation to satisfy the customer at most and convey value addition to the bank. It is also suggested that the banks should offer proper care and support to employees, which would, in turn, encourage them to treat the customers well to reach a "win-win" situation.

As organizational growth is mainly dependent on its manpower, the skilled employee is the key asset to recover the service with the aforementioned justice. Frontline employees need to take the initiation and implement protocols and to develop innovative and new services. Our research suggests that improvements in employee performance lead to decrease in service failures. Hence, it is inevitable for a service provider to spend valued resources on the recruitment and training process to uplift the skills and provide knowledge about the business operations and develop personnel initiation to address service failures. If it fails to do so, it may lead to losing loyalty and at most ruin the business. Therefore, the managers should maintain a cordial relationship with their frontline employees to serve the purpose.

7 Limitations

Our model is not exceptional from the fact that no model can be idealistic. First, this can be realized in the banking sector by considering other variables like failure severity, relationship quality, and firm reputation to analyze customer-provider-related challenges in real-time scenarios. Another limitation stems from the fact that the paper targets a single business sector of banking in a limited geographical area. So, this work can be extended to other geographical areas like rural and semi-urban areas and sectors like airline, hotels, and hospitality. Next, although trust improved over a while, the strength of trust has been assessed at a single point of time only. Assessing perceived justice and customer trust, customer loyalty at different periods would have validated the findings longitudinally. Finally, it is noticed that only customer opinions and perceptions were collected, which is insufficient for justifying the policies related to customer trust for service recovery. As a result, it is recommended that responses be collected from intermediaries such as distributors, promoters, sister concerned service providers, third-party agencies, and customers.

8 Conclusion

Customer trust has become an important factor in fostering customer retention in the banking sector. The importance of trust is to establish a strong customer-provider relationship and reduce the customer switching intention. As a result, the study provides empirical evidence that customer-perceived justice enhances customer loyalty through customer trust in India's banking sector.

References

1. Babin, B. J., Zhuang, W., and Borges, A. (2021) 'Managing service recovery experience: effects of the forgiveness for older consumers', *Journal of Retailing and Consumer Services*, 58:3, pp. 102–222.
2. Baron, R. M., and Kenny, D. A. (1986) 'The moderator–mediator variable distinction in social psychological research: Conceptual, strategic, and statistical considerations', *Journal of personality and social psychology*, 51:6, pp. 173.
3. Blodgett, J. G., Hill, D. J., and Tax, S. S. (1997) 'The effects of distributive, procedural, and interactional justice on post-complaint behavior', *Journal of retailing*, 73:2, pp. 185–210.
4. Chang, Y. W., and Chang, Y. H. (2010) 'Does service recovery affect satisfaction and customer loyalty? An empirical study of airline services', *Journal of air transport management*, 16:6, pp. 340–342.
5. DeWitt, T., Nguyen, D. T., and Marshall, R. (2008) 'Exploring customer loyalty following service recovery: The mediating effects of trust and emotions', *Journal of service research*, 10:3, pp. 269–281.
6. Dunn, J. R., and Schweitzer, M. E. (2005) 'Feeling and believing: the influence of emotion on trust', *Journal of personality and social psychology*, 88:5, pp. 736.
7. Garbarino, E., & Johnson, M. S. (1999) 'The different roles of satisfaction, trust, and commitment in customer relationships', *Journal of marketing*, 63:2, pp. 70–87.
8. Greenberg, J., and McCarty, C. (1990) 'The interpersonal aspects of procedural justice: A new perspective on pay fairness', *Labor Law Journal*, 41:8, pp. 580.
9. Ha, J., and Jang, S. S. (2009) 'Perceived justice in service recovery and behavioral intentions: The role of relationship quality', *International Journal of Hospitality Management*, 28:3, PP. 319–327.
10. Huang, M. H. (2011) 'Re-examining the effect of service recovery: The moderating role of brand equity', *Journal of Services Marketing,* 25:7, pp. 509–516.
11. Hoffman, K. D., and Kelley, S. W. (2000) 'Perceived justice needs and recovery evaluation: a contingency approach', *European Journal of marketing*, 34:3/4, pp. 418–432.
12. Karatepe, O. M. (2006) 'Customer complaints and organizational responses: The effects of complainants' perceptions of justice on satisfaction and loyalty', *International Journal of Hospitality Management*, *25*(1), 69–90.
13. Kau, A. K., and Loh, E. W. Y. (2006) 'The effects of service recovery on consumer satisfaction: a comparison between complainants and non-complainants', *Journal of services marketing,* 20:2, pp. 101–111.
14. Kim, T. T., Kim, W. G., and Kim, H. B. (2009) 'The effects of perceived justice on recovery satisfaction, trust, word-of-mouth, and revisit intention in upscale hotels', *Tourism management*, 30:1, pp. 51–62.
15. Liao, H. (2007) 'Do it right this time: The role of employee service recovery performance in customer-perceived justice and customer loyalty after service failures', *Journal of applied psychology*, 92:2, pp. 475.

16. Mahmoud, M. A., Hinson, R. E., and Adika, M. K. (2018) 'The effect of trust, commitment, and conflict handling on customer retention: the mediating role of customer satisfaction', *Journal of Relationship Marketing*, 17:4, pp. 257–276.
17. Mattila, A. S., and Wirtz, J. (2004) 'Consumer complaining to firms: the determinants of channel choice', *Journal of Services Marketing,* 18:2, pp. 147–155.
18. Maxham, J. G. (2001) 'Service recovery's influence on consumer satisfaction, positive word-of-mouth, and purchase intentions', *Journal of business research*, 54:1, pp. 11–24.
19. Morgan, R. M., and Hunt, S. D. (1994) 'The commitment-trust theory of relationship marketing', *Journal of marketing*, 58:3, pp. 20–38.
20. Nadiri, H. (2016) 'Diagnosing the impact of retail bank customers' perceived justice on their service recovery satisfaction and post-purchase behaviours: An empirical study in financial centre of middle-east', *Economic research-Ekonomska istraživanja*, 29:1, pp. 193–216.
21. Ozkan-Tektas, O., and Basgoze, P. (2017) 'Pre-recovery emotions and satisfaction: A moderated mediation model of service recovery and reputation in the banking sector', *European Management Journal*, 35:3, pp. 388–395.
22. Parasuraman, A., Berry, L. L., and Zeithaml, V. A. (1991), 'Understanding customer expectations of service', *Sloan management review*, 32:3, pp. 39–48.
23. Schoefer, K. (2008) 'The role of cognition and affect in the formation of customer satisfaction judgements concerning service recovery encounters', *Journal of Consumer Behaviour: An International Research Review*, 7:3, pp. 210–221.
24. Smith, A. K., and Bolton, R. N. (2002) 'The effect of customers' emotional responses to service failures on their recovery effort evaluations and satisfaction judgments', *Journal of the academy of marketing science*, 30:1, pp. 5–23.
25. Sparks, B. A., and McColl-Kennedy, J. R. (2001) 'Justice strategy options for increased customer satisfaction in a services recovery setting', *Journal of Business Research*, 54:3, pp. 209–218.
26. Tax, S. S., Brown, S. W., and Chandrashekaran, M. (1998) 'Customer evaluations of service complaint experiences: implications for relationship marketing', *Journal of marketing*, 62:2, pp. 60–76.
27. Van Vaerenbergh, Y., Vermeir, I., and Larivière, B. (2013) 'Service recovery's impact on customers next-in-line', *Managing Service Quality,* 23:6, pp. 495–512.
28. Voorhees, C. M., & Brady, M. K. (2005) 'A service perspective on the drivers of complaint intentions' *Journal of Service Research*, 8:2, pp. 192–204.
29. Wang, Y. S., Wu, S. C., Lin, H. H., and Wang, Y. Y. (2011) 'The relationship of service failure severity, service recovery justice and perceived switching costs with customer loyalty in the context of e-tailing', *International journal of information management*, 31:4, pp. 350–359.
30. Wen, B., and Chi, C. G. Q. (2013) 'Examine the cognitive and affective antecedents to service recovery satisfaction: A field study of delayed airline passengers', *International Journal of Contemporary Hospitality Management*, 25:3, pp. 306–327.
31. Wong, A., and Sohal, A. (2002) 'An examination of the relationship between trust, commitment and relationship quality', *International journal of retail & distribution management,* 30:1, pp. 34–50.
32. Zou, S., and Migacz, S. J. (2020) 'Why service recovery fails? Examining the roles of restaurant type and failure severity in double deviation with justice theory', *Cornell Hospitality Quarterly*.

Performance Analysis of Wind Diesel Generation with Three-Phase Fault

Akhilesh Sharma, Vikas Pandey, Shashikant, Ramendra Singh, and Meenakshi Sharma

1 Introduction

In 1970 was the first time oil crisis occurred. It was repeated again in 1973. Countries which were dependent on oil exporting started diminishing their dependency on oil by exploring other sources of energy to meet their energy demand. In the process of energy exploration, they found that renewable energy resources may serve some alternative ways to overcome the oil crisis [1].

Due to the European power market, the development in the wind industry became fast. In 2005, more than 6180 MW of wind power generators were installed. To generate power in a small scale, the use of the permanent magnet synchronous generator is best suited because the generators provide flexibility to operate at different working conditions and are maintenance free. To fulfill the power demand in the isolated areas, small-scale wind power generator, having the capacity of 0.2–30 kW, may be used. In India, the wind energy generation started in the 1990s. Although the wind power generation had just begun in India as compared to countries like Denmark or the USA, its development took at faster rate, and India became the fourth country having the largest capacity of wind power installation, in the world [2, 3].

In the proposed model, the nonconventional approach has been applied to utilize the energy. The nature of wind is not constant. It is irregular with varying speed, so it becomes difficult to harness the electrical energy continuously in the same manner. There is a need of backup supply with the wind power generation to maintain continuous electrical supply. To do so, an isolated diesel power plant is connected with the wind generation station. The diesel power plant provides electrical energy

A. Sharma (✉) · V. Pandey · Shashikant · R. Singh · M. Sharma
North Eastern Regional Institute of Science and Technology, Electrical Engineering, Naharlagun, Arunachal Pradesh, India

M. Mishra et al. (eds.), *Applications of Computational Intelligence in Management & Mathematics*, Springer Proceedings in Mathematics & Statistics 417,
https://doi.org/10.1007/978-3-031-25194-8_17

when the wind power plant is not able to generate electrical energy due to irregular wind. Thus, the diesel power plant makes a backup for the time intervals when the wind is irregular basically. This makes a hybrid system for the power generation where both conventional and nonconventional approaches for the harnessing energy are applied.

2 Wind Turbine Generation

In the proposed Simulink model, one part is the wind turbine generation which consists of wind turbine conversion system. In the conversion system, the wind, having kinetic energy, is converted into mechanical energy by the turbine or rotor. The wind blowing has constant speed, and the energy extracted by the wind is transmitted to the electrical generator so the generation system involves two main parts, namely, the wind turbine and the asynchronous generator.

2.1 Technology of the Wind Turbine

The commercial wind turbine has been used since the 1980s. In turbine research, there were many designs proposed depending upon the dependency of the wind. These are operated at different speeds because the natural conditions of the wind do not remain constant throughout the day. The natural conditions vary from time to time in a day. The variations in the speed to produce power by the wind turbine range from 3 to 25 m/s.

The aero turbine has the property to convert wind energy into rotating mechanical energy. They require the control. The mechanical system consists of the coupling with the step-up gears. The wind energy converted into mechanical energy is increased by these step-up gears, and this increased mechanical energy is sent to the electrical generator by the coupling.

2.2 Fixed Energy Conversion System

In this type of wind energy conversion system, rotor speed is fixed to a specified speed. The fixed speed energy conversion system is named as the "Danish concept." Maximum energy conversion can be achieved at a specific speed. The electrical generators which are used with the fixed speed wind energy conversion system are induction generators. The fixed speed wind energy conversion system has different properties. They are robust and cheap and require low maintenance. For fixed speed wind energy conversion system, the curve representing the C_p- λ-β at stationary wind speed assumption is satisfactory for study of power system. In the fixed speed

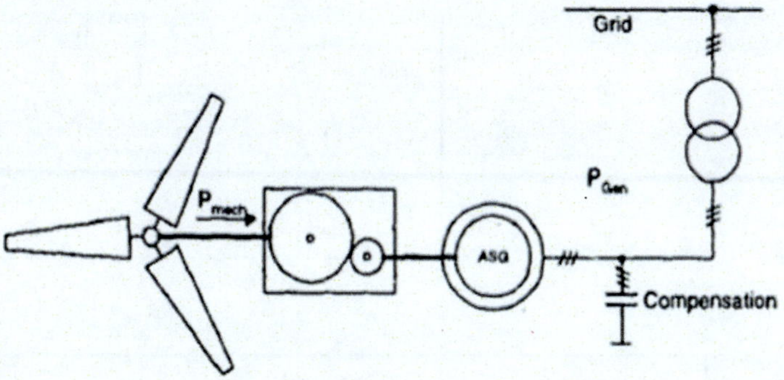

Fig. 1 Fixed speed wind turbine configuration

wind turbine system, relationship between the λ and the C_P can be represented by Equation. [4].

$$C_p(\lambda, \beta) = 0.44\left\{125\left(\frac{1}{\lambda} + 0.002\right) - 6.94\right\} * e^{-16.5\left(\frac{1}{\lambda}+0.002\right)} \quad (1)$$

In Fig. 1, a fixed speed wind turbine configuration is shown. The left side in the figure indicates a fixed speed wind turbine where the kinetic energy of the wind makes the blade of the wind turbine to rotate and develop mechanical power, P_{mech}. The mechanical power is then transferred to mechanical gear system and finally to the asynchronous generator (ASQ). The ASQ receives mechanical energy through the step-up gears and converts it into electrical energy which is then connected to isolated grid system. In this type of system, the preferred generator is a squirrel cage induction generator. It can be used in both single and double speed versions. Constant speed drive has many advantages such as it has simple construction and is robust in nature. It can withstand fluctuating wind load [5, 6].

2.3 Fixed Energy Conversion System

In the present Simulink model, a constant speed wind turbine is used for energy conversion of the wind energy into the mechanical energy. This converted mechanical energy is supplied to the asynchronous generator. The overall system is used for the per unit system. The turbine model consists of a pitch angle. For a constant speed, the pitch angle of the wind turbine may be considered zero.

In the model, a wind turbine, a constant speed of the wind has been considered. It will produce a constant mechanical power. This will produce a motor torque T_m. This torque is provided to the rotor of the asynchronous machine block. The value of this mechanical torque is negative so the permanent magnet synchronous machine will work in the generating mode. The power can be expressed in the pu system as under:

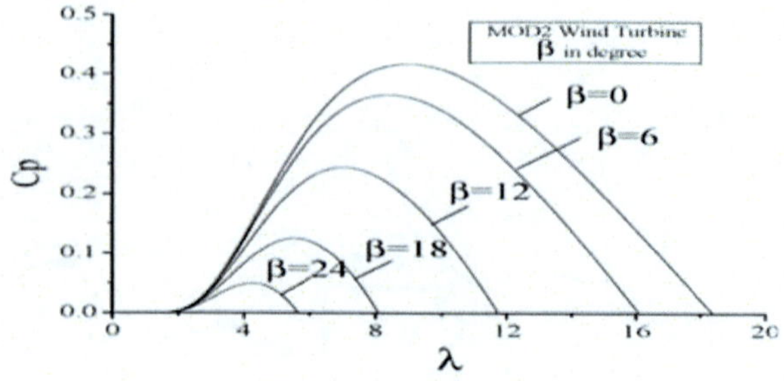

Fig. 2 $C_p - \lambda$ characteristics at different values of β

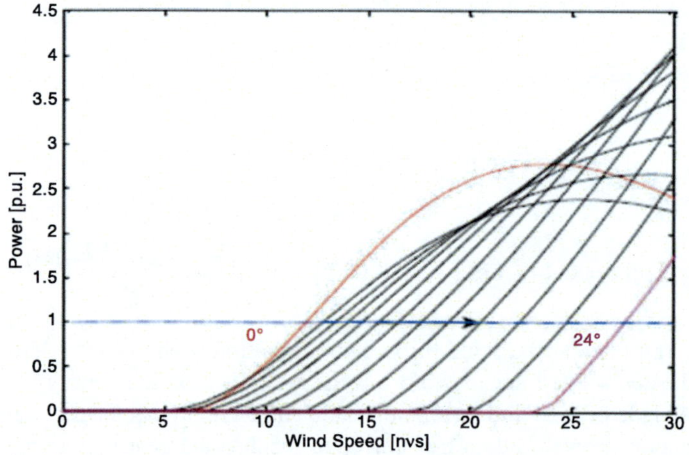

Fig. 3 Power curve at different pitch angles

$$P_{wind\,pu} = V^2_{wind\,pu} * C_{p\,pu} * K_{pw} \tag{2}$$

In Fig. 2, the variation of C_p versus λ at the different pitch angle is presented; as a constant speed wind turbine is used, the pitch angle is zero. In that case, the value of C_p is higher in comparison with the other pitch angles, as seen in the figure. The values of the pitch angle are decreasing from 24 to 0, while the C_p values are improving. It is because the wind power directly depends on the value of C_p as seen in Eq. (2). It helps the wind power extract more energy for zero pitch angle [7, 8].

Generally, the range of pitch angle varies between 0° and 24°. For effective aerodynamic braking blade, the pitch angle may be extended to 90°. In Fig. 3, the power as function of wind speed at the different pitch angle is presented. Pitch angle requirement at different wind speeds can be defined as the intersection of the power curves and the rated power [9].

3 Diesel Engine Generation

Electrical energy in the proposed model is generated by two means: one is the wind energy generation, and the other is the diesel engine generator. In the proposed system, the mechanical power is produced by the burning of the diesel in the diesel engine. Mechanical power is transferred to the synchronous generator. The generator needs the external excitation; the excitation system provides the field supply for the generation of the electrical energy to occur. In the present model, the blocks involved are the diesel engine, diesel engine governor, and the excitation system. So far, the diesel engine governor is used for the controlling of the diesel engine. In the diesel engine, the combustion of the diesel takes place. Governor system is used for the controlling of the diesel generator so that the desired mechanical power and the desired speed is obtained. In a similar fashion, the excitation system is used for the controlling of the field excitation voltage, mechanical power, and the excitation voltage, applied to the synchronous generator [10–16].

4 Electrical Energy Controlling System

The measured voltage is applied to the voltage regulator, and this voltage regulator contains the PI controller. The voltage regulator provides the signal to gate of the IGBTs. The voltage regulator is of discrete type; the reference voltage is present. The comparison to this reference is there, and the PI controller compares the reference voltage and the achieved voltage. If there is error in the voltage measurement, then it sends signals [17, 18]. A PI controller consists of two gains, namely, the proportional gain K_p and the integral gain K_i, whose transfer function is defined as [19]:

$$G(s) = K_p + \frac{K_i}{s} \tag{3}$$

In the voltage regulator, the transformation of the conversion of V_{abc} occurs in the dq0 reference frame. The voltage is changed into the direct axis, quadrature axis, and zero sequence voltage in the rotating two-axis reference frame. The PI control checks the error with respect to reference, and then dq0 reference frame is changed to the V_{abc}. It is then sent to the discrete pulse width modulation generator. With the help of PLL (phase lock loop), phasor voltage of the supply side is synchronized with the controller reference frame [20].

For the conversion of the abc to dq0, we get the different voltages in two-axis rotating field; rotating frame is rotating with the speed, ω, which is in the rad/s. They are the zero sequence voltage, direct axis voltage, and the quadrature axis voltages. The mathematical representation of the different voltages can be represented by the equations: For zero sequence voltage:

$$v_o = \frac{1}{3}(v_a + v_b + v_c) \tag{4}$$

For the direct axis voltage:

$$v_d = \frac{2}{3}\left(v_a \sin \omega t + v_b \sin\left(\omega t - \frac{2\pi}{2}\right) + v_b \sin\left(\omega t + \frac{2\pi}{2}\right)\right) \tag{5}$$

For the quadrature axis voltage:

$$v_q = \frac{2}{3}\left(v_a \cos \omega t + v_b \cos\left(\omega t - \frac{2\pi}{2}\right) + v_b \cos\left(\omega t + \frac{2\pi}{2}\right)\right) \tag{6}$$

5 Simulink Model of Proposed System

The block diagram and Simulink model of the proposed system are shown in Figs. 4 and 5, respectively. In the proposed system, electrical energy is generated with the wind turbine and the diesel generator. Generated electrical energy is rectified with a three-phase uncontrolled rectifier. After the rectification, it is converted to alternating again via IGBT controlled PWM inverter. The inverter is controlled with a voltage regulator. The generated electrical energy is supplied to the R-L-C load, and a fault is made to occur on the load side.

In the proposed model, diesel engine is connected with a permanent magnet synchronous generator, and wind turbine is connected with the asynchronous generator. Both parts are generating the electrical power. The generated power is sent to R-L-C load. The load is also connected with the three-phase fault side. The various steps involved are stated. First of all, the generated electrical power is rectified and filtered with three-phase uncontrolled rectifier that has low pass filters [12] so that a constant DC output voltage is achieved. The generated supply may be synchronized before sending to the rectifier circuit. The DC supply is inverted to an AC using a three-phase inverter which is based on PWM techniques. The inverter contains a low pass filter to smoothen the inverter output voltage so that the harmonics are eliminated from the stepped output of the inverter. The firing of IGBT can be controlled for the desired output so a voltage regulator is used for the purpose.

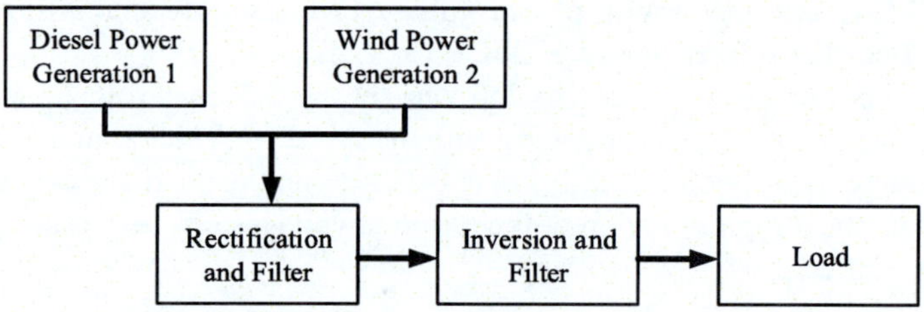

Fig. 4 Block diagram for proposed system

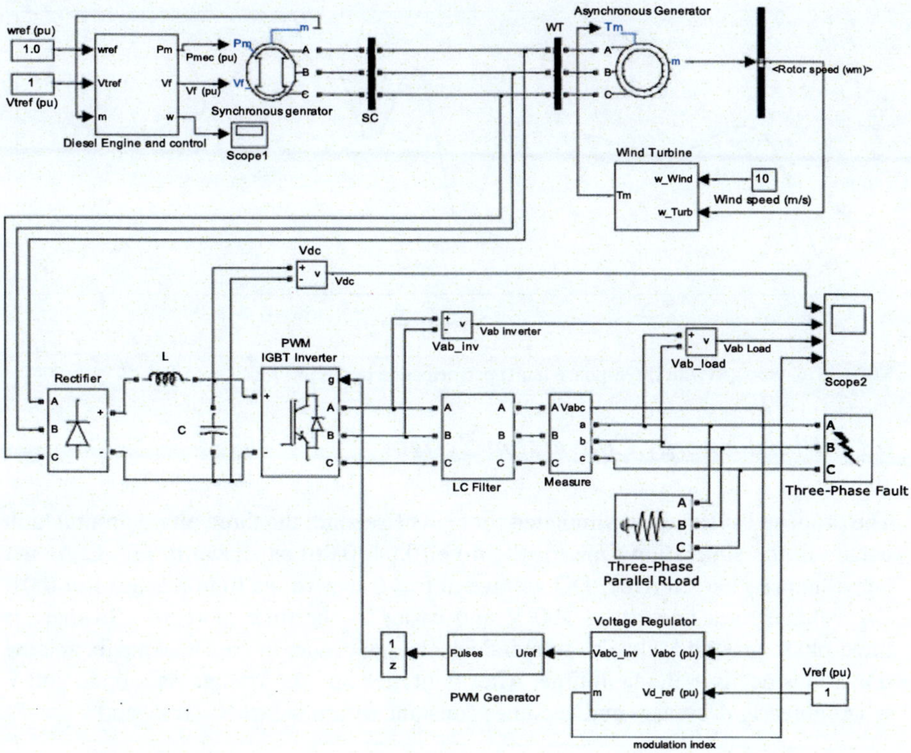

Fig. 5 Simulink model of proposed system

6 Result Analysis

As the proposed model is working properly, in the proposed system, now the performance analysis is done with the load that is working under both normal condition and faulty condition. The fault timing was 0.04–0.06 seconds; it means the three-phase fault starts from 0.04 second and dies out in 0.06 seconds. Effect of the fault will be on the whole system, so performance in the whole system may be checked by considering full load condition.

In the proposed model, the voltage analysis on the three parts is done first; one is at the rectified output voltage, the second one is the inverter output, and the third one is the load end. The output of all three parts is shown in the form of the graphs. Output voltage is on the y-axis and the time is shown on the x-axis. The proposed system in all the three points is being stable after the three-phase fault occurrence, and the desired output is achieved.

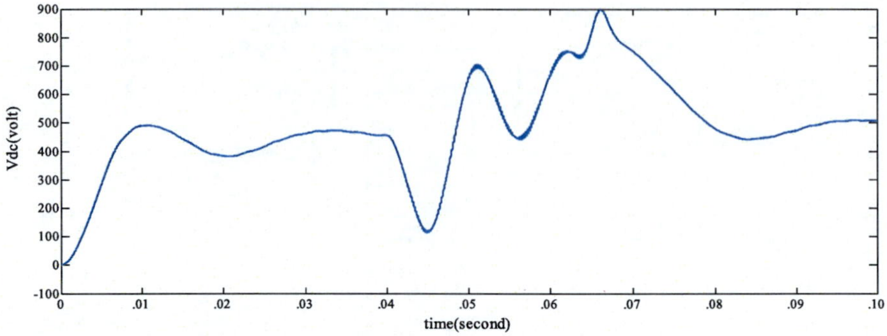

Fig. 6 DC voltage (with three-phase fault) performance in the DC link

6.1 Fixed Energy Conversion System

The proposed system was simulated for t = 0.1 second; the three-phase ground fault occurs in the simulation time for the time (0.04, 0.06) as shown in Fig. 6. As per performance characteristic, DC voltage at t = 0 is zero; as time reaches t = 0.01, DC voltage rises linearly to 500 V and varies as the time increases. So there is three-phase ground fault. The tremendous change occurs in the DC voltage, and the transitions occur in the fault time. After fault dies out, the voltage rises up to 900 V which linearly decreases and becomes constant within the t = 0.10 second.

6.2 Fixed Energy Conversion System

The DC voltage was applied to PWM inverter, which converted the DC voltage into the AC voltage. Simulation time is 0.10 second, and the three-phase fault occurs during 0.04–0.06 second. As per the performance characteristic, voltage of the DC link goes to zero, and the inverter voltage also decreases as seen in Fig. 7, but not zero, and then the random changes in the DC voltage occur. There is also change in the inverter voltage, which is phase-to-phase voltage. At the end of the simulation time, DC voltage of the rectifier becomes constant so is the inverter voltage.

6.3 Fixed Energy Conversion System

The obtained inverter voltage was filtered with a low pass filter, before feeding to the load. The time of simulation was 0.10 second, and the three-phase fault occurs at 0.04 second, which dies out in 0.06. As per performance characteristic, load voltage is zero; as the time increases, the voltage increases and varies sinusoidally. The fault time starts on the completion of the two cycles. So the load voltage becomes zero for

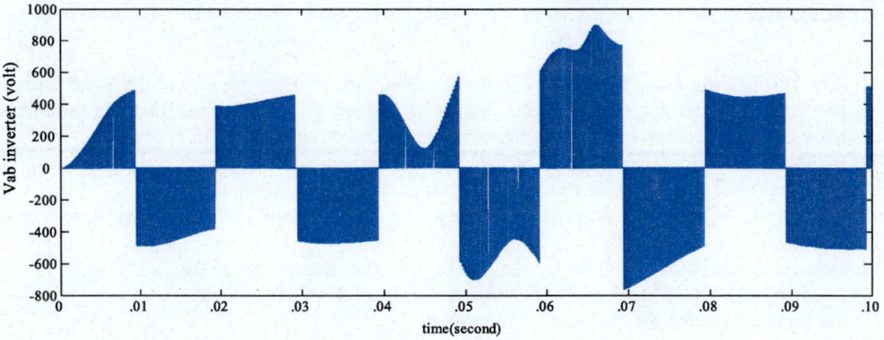

Fig. 7 Inverter (with three-phase fault) voltage performance between two lines

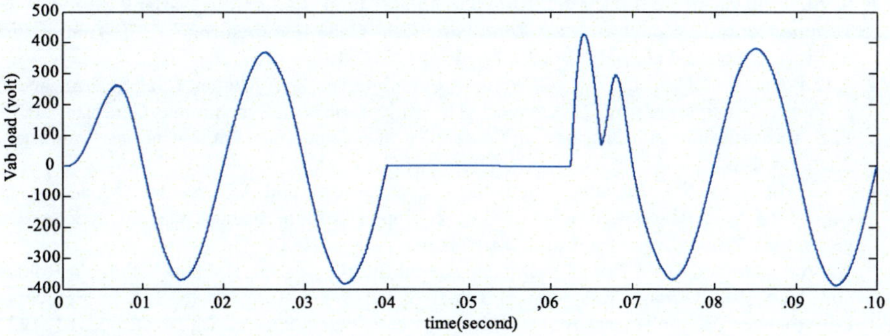

Fig. 8 Load voltage (with three-phase fault) performance between two lines

the time 0.04–0.06 second. After the 0.06 second fault dies out, the load voltage is zero, and after some time, it starts increasing and reaches to its maximum value and varies from time 0.07–0.10 second which comes in the stable state and completes its cycles as shown in Fig. 8.

7 Conclusions

The proposed model is achieving the desired output at all the different conditions. In the beginning, the voltage is normal and stable. When a fault occurs between $t = 0.4$ seconds and $t = 0.6$ seconds, with full load conditions, the voltage in all parts of the system gets disturbed. When the fault is cleared, the voltage becomes stable. So far, the proposed model is working for the fixed speed and using PWM modulation technique to reduce harmonic distortion. Distortions in the system are very less. Performance is observed to be stable and reliable for the wind and diesel-based power generation.

References

1. Jens Vestergaard, Lotte Brandstrup, Robert D. Goddard, “A Brief History of the Wind Turbine Industries in Denmark and the United States”, Academy of International Business (Southeast USA Chapter) Conference Proceedings, November 2004, pp. 322–327.
2. Joanna I. Lewis, “A Comparison of Wind Power Industry Development Strategies in Spain, India and China”, Center for Resource Solutions, China, July 2007, pp. 1–25.
3. Teri Envis, “Centre On Renewable Energy and Environment”, Wind energy information, 2005/2006, pp. 1–56.
4. Quincy Wang, Liuchen Chang, “An Intelligent Maximum Power Traction Algorithm for Inverter-Based Variable Speed Wind turbine systems”, IEEE Transactions on Power Electronics, Sept. 2004, Vol. 19, No. 5, pp. 1242–1249.
5. Gillian Lalor, Alan Mullane, and Mark O’Malley, “Frequency Control and Wind Turbine Technologies”, IEEE Transactions on Power systems, Nov. 2005, Vol. 20, No. 4, pp. 1905–1913.
6. S. M. Muyeen, Rion Takahashi, Toshiaki Murata, Junji Tamura, “A Variable Speed Wind Turbine Control Strategy to Meet Wind Farm Grid Code Requirements” IEEE Transactions on Power systems, Feb. 2010, Vol.25, No. 1, pp. 331–340.
7. A.C. Pinto, B. C. Carvalho, J. C. Oliveira, G. C. Guimarães, A. J. Moraes, C. H. Salerno, and Z. S. Vitório “Analysis of a WECS Connected to Utility Grid with Synchronous Generator” IEEE PES Transmission and Distribution Conference and Exposition Latin America, Venezuela, 2006, pp. 1–6.
8. D.C. Aliprantis, S.A. Papathanassiou, M.P. Papadopoulos, and A.G. Kladas “Modeling and control of a variable-speed wind turbine equipped with permanent magnet synchronous generator”, Proceedings of ICEM 2000, Helsinki, August 2000.
9. Andreas Sumper, Oriol Gomis Bellmunt, Antoni Sudria Andreu, Roberto Villafafila Robles, and Joan Rull Duran, “Response of Fixed Speed Wind Turbines to System Frequency Disturbances”, IEEE Transactions on Power systems, Feb 2009, Vol. 24, No. 1, pp. 181–192.
10. Ruben Pena, Roberto Cárdenas, José Proboste, Jon Clare, and Greg Asher, “Wind–Diesel Generation Using Doubly Fed Induction Machines”, IEEE Transactions on Energy conversion, March 2008, Vol 23, No.1, pp. 202–214.
11. Stephen Drouilhet, “Preparing an Existing Diesel Power Plant for a Wind Hybrid Retrofit: Lessons Learned in the Wales, Alaska, Wind-Diesel Hybrid Power Project”, Wind power conference Washington, D.C., June 2001, pp. 1–10.
12. Alan Mullane, Mark O’Malley, “The Inertial Response of Induction-Machine-Based Wind Turbines”, IEEE Transactions on power systems, August 2005, Vol 20, No.3, pp. 1496–1503.
13. A. Kilk, “Low-Speed Permanent Magnet Synchronous Generator for Small Scale Wind Power Applications”, Tallinn University of Technology, 2007, Vol. 24, No. 2 Special, pp. 318–331.
14. Inigo Garin, Alejandro Munduate, Salvador Alepuz, Josep Bordonau, “Low and Medium voltage Wind Energy conversion systems: Generator Overview and Grid Connection requirements”, 19th International Conference on Electricity Distribution Vienna, May 2007, paper No. 0572, pp. 1–4.
15. Akhilesh sharma, Deepak singh, Vikas Pandey, S Gao, “Selective Harmonic Elimination for Cascaded H-bridge MLI using GA and NR-Method”, 2020 International conference on Electrical and Electronics engineering (ICE3), February 2020, pp. 89–94.
16. Wang Tianxiang, Ho Wanio and Chan Iatneng, “New Simulation Technology in Wind Power System”, University of Macau, 2010.
17. Yan Xu, D. Tom Rizy, Fangxing Li, and John D. Kueck, “Dynamic Voltage regulation using Distributed energy resources” 19th International Conference on Electricity Distribution Vienna, May 2007, Paper no.0736, pp. 1–4.
18. Brahim Berbaoui, Chellali Benachaiba, Rachid Dehini, and Otmane Harici, “Design of DC Link Voltage Controller Using Ant Colony Optimization for Shunt Active Power Filter”, Journal of Electrical Engineering: Theory and Application, 2010, Vol. 1, No. 2, pp. 92–99.

19. B. Ferdi, C. Benachaiba, S. Dib, R. Dehini, "Adaptive PI Control of Dynamic Voltage Restorer Using Fuzzy Logic", Journal of Electrical Engineering: Theory and Application, 2010, Vol. 1, No. 3 pp. 165–173.
20. Marco Liserre, Frede Blaabjerg and Steffan Hansen, "Design and Control of an LCL-Filter-Based Three-Phase Active Rectifier", IEEE Transactions on Industry Applications, Sept./Oct. 2005, Vol. 41, No. 5, pp. 1281–1291.

C-RPI: Cluster-Based Rendezvous Point Identification and Mobile Sink-Based Data Collection in LR-WPAN

S. Jayalekshmi and R. Leela Velusamy

1 Introduction

Low-rate wireless personal area network (LR-WPAN) with the IEEE 802.15.4 standard [1] is an enabling technology for wireless sensor network (WSN) and Internet of Things (IoT) applications [2, 3]. LR-WPAN consists of hundreds to thousands of battery-operated energy-constrained tiny wireless network nodes [4]. These devices are usually placed in a sensing area [region of interest (RoI)—a small geographical region], which can monitor environmental changes such as temperature, light, gas, pressure, motion, moisture, proximity, etc. [5]. These output of LR-WPAN devices are converted into human-readable form and transmitted over a network to data collecting points, known as *sink* or base station (BS), for further processing.

LR-WPAN can be structured or unstructured [6]. In structured deployment, the device positions in the RoI are pre-determined, and they are arranged manually (square, hexagon topologies are used [7]). Random deployment is done in remote areas that makes the network unstructured. The LR-WPAN consists of homogeneous or heterogeneous devices. The random deployment of homogeneous devices with a single BS is considered in this chapter. When there is a change in the environmental condition, the sensor devices will detect it and collect data. The data gathered are passed on to the data collection point (BS or sink) [8]. When sink node is stationary, due to the limited transmit power of LR-WPAN devices, direct communication with the sink is not possible for all the nodes, i.e., single-hop communication to sink is not possible from all the devices. So, the data collected from far away devices are forwarded to the collection point (sink node) through

S. Jayalekshmi (✉) · R. L. Velusamy
NIT Tiruchirappalli, Computer Science and Engineering, Tiruchirappalli, India

M. Mishra et al. (eds.), *Applications of Computational Intelligence in Management & Mathematics*, Springer Proceedings in Mathematics & Statistics 417,
https://doi.org/10.1007/978-3-031-25194-8_18

multi-hop communication. But this will create hot spot problem in the network, since nodes in close proximity to the sink get exhausted quickly and network is partitioned [9]. To resolve the hot spot problem efficiently in low-power wireless devices is to amalgamate sink mobility for data collection in LR-WPAN [10, 11]. The mobile sink (MS) can visit certain points in the ROI either randomly or in a controlled fashion. The locations that are visited by MS are known as *data collection points (DCPs) or rendezvous points (RPs)*. To collect data from the observing region, the MS travels through a selected path and visits RPs. There are various algorithms in the literature to identify the RPs and for optimal path selection of MS.

This chapter addresses the problem of efficient path design of MS using a cluster-based RP identification (C-RPI). The method focuses on full coverage of static wireless devices by selecting a minimum number of RPs. Initially, every device will broadcast its identity (PANiD). Then the MS moves in a predefined path to get the device ids and its neighbor information as a list. C-RPI is a range-based approach for forming rectangular clusters of devices. The center of each rectangular cluster region is taken as RP. Finally, a split and merge operation is performed to reduce the number of clusters by including all devices within "R" distance from RP to the corresponding cluster. The C-RPI is compared with max–min and min–max algorithms. In max–min algorithm, the position of node with maximum neighbors is chosen as first RP by MS. After removing the node and its neighbors from the list, next node having the most number of neighbors is chosen as second RP and so on. This is repeated until the list becomes empty. Min-max algorithm works in reverse order, in which the nodes having a least number of neighbors are selected as first RP. This process is repeated after removing the selected RPs and its neighbors from the list. The iteration is stopped when the list is empty. For each algorithm, after selecting the RPs, traveling sales person (TSP) method is applied to select the optimal path. Thereafter, the MS will travel through the optimal path and data are collected. The implementation is done in NS-3 simulator [12]. The simulation result shows that C-RPI minimizes the number of RPs and tour length of MS, thereby reducing data acquisition delay compared to max–min and min–max algorithms. The contribution of the proposed work is given below:

- A cluster-based RP identification (C-RPI) is proposed.
- TSP using simulated annealing (SA) is done to find the most favorable route of MS.
- Comparison with max–min and min–max algorithms is done.
- Data collection is done through the obtained path, and data collection delay is observed.

The rest of this chapter is organized as follows: In Sect. 2, the related works in the field of MS-based data collection in WSN are presented. The network scenario considered and the proposed algorithms are discussed in Sect. 3. Simulation results are observed and analyzed in Sect. 4 followed by conclusion and future work in Sect. 5.

2 Related Works

An efficient data collection method for low-power devices with the IEEE 802.15.4 standard (LR-WPAN) is very essential for minimizing the energy consumption. There are various applications that require low-data-rate devices with minimum power consumption also. In this section, some of the existing data collection strategies using MS in WSN are discussed.

Two different approaches were proposed by Xing et al., namely RP-CP and RP-UG [13], for minimum energy rendezvous planning in WSN with mobile elements (MEs). In RP-CP, a routing tree is constructed with device id and cost. The rendezvous points and mobile element path are derived from the subtree formed using the sublist of edges. The second is a utility-based greedy approach (RP-UG). In this, virtual nodes are added to the tree edges longer than L_0, an application-dependent parameter. Initially, BS is added to RP list. The utility of a candidate node is calculated, and candidate node is added to the RP list only if its utility value is maximum and the tour length is less than or equal to L. The process is repeated until all source nodes are added to RP list.

A heuristic method for energy-efficient mobile sink path selection in WSN known as weighted rendezvous planning (WRP) was proposed by Salarian et al. in 2014 [14]. The algorithm selects RP as a sensor node having maximum weight. The node weight is the product of the number of packets that it forwards and the number of hops to the closest RP in the path. A node is added to the RPLIST only if tour length is less than the required maximum travel path length.

Kaswan et al. proposed a delay bound data collection in WSN by selecting RPs based on k-means clustering (DBRKM) [15]. In order to reduce the number of potential RP positions, a weight function is derived from network parameters such as the number of one-hop neighbors, average hop count, energy consumption, etc. In each step, RPs with highest weight are removed along with its one-hop neighbors. The process is repeated until the potential RP list is empty. The delay bound path is obtained using TSP with Christofides heuristic.

Yarinezhad et al. [16] have introduced a protocol named as nested routing algorithm in which several virtual rings (closed chains) are formed using router nodes from the center of the deployment region. When sink moves to a new location, it should update its position to router nodes, and the sensor node has to forward the data to the nearest router node. Then, it will perform geographic routing to deliver the data to the sink.

An improved ant colony-based RP selection and sink path estimation was proposed by Praveen et al. [17]. Data forwarding path is constructed using spanning tree. For balancing the energy of RP nodes, RP re-selection and virtual RP concepts are introduced. Minimum spanning-tree-based clustering followed by a geometric algorithm to find the optimal trajectory of MS was proposed by Gutam et al. [18]. Energy of the RP nodes is balanced by applying RP re-selection algorithm. Authors claim that the network lifetime is improved compared to [17].

A hierarchical data collection from dynamic WSN was proposed by Mazumdar et al. [19] in 2021. This is an adaptive method to improve the coverage of wireless nodes when topology changes. It uses multiple data collectors to reduce energy hole problem. Rendezvous node selection and trajectory planning of data collectors is done periodically to address the fault-tolerance issue.

For collecting data in large scale from WSN, a method was proposed by Azar et al. [20]. A multi-interface MS is used to collect data that minimize delay and improve energy efficiency. RoI is partitioned, and MS visits each sub-areas. MS movement path is estimated for each partition.

In 2022, Wu et al. [21] proposed an ant colony optimization strategy for selecting collection points (CP) and for planning MS trajectory. Here a predefined data collection tree is constructed for forwarding the data to the CP. Here isolated nodes are not considered for data collection. Only one data packet is allowed in one round. This may lead to information loss.

Some of the existing data collection models in WSN with MS are presented here. There is no data collection model for LR-WPAN devices in the literature. Most of the methods on WSN focus on energy consumption, network lifetime, and delay. A realistic energy model described in [22] is used in this chapter. The proposed algorithms mainly focus on data collection delay in LR-WPAN devices using MS.

3 Proposed Work

This section describes the network structure, the assumptions made, and the proposed C-RPI algorithm in detail.

3.1 Network Scenario

The set of assumptions used while simulating the LR-WPAN are given below:

- The devices in LR-WPAN are homogeneous.
- Each node has a unique identifier (PANiD).
- Deployed sensor nodes are stationary.
- All sensors have equal initial energy.
- The communication channel is bi-directional.
- Sink does not have any resource constraints.

3.2 Overview of Proposed Algorithms

Wireless devices are randomly deployed in the region of interest (sensing area in WSN) for collecting data. The proposed data collection model for LR-WPAN device comprises five phases:

- *Neighbor Discovery Phase*: All devices collect its neighbor device's information by broadcasting a beacon signal.
- *Initial Sink Path Planning Phase*: The path for sink movement is found out, and the sink moves through the obtained path to collect the neighbor details from each device.
- *RP Selection Phase*: This phase is used to find the data collection points in the observing area using any one of the three proposed algorithms namely min-max, max-min, and cluster-based RP selection (C-RPI).
- *Optimal Path Selection Phase*: The shortest path connecting all RPs is constructed using TSP in this stage.
- *Data Collection Phase*: The sink node moves through the obtained path and collects data periodically from the devices.

Each phase for the data collection from LR-WPAN is shown in Fig. 1, which represents the complete system model.

3.2.1 Neighbor Discovery Phase

Initially, every LR-WPAN device broadcasts its identification number (PANiD) and a message ID (ID2) to inform its neighbors about its presence. On receiving the PANiD with message ID2, the receiving nodes will create a packet with their own PANiD, message ID3, and location. This packet is broadcasted and will be received by all neighbors within the communication range "R." The packet payload created by the receiver j is $\langle PANiD_j, X_j, Y_j \rangle$. Each node i then maintains a list of neighbors, NeighborList ($NL\,[i]$), from the received packet payload of all neighbors.

3.2.2 Initial Sink Path Planning Phase

In order to collect the $NL\,[i]$ from each node i $\in 1 \ldots N$, the sink node has to move from one location to another. While moving, the mobile sink has to cover all the areas in the observing region. The MS path is designed in such a way that it should provide full coverage for devices in LR-WPAN. Assume that the dimension of the sensing region is $D1 \times D2$. This area is divided into "S" square sub-regions. The size of each sub-region is calculated as follows.

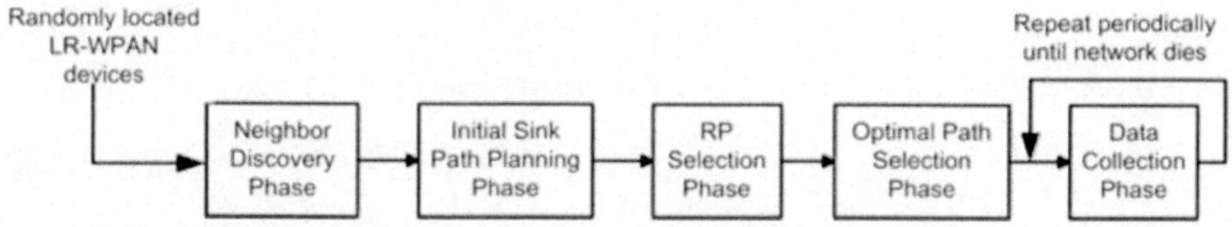

Fig. 1 Proposed system model for data collection from LR-WPAN

Let “R” be the communication range of a wireless node. If a node is placed at the center of a sub-region, it can communicate with all neighbors within “R” distance. If the length of one side of sub-region is 2R, it will create holes in the MS travel path as shown in Fig. 2. So the sub-region is chosen as the square inscribed in a circle of radius “R.” The maximum side length of the square sub-region is $R\sqrt{2}$. When sink is at the center of a sub-region, its sensing range will overlap with other nearby sub-regions.

As shown in Fig. 3, this will eliminate holes and ensures full coverage in the network. The total number of sub-regions $S\,is\left\lceil \frac{D1}{R\sqrt{2}} \right\rceil \times \left\lceil \frac{D2}{R\sqrt{2}} \right\rceil$. Initially, sink is at location (0,0). It will move to the center of sub-region 1 and broadcast a beacon signal with ID1. The neighbors within the communication range “R” will send their NL created in phase 1 as reply. Then sink will move to the center of the next sub-

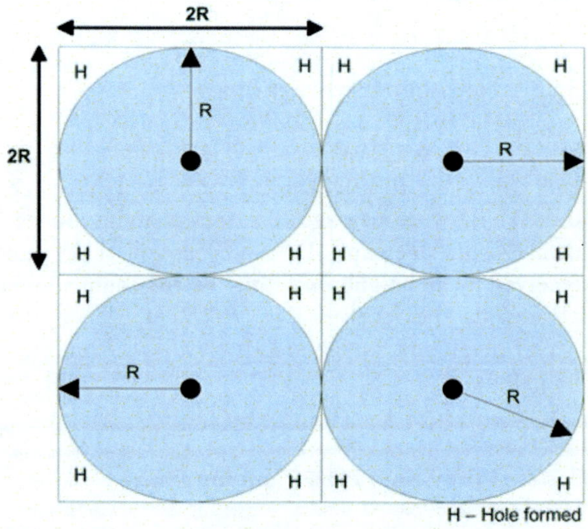

Fig. 2 Square sub-regions with holes (side length 2R)

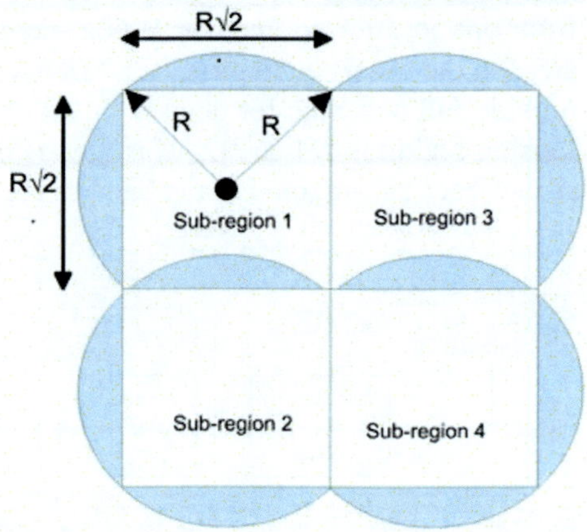

Fig. 3 Square sub-regions formed without holes (side length $R\sqrt{2}$)

region, broadcast beacon, and collect NL from all the nodes in sub-region 2. This process continues for all sub-regions, and the sink gets $NL[i]$ for $i = 1 \ldots N$ nodes after traversing one complete path.

3.2.3 RP Selection Phase

The sink path generated in the previous phase is too long since it passes through all sub-regions. Due to the random deployment, the number of static nodes in the sub-regions is not uniform. Some regions may be densely populated, while others have a less number of nodes or no nodes at all. So, dividing the region of interest into equal sized sub-regions and traversing all sub-regions will cause unnecessary delay for data collection. Hence, it is essential to find out the areas where nodes are present and an appropriate position for the sink that covers maximum nodes in that area. This position is known as rendezvous point (RP). Here, three different algorithms are proposed for the RP selection in LR-WPAN with the objective of minimizing the number of RPs and thereby reducing path length for data collection. The algorithm is designed based on the assumptions mentioned in Sect. 3.1.

Algorithm 1 Cluster-based RP identification (C-RPI)

Require: $NDL[1..N]\, where,\ NDL[i] = \langle X_i, Y_i, NodeId_i \rangle,\ 1 \le i \le N$

Ensure: Set of 'p' Rectangular Clusters $RC\,[1 \ldots p]\, where,\ 1 \le p \le N$

$p = 0$
for $i = 1, \ldots, N$ **do**
    if ($i == 1$) **then**
        $p = 1$
        $RC[p]_{xmin} = RC[p]_{xmax} = X_i$
        $RC[p]_{ymin} = RC[p]_{ymax} = Y_i$
    else
        count = 0
        for $j = 1, \ldots, p$ **do**
            if (($RC[j]_{xmin} \le X_i \le RC[j]_{xmax}$) **then**
                if ($RC[j]_{ymin} \le Y_i \le RC[j]_{ymax}$)) **then**
                    Do Nothing.
                end if
            else
                count ++
                if ($check() == r \&\& r \ge 0$) **then**
                    Update RC[j] co-ordinates for including $Node_i$ in region 'r'
                end if
            end if
        end for
        if ($count == p$) **then**
            p++
            $RC[p]_{xmin} = RC[p]_{xmax} = X_i$
            $RC[p]_{ymin} = RC[p]_{ymax} = Y_i$
        end if
    end if
end for
Return $RC\,[1 \ldots k]$

Cluster-Based RP Identification (C-RPI) This is a cluster formation approach in which rectangular clusters (RCs) are formed based on the distance between the nodes. The center of each cluster is selected as RPs.

At each step, a node is added to some cluster if maximum distance between the node and the cluster members is less than or equal to the communication range "R." For each rectangular cluster, the sink node stores the minimum and maximum values of its X-Y co-ordinates $(RC[j]_{xmin}, RC[j]_{xmax})$ $(RC[j]_{ymin}, RC[j]_{ymax})$, the center point $(RC[p]_{xc}, RC[p]_{yc})$, the number of neighbors $(MemberCount[j])$, and a list of members $(MemberList[j])$.

The steps for cluster formation are as follows. Every node i is having $\langle PANiD_i, X_i, Y_i \rangle$. The procedure starts with the node having $\langle PANiD_1, X_1, Y_1 \rangle$. The first rectangular cluster is formed with X_1 as $RC[1]_{xmin}, RC[1]_{xmax}$ and Y_1 as $RC[1]_{ymin}, RC[1]_{ymax}$. While considering the next node, say i, check whether it is possible to add to any of the existing cluster "j." This is done by dividing the outer space of cluster j into 8 regions and finding the region r in which node i belongs to. Now, try to enlarge the cluster j by including node i in the region r. During this enlarging step, the dimension of the rectangular cluster should not exceed $R\sqrt{2} \times R\sqrt{2}$. If it goes beyond the above said dimension for all the existing clusters, a new cluster is formed with node "i." The example in Fig. 4 shows enlarging of cluster "j" by adding a node "i" in region 1. In this way, all the nodes from 1 to N are added to some cluster. Finally, new RPs are taken as the center of the new rectangular clusters. When the number of data collection points is more, it will increase the path length of MS. This will directly affect the data collection delay. Therefore, as a second step in minimizing the number of RPs, a split–merge mechanism is used.

In this, a circular region with radius R is considered for adding cluster members. The center of the circular region is the position of RP in the rectangular cluster. The split–merge is explained with an example in Fig. 5, with three nearby clusters J, K, and L. While considering the circular region of cluster J and cluster L, it includes all the nodes from cluster K also. So, it is possible to split the nodes in cluster K into two groups and merge them with that of clusters J and L. In this case, cluster K can be removed, and the number of the clusters will be reduced by 1. This will help to include more nodes to the existing cluster and thereby minimizing the number of clusters by splitting and merging them.

3.2.4 Optimal Path Selection Phase

After finding the RPs in the RoI, the next step is to trace the optimal path for data collection. Finding the optimum path is the problem of finding the minimum weight Hamiltonian cycle (a special case of TSP), which is an NP-hard problem. This is done using simulated annealing method.

It is a metaheuristic approach to find the approximate global solution for an optimization problem. In finding the optimal path, an instance of a tour is defined with permutation of the RPs to be visited. The next instances are the set of

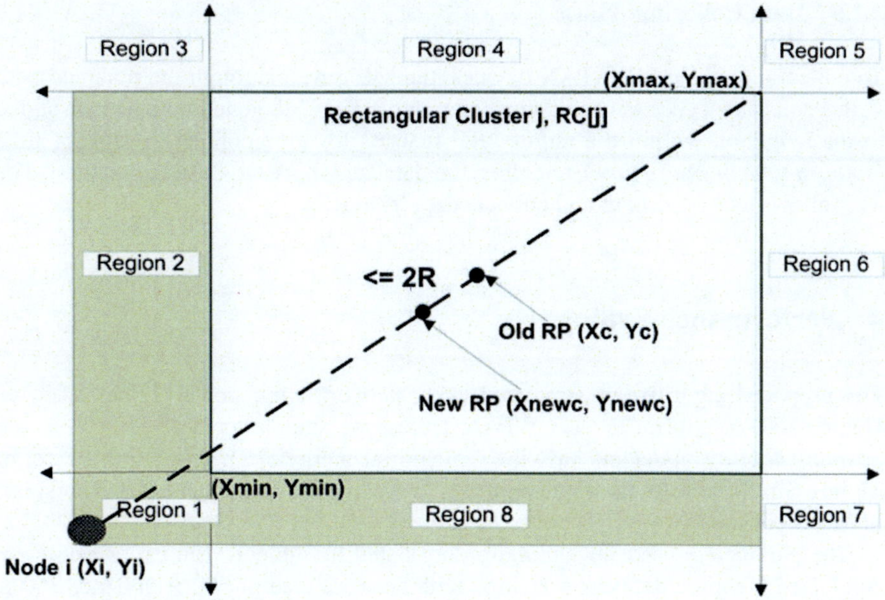

Fig. 4 Enlarging RC[j] by adding a node "i" in Region 1

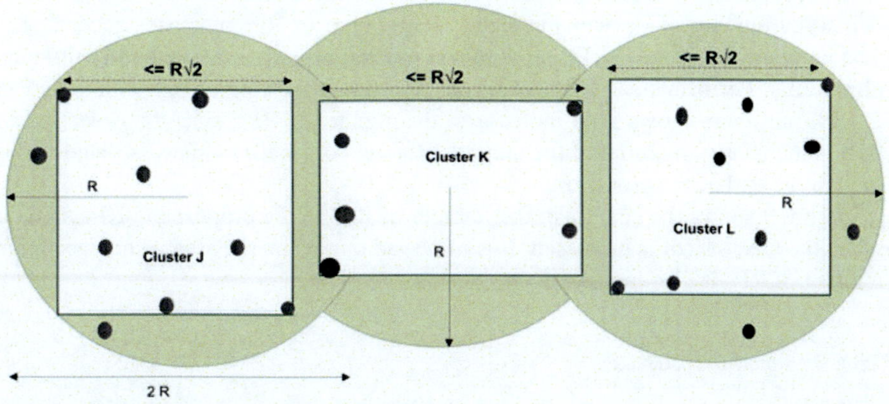

Fig. 5 Reducing the number of clusters by split–merge mechanism

permutations produced by reversing the order of any two successive RPs such that the total travel cost should be less than that of the previous state (progressive improvement). The new instances are created iteratively until the system produces a better result.

3.2.5 Data Collection Phase

To collect data from LR-WPAN devices, the MS travels through the obtained path with a constant speed. At each collection point (RP), a MS generates a beacon signal to the neighbors within the radius "R." It waits for a predefined amount of time *(sojourn time or pause time)*, to collect the data, and moves to the next location. The MS follows a circular path and collects data from RPs.

4 Performance Evaluation

The proposed algorithm is simulated using network simulator-3 [23](ns-3.29) on Ubuntu 16.04 platform. The LR-WPAN devices are randomly placed using *constant position mobility model* and have the same initial energy of 1J. The data collector or Mobile Sink's mobility is implemented using *waypoint mobility model.* The speed of MS is constant.

The parameters used for simulation are listed in Table 1. The experiments are conducted for three network scenarios with network area of $500 \times 500\,\text{m}^2$, $750 \times 750\,\text{m}^2$, and $1000 \times 1000\,\text{m}^2$, by varying the number of nodes between 50 and 250.

Figures 6, 7, and 8 show the number of rendezvous points (RPs) generated for the proposed algorithms against a varying number of nodes in three different network scenarios (network area $500 \times 500\,\text{m}^2$, $750 \times 750\,\text{m}^2$, and $1000 \times 1000\,\text{m}^2$). When the number of devices increases, the number of RPs increases accordingly. The number of RPs in C-RPI is low when compared with max–min and min–max algorithms. Performance of max–min is 10% better than min–max. Compared to C-RPI, min–max shows 30% increase in the number of RPs, since the selection of RPs in the increasing order of the number of neighbors leads to many isolated nodes toward the end of the procedure.

Figure 9 shows the change in the number of RPs in the proposed C-RPI method when the area of RoI is increased. It is observed that when the area is increased, the number of RPs is also increased.

Table 1 Simulation parameters

Parameter	Value
Target area	$500 \times 500, 750 \times 750, 1000 \times 1000\,\text{m}^2$
The number of LR-WPAN devices	50, 100, 150, 200, 250
Initial energy of devices	1.0 J
Transmit power	0 dBm to 5 dBm
Receiver sensitivity	−106.58 dBm
Data packet size	64 Byte
Speed of MS	5 m/s
Pause time at RPs	5 s

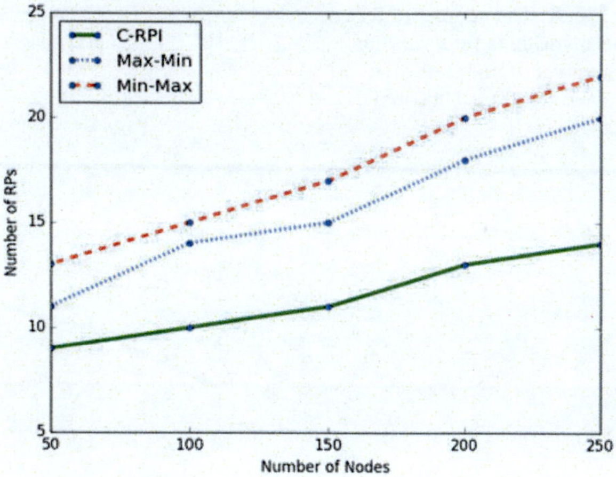

Fig. 6 The number of RPs in 3 algorithms for a varying number of nodes. [Area: $500 \times 500\,\text{m}^2$]

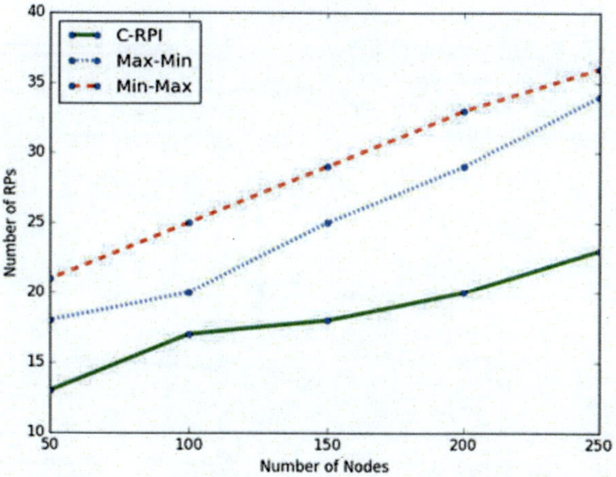

Fig. 7 The number of RPs in 3 algorithms for a varying number of nodes. [Area: $750 \times 750\,\text{m}^2$]

The packet success rate for a packet size of 64 Byte with 2.4 GHz clock with transmit power 0 dBm and 5 dBm is estimated. It is observed that the reliable communication range for 0 dBm is around 90 m and that of 5 dBm is 133 m. The number of rendezvous points is estimated against varying transmit power ranging from 0 to 5 dBm. The C-RPI shows better performance over the other two algorithms because the simulation results indicate that the number of RPs is reduced by 17% and 23% from max–min and min–max, respectively. It is also observed that there is an average decrease of 40% in the number of RPs, when the transmit power is increased as indicated in Fig. 10.

The TSP with simulated annealing gives optimal path for the given set of RPs. The proposed algorithms' path cost is calculated as distance of the travel path. Since C-RPI is having a less number of RPs, the average MS path length is reduced by 12% and 16% when compared to max–min and min–max algorithms, respectively. The

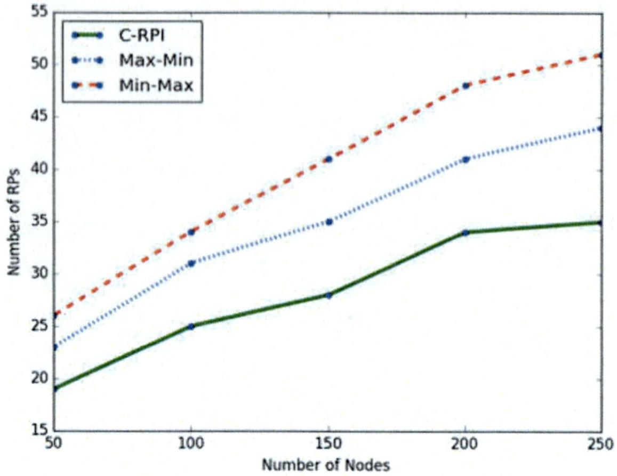

Fig. 8 The number of RPs in 3 algorithms for a varying number of nodes. [Area: $1000 \times 1000\,m^2$]

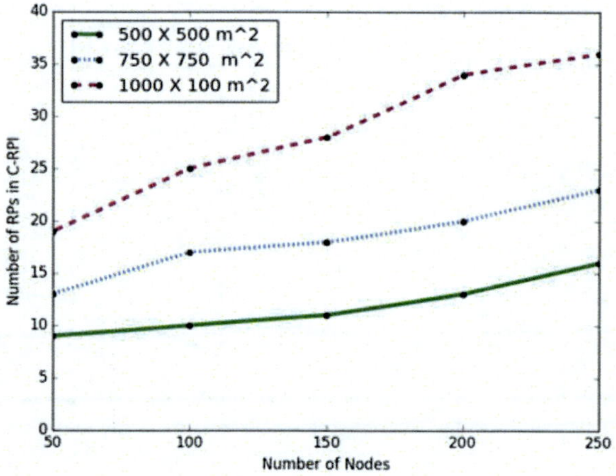

Fig. 9 The number of RPs in C-RPI for 3 network scenarios against a varying number of nodes

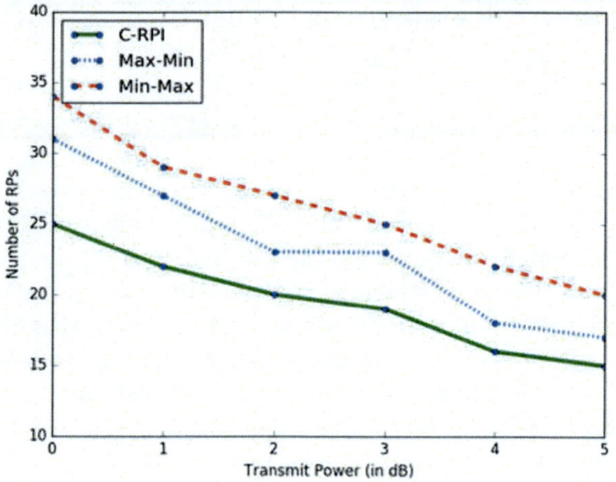

Fig. 10 The number of RPs for 3 algorithms against varying transmit power [No of nodes = 100 and Network area $1000 \times 1000\,m^2$]

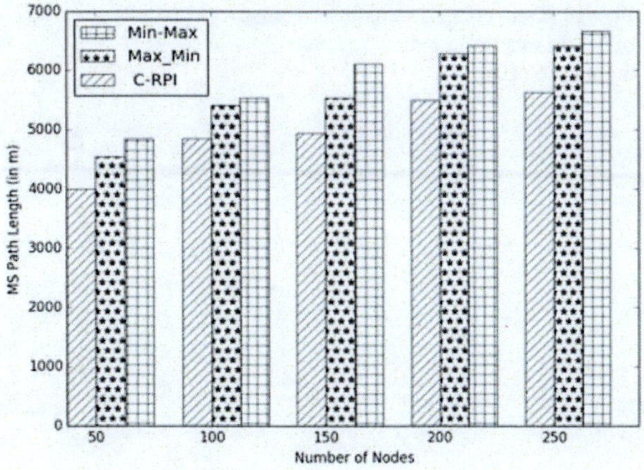

Fig. 11 Average MS path length for 3 algorithms against a varying number of nodes (using SA)

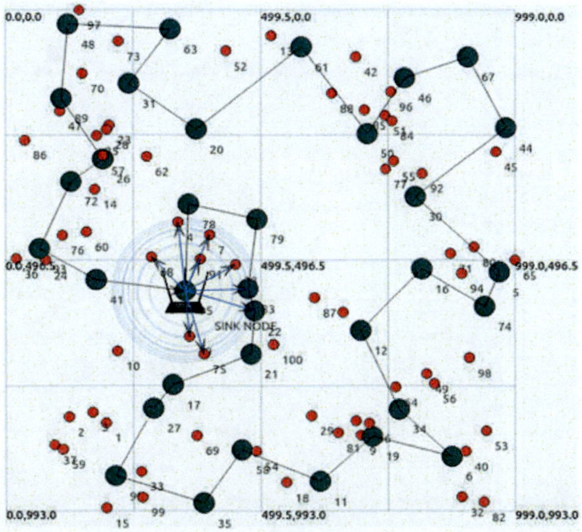

Fig. 12 Data collection path of MS in min–max path length = 5634 m

performance of max–min algorithm is in between C-RPI and min–max, as shown in Fig. 11.

After executing each one of the three algorithms mentioned in Sect. 3.2.3, the data collection path is obtained using SA in phase 4. The outputs observed are shown in Figs. 12, 13, and 14 for min–max, max–min, and C-RPI, respectively.

The polygon (circular zigzag path) shows the data collection trajectory with N_{RP} vertices, where N_{RP} is the number of RPs. The static nodes are represented in red color, and the nodes in the RP location are highlighted with green color for min–max (Fig. 12) and max–min (Fig. 13). For C-RPI (Fig. 14), the RPs are not highlighted because these points are the center of the rectangular clusters but not the node positions as it is considered in min-max and max-min. In this case, the RPs are vertices of the polygon.

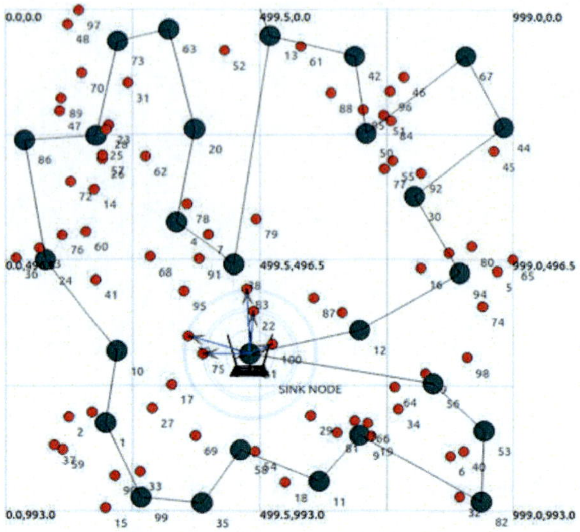

Fig. 13 Data collection path of MS in max–min path length = 5219 m

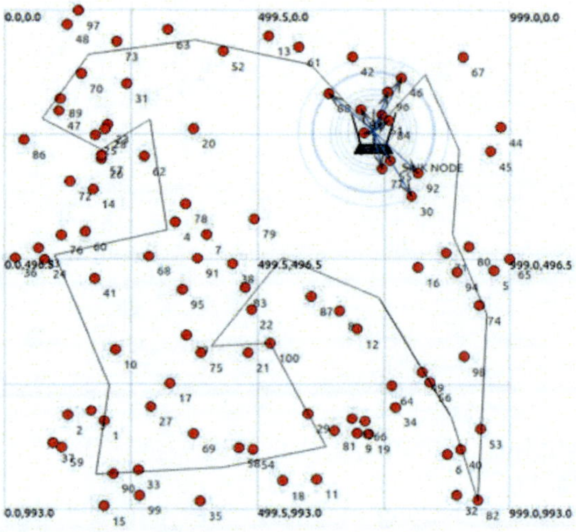

Fig. 14 Data collection path of MS in C-RPI path length = 4378 m

Next, the data collection delay is analyzed for the proposed algorithms. The delay is defined in Eqn. (1) as the sum of the time taken to travel through the data collection path and the pause time at RPs.

$$Delay = \frac{P_l}{S_{MS}} + (N_{RP} \times P_t) \tag{1}$$

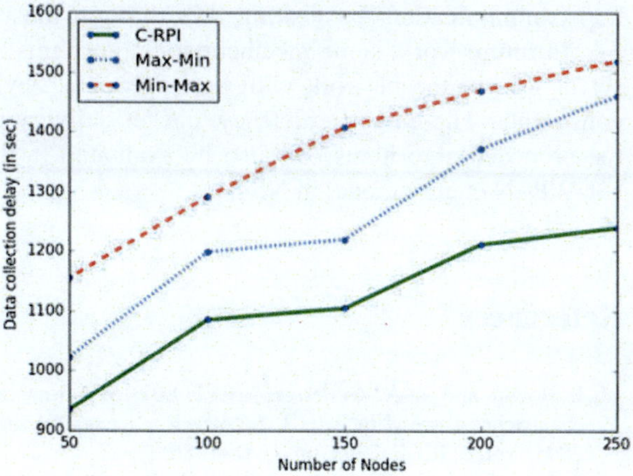

Fig. 15 Data collection delay vs. the number of nodes [Area $1000 \times 1000\,m^2$]

$$where, \quad P_l = Length\ of\ the\ data\ collection\ path\ (m)$$

$$S_{MS} = Speed\ of\ MS\ (m/s)$$

$$N_{RP} = Number\ of\ RPs$$

$$P_t = Pause\ Time\ (s)$$

The results obtained for delay estimation are shown in Fig. 15. The graph shows that the delay is directly proportional to the path length and the number of RPs. The delay obtained for min-max is 18% more when compared to C-RPI. Since P_l and N_{RP} is high for min-max algorithm, its delay is also high. The percentage increase in delay for max–min is 14% with respect to C-RPI. So C-RPI method performs better with minimum delay, and hence, it is suitable for delay-sensitive applications. From the above observations, out of the three algorithms proposed, the performance of C-RPI is better in terms of the number of RPs, path length, and delay.

5 Conclusion

IEEE 802.15.4 is a promising standard for short distance networks such as LR-WPAN. The cluster-based rendezvous point identification strategy viz. C-RPI for data collection from LR-WPAN devices is proposed in this chapter. Network simulator-3 is used to test the proposed work for three network scenarios by varying the area of RoI, the number of nodes, and transmitting power. From the simulation results, it is observed that in all the cases, the performance of C-RPI is better in terms of the number of RPs, average path length, and data collection delay. The simulation is done for applications with uniform data rate in homogeneous networks. The data collection is performed in every 30 min interval using single-hop communication, by visiting RPs through the path obtained using TSP with SA. In future work, some metaheuristic algorithms can be used to find an optimal set of RPs in the network with heterogeneous devices having non-uniform data constraints. The energy efficiency, packet delivery ratio, and network lifetime of the proposed algorithms can also be evaluated in the future, after modifying the LR-WPAN energy model in NS-3.

References

1. I. Howitt and J. A. Gutierrez, "IEEE 802.15. 4 low rate-wireless personal area network coexistence issues," in *2003 IEEE Wireless Communications and Networking, 2003. WCNC 2003.*, vol. 3. IEEE, 2003, pp. 1481–1486.
2. J. A. Gutierrez, E. H. Callaway, and R. L. Barrett, *Low-rate wireless personal area networks: enabling wireless sensors with IEEE 802.15. 4*. IEEE Standards Association, 2004.
3. S.-H. Yang, "Internet of Things," in *Wireless Sensor Networks*. Springer, 2014, pp. 247–261.
4. D. Mirzoev *et al.*, "Low rate wireless personal area networks (LR-WPAN 802.15. 4 standard)," *arXiv preprint arXiv:1404.2345*, 2014.
5. I. F. Akyildiz, W. Su, Y. Sankarasubramaniam, and E. Cayirci, "Wireless sensor networks: a survey," *Computer Networks*, vol. 38, no. 4, pp. 393–422, 2002.
6. J.-W. Lee and J.-J. Lee, "Ant-colony-based scheduling algorithm for energy-efficient coverage of WSN," *IEEE Sensors Journal*, vol. 12, no. 10, pp. 3036–3046, 2012.
7. N. Akshay, M. P. Kumar, B. Harish, and S. Dhanorkar, "An efficient approach for sensor deployments in wireless sensor network," in *Emerging Trends in Robotics and Communication Technologies (INTERACT), 2010 International Conference on*. IEEE, 2010, pp. 350–355.
8. M. Sujeethnanda, S. Kumar, and G. Ramamurthy, "Mobile wireless sensor networks: A cognitive approach."
9. M. Perillo, Z. Cheng, and W. Heinzelman, "On the problem of unbalanced load distribution in wireless sensor networks," in *IEEE Global Telecommunications Conference Workshops, 2004. GlobeCom Workshops 2004*. IEEE, 2004, pp. 74–79.
10. R. Jaichandran, A. A. Irudhayaraj *et al.*, "Effective strategies and optimal solutions for hot spot problem in wireless sensor networks (WSN)," in *10th International Conference on Information Science, Signal Processing and their Applications (ISSPA 2010)*. IEEE, 2010, pp. 389–392.
11. M. I. Khan, W. N. Gansterer, and G. Haring, "Static vs. mobile sink: The influence of basic parameters on energy efficiency in wireless sensor networks," *Computer Communications*, vol. 36, no. 9, pp. 965–978, 2013.
12. K. El Ghomali, N. Elkamoun, K. M. Hou, Y. Chen, J.-P. Chanet, and J.-J. Li, "A new WPAN model for NS-3 simulator," in *NICST'2103 New Information Communication Science and Technology for Sustainable Development: France-China International Workshop*, 2013, pp. 8–p.
13. G. Xing, T. Wang, Z. Xie, and W. Jia, "Rendezvous planning in wireless sensor networks with mobile elements," *IEEE Transactions on Mobile Computing*, vol. 7, no. 12, pp. 1430–1443, 2008.
14. H. Salarian, K.-W. Chin, and F. Naghdy, "An energy-efficient mobile-sink path selection strategy for wireless sensor networks," *IEEE Transactions on Vehicular Technology*, vol. 63, no. 5, pp. 2407–2419, 2013.
15. A. Kaswan, K. Nitesh, and P. K. Jana, "Energy efficient path selection for mobile sink and data gathering in wireless sensor networks," *AEU-International Journal of Electronics and Communications*, vol. 73, pp. 110–118, 2017.

16. R. Yarinezhad, "Reducing delay and prolonging the lifetime of wireless sensor network using efficient routing protocol based on mobile sink and virtual infrastructure," *Ad Hoc Networks*, vol. 84, pp. 42–55, 2019.
17. P. K. Donta, T. Amgoth, and C. S. R. Annavarapu, "An extended ACO-based mobile sink path determination in wireless sensor networks," *Journal of Ambient Intelligence and Humanized Computing*, vol. 12, no. 10, pp. 8991–9006, 2021.
18. B. G. Gutam, P. K. Donta, C. S. R. Annavarapu, and Y.-C. Hu, "Optimal rendezvous points selection and mobile sink trajectory construction for data collection in WSNs," *Journal of Ambient Intelligence and Humanized Computing*, pp. 1–12, 2021.
19. N. Mazumdar, S. Roy, A. Nag, and S. Nandi, "An adaptive hierarchical data dissemination mechanism for mobile data collector enabled dynamic wireless sensor network," *Journal of Network and Computer Applications*, vol. 186, p. 103097, 2021.
20. S. Azar, A. Avokh, J. Abouei, and K. N. Plataniotis, "Energy-and delay-efficient algorithm for large-scale data collection in mobile-sink WSNs," *IEEE Sensors Journal*, vol. 22, no. 7, pp. 7324–7339, 2022.
21. X. Wu, Z. Chen, Y. Zhong, H. Zhu, and P. Zhang, "End-to-end data collection strategy using mobile sink in wireless sensor networks," *International Journal of Distributed Sensor Networks*, vol. 18, no. 3, p. 15501329221077932, 2022.
22. V. Rege and T. Pecorella, "A realistic MAC and energy model for 802.15. 4," in *Proceedings of the Workshop on NS-3*. ACM, 2016, pp. 79–84.
23. "Low-rate wireless personal area network (LR-WPAN)." nsnam-ns3-A Discrete-Event Network Simulator Release 3.30.1, 2013. [Online]. Available: https://www.nsnam.org/docs/models/html/lr-wpan.html/

Effect of Weather on the COVID-19 Pandemic in North East India

Piyali Das, Ngahorza Chiphang, and Arvind Kumar Singh

1 Introduction

As the world is infested with the coronavirus and India is stricken with fear of the pandemic, we are trying to focus on some research related to the correlation between weather parameters and the increasing cases. After the initial detection in Wuhan, China, during December 2019, the virus has spread like wildfire in just 10 months and has infected more than 44 million, with ten lakhs diseased as reported by the World Health Organization (WHO) [1]. After the USA, India has grabbed the second position with 8.04 M infected cases and almost 12 lakh fatalities. In the North East India, after taking 41 days to reach 100 cases, the positive cases have surged nearly about 3.5 lakhs in the recent few days during Aug.–Oct. 2020 (Govt. of India, ArogyaSetu) [2], with a sharp rise in cases in all the states.

The review research published in the *Chemical Engineering Journal* [3] noted that each of these phases was characterized by a different type of biological interaction. SARS-CoV-2, the novel coronavirus that caused COVID-19, is transmitted through droplets expelled from an infected person's nose or mouth when they cough or talk as claimed by WHO. During the early infection phase which was referred to as phase I, WHO said that the virus multiplied inside the body and is likely to cause mild symptoms that may be confused with common cold or flu. There are a total of three stages of the disease as claimed by the *London Times*, June 13, 2020.

P. Das (✉)
Department of Electrical Engineering, North Eastern Regional Institute of Science and Technology, Nirjuli, Arunachal Pradesh, India

N. Chiphang · A. K. Singh
North Eastern Regional Institute of Science and Technology, Nirjuli, Arunachal Pradesh, India

M. Mishra et al. (eds.), *Applications of Computational Intelligence in Management & Mathematics*, Springer Proceedings in Mathematics & Statistics 417,
https://doi.org/10.1007/978-3-031-25194-8_19

2 NE Indian History of COVID-19

The first NE India COVID-19 case was detected in Manipur as per the [4] records, where the patient was recovered within 14–16 days. The objective of this study is to find how the active cases vary with a change in meteorological conditions for the NE states of India. In mainlands of India, COVID-19 positive scenario is becoming worse with the passage of time, whereas in NE, perhaps the people are more immune as per an article published in [5]. Several states are badly affected, and among them top are Assam (AS), Tripura (TR), and Manipur (MN). Weather parameter survey has been done for Meghalaya (ML) and Nagaland (NL) too. The average temperature (oC), wind speed (mph), heat index, and relative humidity (%) are the major parameters among the meteorological factors undertaken for analysis.

The weather data has been collected from visual crossing [6] website. In this paper, the authors focused on the correlation between the meteorological data and total COVID-19 positive cases affected in the abovementioned ten states of India. The results were analyzed from April 15 to October 18, 2020.

3 Methodology

The Spearman rank correlation was calculated for different meteorological parameters and COVID-19 positive cases. The correlation coefficient (S_r) signified the relationship between two different variables. Here how the active cases depend on the metrological data is shown. The correlation coefficient is defined as.

$$S_r = 1 - 6\frac{\sum D_i^2}{n\left(n^2 - 1\right)} \tag{1}$$

where D_i represents the rank difference of two parameters and n is the number of choices. These different parameters are chosen from the data surveyed for the study.

Hypothetically, if the correlation coefficient S_r varies between -1 and $+1$, then $+1$ indicates a strong positive relation between the two variables, and close to -1 indicates that there is a strong negative relation between the two variables. $+1$ implies both the variables will increase together, and -1 implies one of the parameters increases inversely with respect to the other parameter. Now the S_r value implies probability of being wrong in concluding that there is a true association between the variables such as probability of falsely rejecting the null hypothesis or committing an error. If the S_r value is smaller, the probability that the variables are correlated will become high; conventionally it depends on $S_r < 0.05$.

4 Result and Discussion

Table 1 shows the result of Spearman's correlation analysis for ten selected states in India. In context with the methodology, the data analysis showed that among all the ten states, the highest correlation was possessed by Sikkim for average temperature parameter with respect to the number of confirmed cases. This also implied that the confirmed cases were highly dependent on the average temperature. In [7] Meraj, G. et al. have described about the Indian scenario of the temperature dependency for the confirmed cases. It has been also mentioned in the research that to some extent of 25.8 °C, the spread relation may be inverse, but below the said temperature, the spread depends on various aspects of weather parameters. In this present data analysis approach, it has been observed that Maharashtra and Gujarat were having the highest negative correlation and the S_r value was less than 0.05 in both the cases. So it can be concluded that with the increase of temperature the active cases will decrease (Table 2). In the second case, the relative humidity was directly proportional for all the states, and the maximum correlation was occupied by the state Gujarat (Fig. 1). The relative humidity and confirmed case result analysis is shown in Fig. 2. The third parameter was wind speed, which as per the survey data showed an inversely proportional relation with the confirmed cases as shown in Table 3, and the analysis is illustrated in Fig. 3. The total case survey report was captured in Fig. 4 which entailed an exponentially increasing rate for all the states.

The relation between the cloud coverage and weather is discussed in [8], where it is clearly mentioned that due to the lockdown the cloud coverage becomes less which is with negative relation to the total number of affected cases. In this analysis it has been observed that only few states were able to alleviate the air pollution which generated a negative correlation; majority of the states are having positive correlation that implies the pollution rate is on the higher side (Table 4). As per Table 5, the heat index possesses the same pattern with Figs. 5 and 6.

Table 1 Correlation of avg. temp/active case

	Tripura	Manipur	Assam	Meghalaya	Nagaland
Correlation coefficient	0.259	0.376	0.581	0.224	0.21
S_r value	0.000362	0.00000015	0.0000002	0.000062	0.0000002
Number of samples	201	201	201	201	201

Table 2 Correlation between relative humidity (%)/active case

	Tripura	Manipur	Assam	Meghalaya	Nagaland
Correlation coefficient	0.313	0.0685	0.439	0.121	0.31
S_r value	0.0000139	0.353	5.11E–10	0.00000462	0.0000002
Number of samples	183	183	183	183	183

Table 3 Correlation between wind speed (mph)/active case

	Delhi	Gujarat	Maharashtra	Rajasthan	Tamil	Mizoram	Sikkim	Tripura	Assam	Manipur
Correlation coefficient	−0.747	−0.763	−0.786	−0.732	−0.738	−0.604	−0.786	−0.555	−0.35	−0.392
P value	0.0000002	0.0000002	0.0000002	0.0000002	0.0000002	2.955E–07	0.0000002	0.0000002	0.00000113	4.11E–08
Number of samples	183	183	183	183	183	183	183	183	183	183

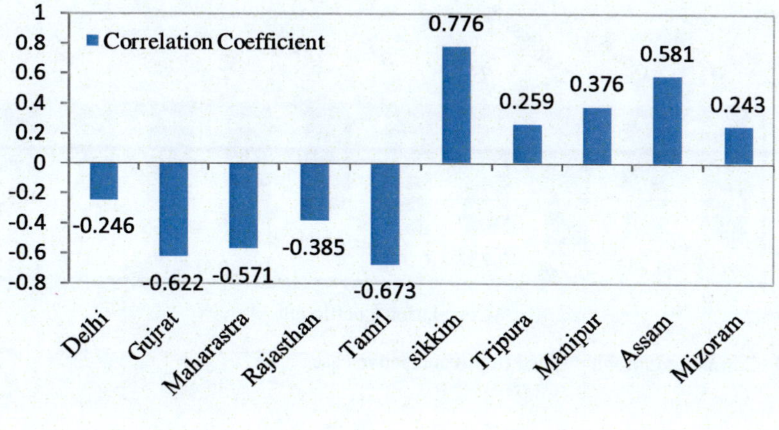

Fig. 1 Correlation for avg. temp and active case

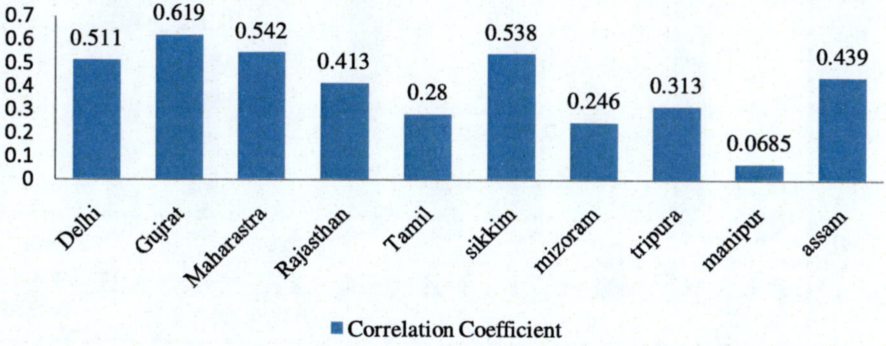

Fig. 2 Correlation for relative humidity (%) and active case

5 Conclusion

The overall impact of the weather with respect to the number of total confirmed cases was analyzed in this study. The number of confirmed cases across the planet has surpassed 40 million, but expertise suggestion is that it is only the tip of the iceberg when it comes to the true impact of the pandemic that has upended life and work around the world. This analysis tried to show the impact of average temperature, relative humidity, and wind speed in terms of confirmed cases in India. In different states of India, it has been observed that the wind speed has negative correlation which implies that while the speed increases the number of cases will decrease. In the case of relative humidity, Gujarat and Maharashtra

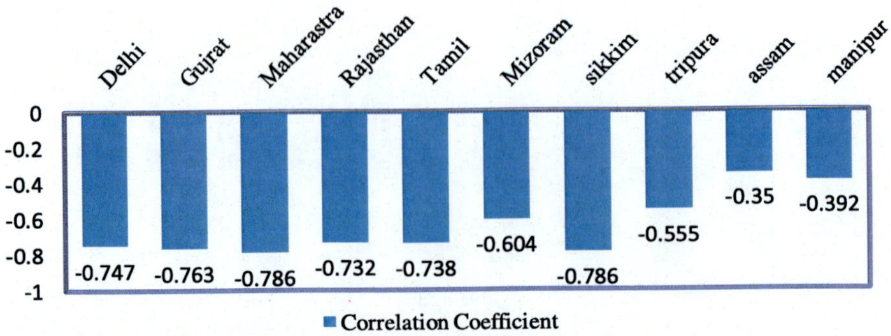

Fig. 3 Correlation for wind speed (mph) and active case

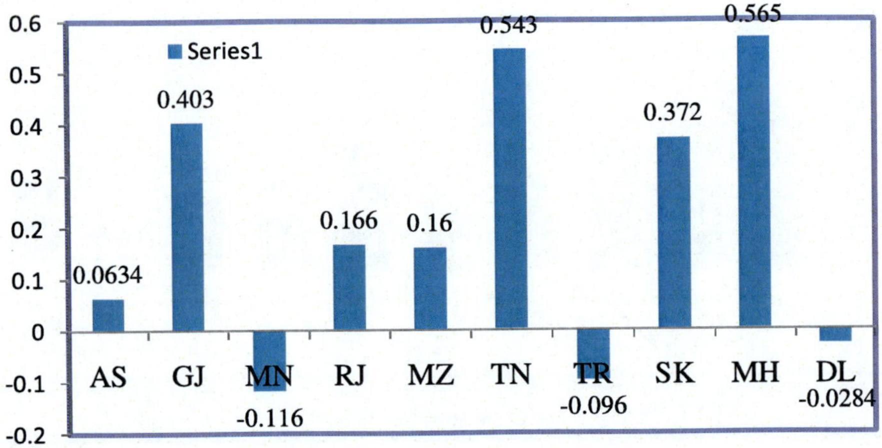

Fig. 4 Correlation for cloud coverage and active case

took the highest places with respect to the Spearman correlation coefficient, which implied that if it increases the cases also increases. The average temperature has a different relation with different states. In northeastern states, the correlation was found to be positive, whereas in the eastern, western, and southern zones, the correlation relation was negative. In the midst of this pandemic situation, weather and metrological parameters are also one of the major parts to influence the rate of spreading infection.

In future analysis, the returnee's rate for different states, the rate at which activities were unlocked, and recovery rate of infected people may also be included in this analysis. The weather parameters such as dew point, heat aspect, etc. may also be included in this analysis.

Table 4 Correlation between wind cloud coverage/active case

	Delhi	Gujarat	Maharashtra	Rajasthan	Tamil	Mizoram	Sikkim	Tripura	Assam	Manipur
Correlation coefficient	−0.0284	0.403	0.565	0.166	0.543	0.16	0.372	−0.096	0.0634	−0.116
P value	0.7	1.59E–08	2E–07	0.0236	2E–07	0.0296	2.1E–07	0.192	0.39	0.115
Number of samples	186	186	186	186	186	186	186	186	186	186

Table 5 Correlation between wind heat index/active case

	Delhi	Gujarat	Maharashtra	Rajasthan	Tamil	Mizoram	Sikkim	Tripura	Assam	Manipur
Correlation coefficient	−0.373	0.23	0.0979	−0.0328	−0.64	0.16	0.604	0.309	0.705	0.00727
P value	1.91E–07	0.00169	0.184	0.657	2E–07	0.0296	2E–07	1.88E–05	0.0000002	0.933
Number of samples	186	186	186	186	186	186	186	186	183	186

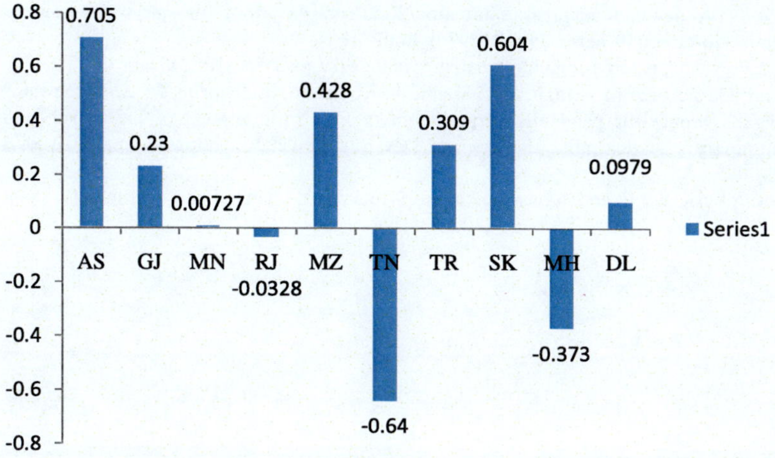

Fig. 5 Correlation for heat index and active case

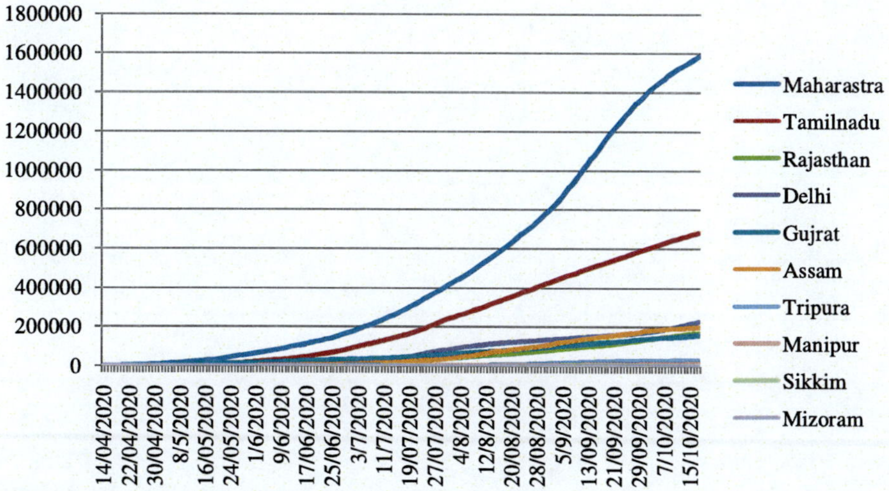

Fig. 6 Total cases for ten states since April 14 to Oct. 18, 2020

References

1. https://covid19.who.int/
2. https://www.aarogyasetu.gov.in/, https://www.mygov.in/corona-data/covid19-statewise-status/
3. Mohan SV, Hemalatha M,Kopperi H, Ranjith I, Kumar AK, SARS-CoV-2 in environmental perspective: Occurrence, persistence, surveillance, inactivation and challenges, Chem Eng J. 2021 Feb 1;405:126893. https://doi.org/10.1016/j.cej.2020.126893.
4. arunachaltimes.in

5. https://www.newindianexpress.com/nation/2020/apr/14/northeast-people-will-be-more-immune-to-covid-19-heres-why-2129749.html
6. https://www.visualcrossing.com/weather/weather-data-services#/viewData
7. Meraj, G., Farooq, M., Singh, S.K. Shakil A. Romshoo, Sudhanshu, M. S. Nathawat & Shruti Kanga, Coronavirus pandemic versus temperature in the context of Indian subcontinent: a preliminary statistical analysis. Environ Dev Sustain (2020). https://doi.org/10.1007/s10668-020-00854-3
8. https://physicsworld.com/a/has-the-covid-19-lockdown-changed-earths-climate/

A Comparative Study on the Performance of Soft Computing Models in the Prediction of Orthopaedic Disease in the Environment of Internet of Things

Jagannibas Paul Choudhury and Madhab Paul Choudhury

1 Introduction

In the field of orthopaedics, due to the advancement in the treatment and surgery, the gathering of information and education, advanced research, and development has become possible with the Internet of things. After orthopaedics surgery, it is necessary to do effective conversation between doctors and the patient. In that respect, the chances of recovery of the patient become better. Due to the implementation of IOT technology, the operational cost and errors are reduced. Necessary drugs can be applied to improve the patient health. Under IOT, all physical orthopaedics devices remain connected with the Internet.

2 Literature Review

Farahnaz Sadoughi, Ali Behmanesh, and Nasrin Sayfouri [1] have made an effort to make a study with an objective to find out IOT developments in medicine with the help of graphical/tabular classifications. It has been found that India, China, and the United States are the countries who are using medical research using IOT.

Bikash Pradhan, Saugat Bhattacharyya, and Kunal Pal [2] have opined that the Internet of things (IOT) has been used in manufacturing of various medical devices and sensors. Healthcare professionals can use quality medical services in a remote

J. P. Choudhury · M. P. Choudhury (✉)
Narula Institute Technology, CSE, Agarpara, Kolkata, West Bengal, India

NIT Jamshedpur, CSE, Jamshedpur, Jharkhand, India

M. Mishra et al. (eds.), *Applications of Computational Intelligence in Management & Mathematics*, Springer Proceedings in Mathematics & Statistics 417,
https://doi.org/10.1007/978-3-031-25194-8_20

location. As a result, patient safety has been improved, healthcare costs have been reduced, the healthcare services have been improved, and operational efficiency has been enhanced.

Amani Aldahiri, Bashair Alrashed, and Walayat Hussain [3] have used machine learning techniques in the health management system for prediction. The authors have made an effort to analyse the most well-known machine learning (ML) algorithms for the purpose of classification and prediction of IOT data to choose requisite algorithms for developing an efficient model of prediction.

Abid Haleem, Mohd Javaid, and Ibrahim Haleem Khan [4] have discussed the role of Internet of things in orthopaedics. With the help of different IOT technologies, proper information becomes available in the field of orthopaedics, and as a result, proper information has been gathered. Big data, cloud computing, smart sensors, artificial intelligence, and actuators have been used. It can detect mistakes during orthopaedics surgery.

Deepak Chahal and Latika Kharb [5] have analysed the data set of orthopaedic patients with the help of computer-aided diagnostic systems with an objective to decide regarding the health of a patient. They have tried to get a solution in the vertebral column disorders of a patient.

Vatan and Sandip Kumar Goyal [6] have made an effort to make the analysis of the clustering protocols using the method of soft computing in the field of agriculture in order to increase the lifetime of wireless sensor network.

Nicola Maffulli, Hugo C. Rodriguez, Ian W. Stone, Andrew Nam, Albert Song, Manu Gupta, Rebecca Alvarado, David Ramon, and Ashim Gupta [7] have tried to compare the performance of AI (artificial intelligence) and ML (machine learning) models in making a clinical diagnosis to forecast post-operative outcomes in orthopaedic surgery.

3 Content and Problem Statement

From the literature review it is evident that the authors have shown the importance of models of Internet of things in clinical and surgical fields [1, 2, 4, 6, 7]. Discussion on machine learning algorithms in surgical fields has been made in [3]. The supremacy of machine learning models and artificial intelligent has been discussed in [7]. From the literature review, it has been found that a lot of authors have expressed their views that the models of Internet of things should be used in clinical and surgical fields. But they have not worked on any data of the surgical field where the models of Internet of things can be applied. The importance of machine learning algorithms has been discussed in [3]. But they have not worked on any surgical data. For this purpose, here an effort is being made to work on vertebral data which is available from UCI Machine Repository [8]. Similar research work has been carried using soft computing models [9–11].

In this paper an effort has been made to apply multivariate statistical tools on the available data items for eliminating redundant items and to form a cumulative data item. The particular multivariate statistical tool has to be chosen using the minimum value of standard deviation and coefficient of variation for both the methods. Thereafter the available data has to be fuzzified, and neural network has to be applied on these fuzzy data. The output data from neural network has to be defuzzified, and operators of evolutionary algorithm, particle swarm optimization, and harmony search algorithm are necessary to be applied. The performance of these models has to be compared based on the value of average error, parameters of residual analysis, Atkinson's Information Criterion (AIC), and Bayesian's Information Criterion (BIC). Under residual analysis the absolute residual, summation of absolute residual, summation of mean of absolute residual, summation of residual error, mean of mean of absolute residual, and standard deviation of absolute residual have to be computed. The value of absolute residual is the positive difference of input data value and estimated data value based on any particular model. Clustering algorithms (K-means, hierarchical) have to be used on the estimated data based on selected model to get optimum number of clusters. The selection of clustering algorithm has to be made based on the value of Dunn index, Davies-Bouldin index, and Silhouette index. Using the clusters as formed using orthopaedic disorder data, confidence matrix has to be formed which can indicate the accuracy of the proposed system. Finally certain test data (containing vertebral data parameters) from new persons have to be taken, and the proposed model has to be applied with that test data. The difference of the test data has to be computed with the cluster centres as formed. The minimum distance value with particular centre indicates that data is similar to that cluster which signifies the deformity of the person. This procedure indicates whether the prediction made by the proposed model is correct or not.

4 Methodology

4.1 Multivariate Statistical Tool

Multivariate statistical analysis [7] is used as a necessary tool for finding out the unnecessary variable among other variables in mutual influence of variables with each other [7].

4.1.1 Factor Analysis

Step 1: The correlation matrix has to be calculated using all the features in the data set.

Step 2: The eigenvalues and eigenvectors have to be calculated for the correlation matrix.

Step 3: To consider those eigenvalues whose contribution is more than 2.5%.

Step 4: To pick the eigenvectors corresponding to the selected eigenvalues and to form cumulative item set for a particular data set.

4.1.2 Principal Component Analysis (PCA)

Principal component analysis (PCA) is a procedure using the method of statistics with an objective to reduce the dimensionality.

The covariance matrix has to be calculated using the features of the data set. The steps as narrated in Sect. 4.1.1 have to be applied.

4.2 Fuzzy Logic

Under fuzzy logic a variable handles multiple truth values for the same variable. Fuzzy logic handles imprecise spectrum of data. The law of heuristics can be used to get an array of accurate solutions. The membership functions are used. These are Gaussian, triangular, trapezoidal, generalized bell, sigmoid, etc.

4.3 Neural Network

A neural network makes an effort to approximate the predefined relationship in a set of data which operates like the human brain. The neural networks handle systems of artificial neurons. There are three main components: an input layer, a processing layer, and an output layer. The inputs may be weighted based on certain criteria. The processing layer is kept as hidden, which consists of nodes and connections between these nodes, which are analogous to the neurons and synapses of an animal brain.

4.4 Evolutionary Algorithm

An evolutionary algorithm (EA) is an algorithm that works and follows the procedure equivalent to the behaviour of living organisms (reproduction, mutation, recombination, and selection). The quality of the solutions that the algorithm produces is determined by fitness function. The evolutionary algorithm is a combination of evolutionary computing and bio-inspired computing. EAs are following the decision of Darwinian evolution.

4.5 Particle Swarm Optimization

Particle swarm optimization (PSO) is a method that tries to increase the efficiency of the problem repeatedly with respect to a specified target, decided early. It tries to get a solution of a problem by displacing these particles around the search-space and changing the particle's position and velocity. Each particle's movement is guided by its local best position and global best position.

4.6 Harmony Search (HS) Algorithm

The HS algorithm follows the behaviour of a singer who sings, where each singer produces a pitch following only one operation out of random selection, memory consideration, and pitch adjustment with an objective of increased efficiency of the problem. This process converts into a mathematical solution.

4.7 Average Error, Residual Analysis, AIC, and BIC

If a_i is the input data value and b_i is the estimated data value based on any particular model,

Total number of terms = n
Absolute residual abres = $(|b_i - a_i|)$
Sum of absolute residual = $\sum$abrs = $\sum (|b_i - a_i|)$
Residual error = abres/a_i where abres is value of absolute residual and a_i is input data value
Summation of residual error = Σ residual error
Mean of mean of absolute residual = sum of residual error/total number of terms
Mean of absolute residual, mean abres = summation of absolute residual/n
Standard deviation of absolute residual = $\sum$ (value of absolute residual – mean of absolute residual$)^2$/n
Average error = $(\sum(|b_i - a_i|)/a_i) * 100/n$
Likelihood = sum of absolute residual/n
Term = log(likelihood)
k is number of variables. Here the value of k = 2 since there is 1 input variable and 1 output variable.
Atkinson's Information Criterion (AIC) = $(-1) * 2 *$ term $+ 2 * k + (2 * k * (k+1))/(n - k - 1)$
Bayesian's Information Criterion (BIC) = $(-1) * 2 *$ term $+ k * \log(n)$

4.8 Dunn Index, Davies-Bouldin (DB) Index, and Silhoutte Index

4.8.1 Dunn Index

$$\text{Dunn Index } (U) = \min\left\{\min\left\{\partial\left(X_i, X_j\right) / \max\left(\Delta\left(X_k\right)\right)\right\}\right\}$$

where $\partial(X_i, X_j)$ is the inter-cluster distance of cluster X_k, i.e. distance within the cluster X_k.

$\Delta(X_k)$ is the intra-cluster distance of cluster X_k, i.e. distance within the cluster X_k.

4.8.2 Davies-Bouldin (DB) Index

$$\text{DB index } (U) = 1/k\ \Sigma \max\left\{\Delta\left(X_i\right) + \Delta\left(X_j\right) / \partial\left(X_i, X_j\right)\right\}$$

where $\partial(X_i, X_j)$ is the inter-cluster distance of cluster X_k, i.e. distance within the cluster X_k. $\Delta(X_k)$ is the intra-cluster distance of cluster X_k, i.e. distance to be calculated with respect of cluster centre within the cluster X_k.

4.8.3 Silhouette Index

$$\text{Silhouette Index } S(i) = (b(i) - a(i)) / \left(\max\{(a(i), b(i))\}\right.$$

where a(i) is the average distance of the ith object to other objects within the same cluster and b(i) is the average dissimilarity of the ith object with all other objects placed in the cluster which is the nearest cluster of that object.

5 Implementation

The vertebral data set have been used which are available in UCI machine repository [8] which have been used as input data set. Methods of factor analysis and principal component analysis have to be worked on the available data items to get cumulative data item.

Table 1 Comparison of factor analysis and principal component analysis

Model	Factor analysis	Principal component analysis
Mean	271.063	564,830
Standard deviation	*64.969*	139,590
Coefficient of variation	*0.239*	*0.24*

5.1 *Comparison on the Performance of Factor Analysis and Principal Component Analysis*

A comparative study has been made between estimated data based on factor analysis and on principal component analysis on the basis of the value of standard deviation and coefficient of determination. Lesser value of standard deviation and coefficient of variation indicates the superiority of the model. The value of coefficient of variation becomes almost the same for both cases. However the value of standard deviation is less in factor analysis than principal component analysis. Therefore, factor analysis can be used instead of principal component analysis (Table 1).

5.2 *Contribution of Neural Network, Evolutionary Algorithm, Particle Swarm Optimization, and Harmony Search (HS) Algorithm*

The estimated data based on factor analysis has been fuzzified using Gaussian membership function. Gaussian membership function has been used because the performance of that Gaussian membership function is better than other membership functions (triangular, trapezoidal, etc.).

The estimated data as formed using fuzzy logic has been applied to feed-forward back propagation neural network (BPNN). The output data from BPNN is fuzzy, and it has to be defuzzified to get the real numbers.

Evolutionary algorithm operators have been employed on the estimated data using neural network.

Thereafter particle swarm optimization algorithm has been employed to the estimated data using evolutionary algorithm.

Harmony search algorithm has been employed on the estimated data using particle swarm optimization. The proposed models have been made so that the performance of the proposed system is increased.

Now error analysis (computation of average error) and computation of AIC (Atkinson's Information Criterion) coefficient and BIC (Bayesian's Information criterion) coefficient has been made. The result has been provided in Table 2.

5.3 Contribution of Clustering Algorithm

From Table 2, it has been found that a lesser value of summation of absolute residual, summation of mean of absolute residual, summation of residual error, mean of mean of absolute residual, standard deviation of absolute residual, and average error and a greater value of Atkinson's Information Criterion (AIC) and Bayesian's Information Criterion(BIC) give more preference to the model. Thus the estimated data using harmony search algorithm has to be used for next processing. Clustering algorithms have been used on the estimated data value based on harmony search algorithm as narrated in the previous section.

Step 1
K-means clustering algorithm and hierarchical clustering algorithm have been employed using the estimated data value using harmony search algorithm as narrated in Sect. 5.2.

Step 2
The evaluation of clustering algorithms has been based on the value of Dunn index, DB index, and Silhouette index. The index values have been provided in Table 3.

Step 3
It is known that that with the lesser value of DB index and more value of Dunn Index and Silhouette index show more preference to that particular clustering algorithm than other clustering algorithm. From Table 3, it has been noticed that K-means clustering algorithm is preferable than hierarchical clustering algorithm.

Table 2 The performance of Neural Network and Evolutionary Algorithm, Particle Swarm Optimization and Harmony Search

Model	Neural network	Evolutionary algorithm	Particle swarm optimization	Harmony search
Sum of absolute residual	288.688	166.059	45.479	*30.608*
Sum of mean of absolute residual	1.241	0.6916	0.1745	*0.123*
Sum of residual error	9.622	5.535	1.5159	*1.02*
Mean of mean of absolute residual	0.041	0.023	0.0058	*0.0041*
Standard deviation of absolute residual	62.70	10.397	1.167	*0.0244*
Average error	4.137%	2.305%	0.582%	*0.409%*
Atkinson's Information Criterion (AIC)	−0.083	1.022	3.612	*4.404*
Bayesian's Information Criterion (BIC)	2.274	3.38	5.9703	*6.762*

Table 3 Clustering indexes vs. clustering algorithms

Name of clustering index	K-means clustering algorithm	Hierarchical clustering algorithm
Dunn index	*0.298*	0.0714
DB index	*0.541*	1.407
Silhouette index	0.526	0.347

Table 4 Confusion matrix

Orthopaedic data level	Accuracy	Recall	Precision	Specificity	F-measure
Disk hernia	0.6051	0.983	0.279	0.8296	0.3427
Spondylolisthesis	0.6497	0.433	0.48	1	0.6486
Normal	0.4796	0.24	0.6875	0.9338	0.529

Step 4
Further it has been noticed that the number of clusters as 3 gives minimum intra-cluster distance. Therefore the said number of clusters as 3 has been used for further processing. Here it is to note that Euclidean distance has been used as intra-cluster distance.

Step 5
Based on the maximum number of elements in cluster allocation, the selected cluster number has been mapped with the orthopaedic disorder level data types. Therefore it can be mentioned that cluster 1 is meant for 'disk hernia' disorder level, cluster 2 for 'spondylolisthesis' disorder level, and cluster 3 for 'normal' order level.

Step 6
Based on mapping of orthopaedic data items along with cluster number, confusion matrix has been formed based on certain parameters which has been provided in Table 4. The parameters used are as follows:

$$\begin{aligned}
&\text{Accuracy} = (\text{TP} + \text{TN}) / (\text{TP} + \text{FP} + \text{TN} + \text{FN}), \\
&\text{Recall} = \text{TP}/ (\text{TP} + \text{FN}), \\
&\text{Precision} = \text{TP}/ (\text{TP} + \text{FP}), \\
&\text{Specificity} = \text{TN}/ (\text{FP} + \text{TN}), \\
&\text{F} - \text{measure} = \left(2^*\text{precision}^*\text{recall}\right) / (\text{precision} + \text{recall})
\end{aligned}$$

where TP means true positive, TN means true negative, FP means false positive, and FN means false negative, respectively.

5.4 *Testing of Orthopaedic Disorder Using New Data Comprising Vertebral Parameter Values*

Step 1
A set of new vertebral parameter values has been taken. The steps narrated in Sect. 5.2 have been applied on the vertebral parameter values to get the estimated data using harmony search algorithm.

Table 5 Data set versus selected cluster no

Dat a set no	Data set value	Original data set type	Distance within cluster 1	Distance within cluster 2	Distance within cluster 3	Cluster selected	Predicted type
1	222.5	Disk hernia	*11*	103.27	30.3	1	Disk hernia
2	226.67	Disk hernia	*15.17*	99.107	26.13	1	Disk hernia
3	242	Disk hernia	30.485	83.77	*10.8*	3	Normal
4	254	Disk hernia	42.49	71.77	*1.2*	3	Normal
5	312.11	Spondylolisthesis	100.58	13.66	59.31	2	Spondylolisthesis
6	312.64	Spondylolisthesis	101.124	*13.13*	59.839	2	Spondylolisthesis
7	320.571	Spondylolisthesis	109.06	*5.21*	67.77	2	Spondylolisthesis
8	365.69	Spondylolisthesis	154.18	*39.92*	112.897	2	Spondylolisthesis
9	250.313	Normal	38.79	75.46	*2.487*	3	Normal
10	255.927	Normal	44.4	69.85	*3.127*	3	Normal
11	243.09	Normal	31.578	82.68	*9.707*	3	Normal
12	239.527	Normal	28.012	86.25	*13.273*	3	Normal

Step 2
The distance vector value has been computed between the estimated data using harmony search algorithm with three cluster centres as formed in step 3 of Sect. 5.3. The minimum distance indicates the data value is similar to the characteristic of that selected cluster. The selection of cluster centre has been provided in Table 5.

Step 3
From Table 5 it is evident that data no. 1 with data set value as 222.5 belongs to disk hernia orthopaedic disorder category; selected cluster is cluster 1 which belongs to disk hernia. Similarly data no. 4 with data value 254 belongs to disk hernia disorder category; selected cluster is cluster no. 3 which belongs to normal category. Further it can be mentioned that data no. 10 with data value as 255.927 belongs to normal category; selected cluster is cluster no. 3 which belongs to normal category. Accordingly for all data values, predicted cluster no. with predicted cluster data type has been found out.

6 Result

From Table 5, original data type and predicted data type has been ascertained. Table 6 has been created based on the original data type and predicted data type which has indicated whether correct prediction has been made or not.

From Table 6 it is evident that out of 12 data items, 10 cases have been predicted correctly, whereas 2 cases could not be predicted correctly. Accordingly based on

Table 6 Initial data type, predicated data type, and evaluation of prediction (correct or not)

Data no	Initial data type	Predicted data type based on cluster allocation	Evaluation (correct predication yes/no)
1	Disk hernia	Disk hernia	Yes
2	Disk hernia	Disk hernia	Yes
3	Disk hernia	Normal	No
4	Disk hernia	Normal	No
5	Spondylolisthesis	Spondylolisthesis	Yes
6	Spondylolisthesis	Spondylolisthesis	Yes
7	Spondylolisthesis	Spondylolisthesis	Yes
8	Spondylolisthesis	Spondylolisthesis	Yes
9	Normal	Normal	Yes
10	Normal	Normal	Yes
11	Normal	Normal	Yes
12	Normal	Normal	Yes

any data set, type of orthopaedic disorder can be ascertained in advance. In that case any preventive measure can be taken by the doctor/physiotherapist so that type of orthopaedic disorder can be reduced/eliminated.

7 Conclusion

Vertebral data set has been used which is available from UCI machine repository data [8]. Here in this data set, the type of orthopaedic disorder with respect to orthopaedic parameters has been furnished. It is to note that if there is no orthopaedic disorder, type of orthopaedic is assigned as normal. Here three types of orthopaedic disorder data are available; based on the orthopaedic parameter values, any orthopaedic disorder of the patient can be ascertained.

8 Theoretical/Managerial Implications

The orthopedic disorder of any person can be detected by taking Image of defected organ of the person and accordingly the value of orthopedic parameters can be found out and the said values can be processed for detection of orthopedic disorder using the proposed methodology. The data can be collected from persons in a camp located at distance from processing laboratory, send vide internet to the processing laboratory for examination and decision.

The healthcare sector is one of the most important sector which plays a major and vital role in the society. Due to the advancement of the Internet of things (IoT) methodology, a lot of problems of the society can be solved. Orthopedic field is a major medical area where patients' remedy can be made possible in quickest possible time with the help of the Internet of things methodology and procedures. Remote monitoring of patients' condition can be made possible through the implementation of these procedures now a days. Through the implementation of the proposed procedures, patient from distant locations can get treatment for their remedy and cure.

References

1. Farahnaz Sadoughia, Ali Behmaneshb, Nasrin Sayfouric, "Internet of things in medicine: A systematic mapping study", Journal of Biomedical Informatics, www.elsevier.com/locate/yjbin
2. Bikash Pradhan, Saugat Bhattacharyya, and Kunal Pal, "IoT-Based Applications in Healthcare Devices", Hindawi Journal of Healthcare Engineering, Volume 2021, Article ID 6632599, 18 pages. https://doi.org/10.1155/2021/6632599
3. Amani Aldahiri, Bashair Alrashed and Walayat Hussain, "Trends in Using IoT with Machine Learning in Health Prediction System", Forecasting 2021, 3, 181–206. https://doi.org/10.3390/forecast3010012
4. Abid Haleem, Mohd Javaid, Ibrahim Haleem Khan, "Internet of Things (IoT) applications in orthopaedics", Journal of Clinical Orthopaedics and Trauma, (2019). https://doi.org/10.1016/j.jcot.2019.07.003
5. Deepak Chahal, Latika Kharb, "Smart Diagnosis of Orthopaedic Disorders using Internet of Things (IoT)", International Journal of Engineering and Advanced Technology (IJEAT) ISSN: 2249–8958, Volume-8 Issue-6, August 2019. https://doi.org/10.35940/ijeat.7191.088619
6. Vatan, Sandip Kumar Goyal, "Soft Computing based Clustering Protocols in IoT for Precision and Smart Agriculture: A Survey", Proceedings of ISIC'21: International Semantic Intelligence Conference, February 25–27, 2021, New Delhi, India. 2020 Copyright for this paper by its authors. Use permitted under Creative Commons License Attribution 4.0 International (CC BY 4.0). CEUR Workshop Proceedings (CEUR-WS.org). https://ceur-ws.org/Vol-2786/
7. Nicola Maffulli, Hugo C. Rodriguez, Ian W. Stone, Andrew Nam, Albert Song, Manu Gupta, Rebecca Alvarado, David Ramon and Ashim Gupta, "Artificial intelligence and machine learning in orthopedic surgery: a systematic review protocol", Journal of Orthopaedic Surgery and Research (2020) 15:478. https://doi.org/10.1186/s13018-020-02002-z
8. Vertebral data set http://archive.ics.uci.edu/ml/datasets/vertebral+column
9. Dharmpal Singh, J. Paul Choudhury, Mallika De, "A Comparative Study of Meta Heuristic model to assess the type of Breast Cancer disease", IETE Journal of Research, June 2020, https://doi.org/10.1080/03772063-2020.1775139, Taylor & Francis Online.
10. Manisha Burman, J. Paul Choudhury, Susanta Biswas, "Automated Skin disease detection using multiclass PNN", International Journal of Innovations in Engineering & Technology (ISSN 2319-1058), https://doi.org/10.21172/ijet144.03. vol 14, issue 4, November 2019, pages 19-24.
11. Md. Iqbal Quraishi, Abul Hasnat, J. Paul Choudhury, "Selection of Optimal Pixel resolution for Landslide Susceptibility analysis within Bukit Antarabangsa, Kuala Lampur, by using Image Processing and Multivariate Statistical tools", EURASIP Journal on Image & Video Processing, Springer Open (2017) 2017-2. https://doi.org/10.1186/S13640-01S-0169-2, pp 1–12. Thompson Reuter.

Maximization of Active Power Delivery in WECS Using BDFRG

Manish Paul and Adikanda Parida

Nomenclature

P_r	Number of poles on the rotor side
P_p	Number of pole pairs in the primary winding
P_s	Number of pole pairs in the control winding
ω_r	Angular velocity of the rotor
ω_p	Speed of magnetic field of the primary winding
ω_s	Speed of magnetic field of control winding
V_{pd}	Direct voltage component of the primary winding
V_{pq}	Quadrature voltage component of the primary winding
V_{sd}	Direct voltage component of control winding
V_{sq}	Quadrature voltage component of control winding
I_{pd}	Direct current component of the primary winding
I_{pq}	Quadrature current component of the primary winding
I_{sd}	Direct current component of control winding
I_{sq}	Quadrature current component of control winding
R_p	Primary winding resistance
R_s	Control winding resistance
L_p	Primary winding self-inductance
L_s	Control winding self-inductance
L_m	Mutual inductance between primary and control winding
T_e	Electromagnetic torque
T_L	Load torque
λ_p	Primary rotor flux

M. Paul (✉) · A. Parida
Department of Electrical Engineering, NERIST, Nirjuli, India

M. Mishra et al. (eds.), *Applications of Computational Intelligence in Management & Mathematics*, Springer Proceedings in Mathematics & Statistics 417,
https://doi.org/10.1007/978-3-031-25194-8_21

λ_s Secondary rotor flux
ρ *d*/*dt*

1 Introduction

Power generations from renewable sources play a very vital role to address the issues like increasing energy demand, cost of fossil fuels, and climatic change. Power generations from wind have become popular nowadays [1]. However, the efficiency and the active and reactive power delivery using existing wind generators are always an issue [2]. WECS using doubly fed induction generators (DFIG) is a natural option [3]. Due to features like a higher saliency ratio and efficiency, BDFRG can be a better option for WECS [4]. However, regarding the active and reactive power control, BDFRG requires an almost similar controller to DFIG [5–8]. In this paper, the control mechanism of BDFRG has been highlighted, and the utility of BDFRG for the improved power generation capability of WECS has also been discussed. The presented control mechanism for BDFRG has some advantages like fixed switching frequency operation, which results in a decrease in power ripple in the output. In addition, it uses space vector pulse width modulation (SVPWM) control, which replaces the hysteresis controllers as hysteresis controllers have limitations like saturation.

2 Constructional Features of BDFRG

The BDFRG has two three-phase windings on the stator, which are sinusoidally distributed. The primary winding is the power winding which handles load power at supply frequency. Similarly, the secondary winding is the control winding which operates at a different frequency as shown in Fig. 1 [9]. Both the mentioned windings are electrically coupled to each other through a back-to-back converter. Both the converters are controlled using control signals generated by SVPWM in the case of the proposed scheme. It can be observed from Fig. 1 that the converter needs to handle only a fraction (almost 30%) of the total generated power. Therefore, unlike the self-excited induction generators, the converter system in the presented scheme is less costly. The number of salient pole pairs in the rotor can be computed from the number of pole pairs of both the primary and control windings. The same can be expressed as

$$P_r = \left(P_p + P_s\right)/2 \tag{1}$$

There is no magnetic coupling between both the windings as the number of pole pairs is different. Similarly, the rotor speed in rad./sec. of the BDFRG can

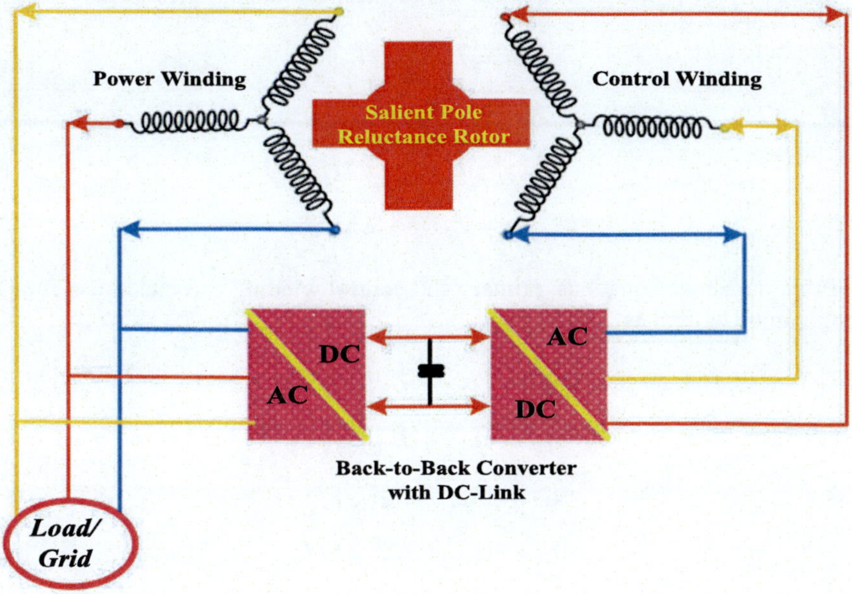

Fig. 1 BDFRG model

be computed as

$$\omega_r = \left(\omega_p + \omega_s\right)/2 \tag{2}$$

3 Dynamic Modeling of BDFRG

Unlike the doubly fed induction generators, in the case of BDFRG, the control winding is shifted to the stator side, and the slip ring in the rotor side is absent. Therefore, the rotor iron loss and the losses in the slip rings are absent in the case of BDFRG. This is the reason for the improved efficiency of the BDFRG-based WECS [10]. The d-q axis primary and control winding voltage equations in terms of primary and control winding currents can be expressed as [11]

$$\begin{bmatrix} V_{pd} \\ V_{pq} \\ V_{sd} \\ V_{sq} \end{bmatrix} = \begin{bmatrix} R_p + \rho L_p & -\omega L_p & \rho L_m & \omega L_m \\ \omega L_p & R_p + \rho L_p & \omega L_m & -\rho L_m \\ \rho L_m & (\omega r - \omega) L_m & R_s + \rho L_s & (\omega_r - \omega) L_s \\ (\omega_r - \omega) L_m & -\rho L_m & (\omega_r - \omega) L_s & R_s + \rho L_s \end{bmatrix} \begin{bmatrix} i_{pd} \\ i_{pq} \\ i_{sd} \\ i_{sq} \end{bmatrix} \tag{3}$$

Similarly, the primary and control winding fluxes can be expressed as

$$\lambda_{pd} = L_p i_{pd} + L_m i_{sd} \tag{4}$$

$$\lambda_{pq} = L_p i_{pq} - L_m i_{sq} \tag{5}$$

$$\lambda_{sd} = L_s i_{sd} + L_m i_{pd} \tag{6}$$

$$\lambda_{sq} = L_s i_{sq} - L_m i_{pq} \tag{7}$$

From the above equations, primary and control winding currents in d-q components can be written as

$$i_{pd} = \left(\lambda_{pd} - L_m i_{sd}\right)/L_p \tag{8}$$

$$i_{pq} = \left(\lambda_{pq} + L_m i_{sq}\right)/L_p \tag{9}$$

$$i_{sd} = \left(\lambda_{sd} - L_m i_{pd}\right)/L_s \tag{10}$$

$$i_{sq} = \left(\lambda_{sq} + L_m i_{pq}\right)/L_s \tag{11}$$

From the above equations, torque equation for the machine can be derived as [12]

$$T_e = 3\mathrm{P_r} L_m \left(\lambda_{pd} i_{sq} - \lambda_{pq} i_{sd}\right)/2L_p \tag{12}$$

Rotor speed expression can be derived from the above equations as

$$\omega_r = \int \left[\left(T_e - T_L - K_d \omega\right)/J\right].dt \tag{13}$$

The three-phase active and reactive power can be expressed as [7] [12]

$$P = 3\left(V_{pd} i_{pd} + V_{pq} i_{pq}\right)/2 \tag{14}$$

$$Q = 3\left(V_{pq} i_{pd} - V_{pd} i_{pq}\right)/2 \tag{15}$$

4 Simulation of Active and Reactive Power Control

The proposed control scheme for BDFRG-based WECS with active and reactive power control mechanisms is shown in Fig. 2. An SVPWM controller is used to control the output of the converters by controlling the DC link voltage. In this control scheme, the parameters like V_{pd}, I_{pd}, V_{pq}, I_{pq}, P, and Q are estimated in the stationary reference frame. The details of the parameters considered for simulations are shown in Table 1.

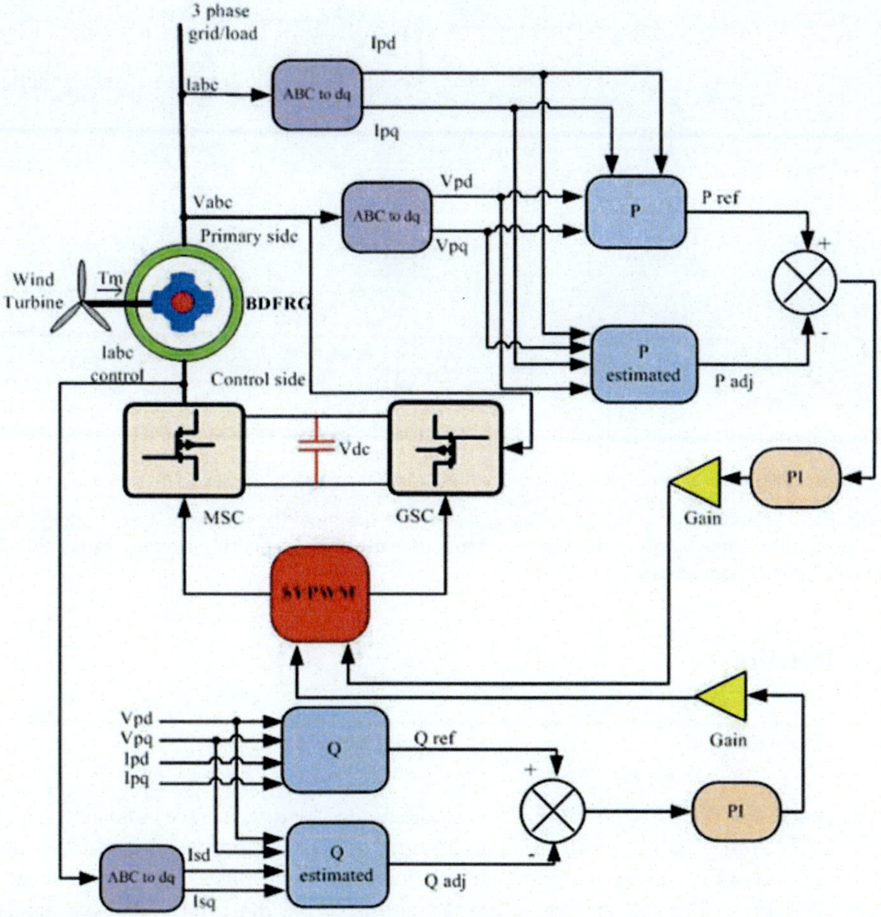

Fig. 2 Schematic block diagram of the proposed system

Table 1 System parameters

Parameter	Values	Parameter	Values
Machine power	2.5 kW	L_s	87 mH
P_s	2	L_m	62 mH
P_p	6	R_s	0.9 Ω
R_p	1.2 Ω	L_p	46 mH

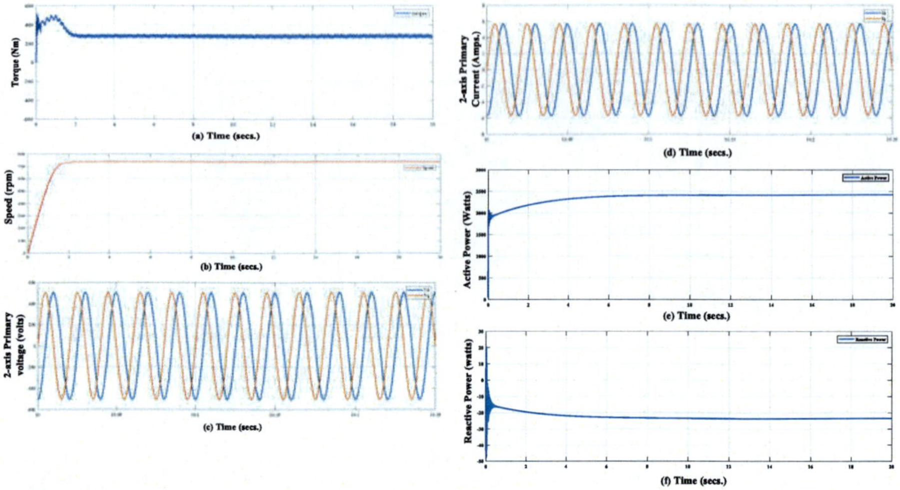

Fig. 3 Simulation results: (**a**) torque generated by the machine, (**b**) rotor speed, (**c**) primary side voltage in the d-q axis, (**d**) primary side current in the d-q axis, (**e**) active power delivered, and (**f**) reactive power compensated

5 Results

6 Discussions

The proposed WECS with BDFRG is implemented with a 2.5 kW machine using the MATLAB/Simulink platform. The performance of the proposed controller can be observed in Fig. 3. The experimental results from the proposed scheme are shown in Fig. 3. From Fig. 3a, it is seen that the torque of the machine is approximately 250 N-m and stabilizes in 2 seconds. Figure 3b shows the speed of the machine, which runs at 750 r.p.m., which is also the rated speed. The primary winding voltage and current in the d-q component are shown in Fig. 3c, d which are approximately 510 V and 5.75A, respectively. Figure 3e shows the active power delivered by the machine is 2450 W. This is also compared to the reference power to generate the signal for the controller circuit. Figure 3f shows a low compensation of reactive power of 22 W.

7 Conclusion

As proposed the improved controlling technique for controlling the active and reactive power of BDFRG has been successfully implemented in this paper. The monitored electrical power (P) and the estimated power (P) depict the BDFRG input conditions which are represented by Eqs. (14) and (15). It has been observed that the proposed controller appropriately works to estimate the torque and active power by sensing the voltage and current waves of the primary winding of the BDFRG. Moreover, this control technique does not require any flux estimation, as it is directly based on the maximization of active power delivery by taking power as the reference signals.

References

1. Mohammadi J, Afsharnia S, Vaez-Zadeh S. Efficient fault-ride-through control strategy of DFIG-based wind turbines during the grid faults. Energy Convers Manage 2014;78:88–95.
2. R.E. Betz, M.G. Jovanovic, The brushless doubly fed reluctance machine and the synchronous reluctance machine – a comparison, IEEE Trans. Ind. Appl. 36 (4) (2000) 1103–1110.
3. R.E. Betz, M.G. Jovanovic, Theoretical analysis of control properties for the brushless doubly fed reluctance machine, IEEE Trans. Energy Convers. 17 (3) (2002) 332–339.
4. M.G. Jovanovic, R.E. Betz, Y. Jian, the use of doubly fed reluctance machines for large pumps and wind turbines, IEEE Trans. Ind. Appl. 38 (6) (2002) 1508–1516.
5. M. Jovanovic, Sensored and sensorless speed control methods for brushless doubly fed reluctance motors, IET Electric Power Applications 3 (6) (2009) 503–513.
6. H. Chaal, M. Jovanovic, toward a generic torque and reactive power controller for doubly fed machines, IEEE Transactions on Power Electronics 27 (1) (2012) 113–121.
7. H. Chaal, M. Jovanovic, Power control of brushless doubly-fed reluctance drive and generator systems, Renewable Energy 37 (1) 2012) 419–425.
8. M. G. Jovanovic, J. Yu, E. Levi, Encoderless direct torque controller for limited speed range applications of brushless doubly fed reluctance motors, IEEE Transactions on Industry Applications 42 (3) (2006) 712–722.
9. A.S. Abdel-khalik, M.I. Masoud, M.M. Ahmed, Generalized theory of mixed pole machines with a general rotor configuration, Alexandria Eng. J. 52 (1) (2013) 19–33.
10. M. Paul and A. K. Das, “Improved Modeling of BDFRG in Wind Power Generation Application,” 2021 IEEE 2nd International Conference On Electrical Power and Energy Systems (ICEPES), 2021, pp. 1-4, https://doi.org/10.1109/ICEPES52894.2021.9699561.
11. Song William K., Dorrell David G., “Modeling and simulation study for dynamic model of brushless doubly fed reluctance machine using Matlabsimulink”, In the proceedings of 2015 IEEE 3rd International Conference on Artificial Intelligence, Modelling, and Simulation.
12. R. E. Betz and M. G. Jovanovic, “Introduction to the Space Vector Modeling of the Brushless Doubly Fed Reluctance Machine”, Electric Power Components and Systems, 2003, vol. 31, no. 8, pp. 729-755.

Identification and Detection of Credit Card Frauds Using CNN

C. M. Nalayini, Jeevaa Katiravan, A. R. Sathyabama, P. V. Rajasuganya, and K. Abirami

1 Introduction

Credit card usage is touching its peak in recent years for its ease of making payments and efficiency in almost every parts of the world in most of the fields like trading, marketing, E-commerce sites, business transactions, and even for the purchasing needs of the individuals [18]. As long as the need for using credit card increases, the need for protection and secured transaction also increases. There are two types of credit card frauds [13]. A transaction in which the fraudulent party shows the duplicate credit card to the agent in person is known as card-present fraud. This way of deception has become less common in the present as the fraudsters has hauled towards the sophisticated online modes of frauds. Users won't do physical transactions; instead they will go for online transactions [15]. Though many of the transactions are processed via more secure authentication tools and methodologies [6], credit card frauds are still a big challenge in the real time scenario. A model can be trained in such a way that it can predict the fraudulent cases at the earliest is the scope of this work.

C. M. Nalayini (✉) · A. R. Sathyabama · K. Abirami
Information Technology, Velammal Engineering College, Chennai, India
e-mail: nalayini@velammal.edu.in; sathyabama@velammal.edu.in; abirami@velammal.edu.in

J. Katiravan
Department of Information Technology, Velammal Engineering College, Chennai, India

P. V. Rajasuganya
Artificial Intelligence and DataScience, Velammal Engineering College, Chennai, India
e-mail: rajasuganya@velammal.edu.in

M. Mishra et al. (eds.), *Applications of Computational Intelligence in Management & Mathematics*, Springer Proceedings in Mathematics & Statistics 417,
https://doi.org/10.1007/978-3-031-25194-8_22

1.1 Credit Card Frauds

Credit card transaction has only two cases. Credit card transaction categorization is often a binary classification problem [1]. A credit card transaction is classified as either a valid (negative class) or a fraudulent (positive class) transaction. According to the Federal Trade Commission's Fraud Detection Data Book 2020 Report on Consumer Sentinel Network, 93,408 reports have been anticipated for the year 2020, ranking second in identity theft instances [14]. This statistical report gives information of credit card frauds that occurred in all the 4 quarters of the fiscal year 2020 in the United States. Attackers launch volumetric attacks like DDoS and its variations to deny the services. Credit card cracking involves several bots which run automated programs over www. Cracking helps the attacker to obtain the card number and PAN number through various guessing patterns to perform credit card frauds conveniently. For that they involve DDoS attack [8, 20, 21] as one of the weapons to deny the services of the bank server thereby gaining the unauthorized remote access of the customer's concern and commit credit card frauds successfully. To address this issue and prohibit such fraudulent actions all over the world, an efficient model using convolutional neural network of deep learning is proposed in this paper. This algorithm is much suitable and accurate for detecting credit card fraud transactions that occur less frequently but with huge loss and impact in the financial sector. The main advantage of using CNN is that it automatically detects the important features in the initial layers of inputting into the model using convolutional feature sequencing.

There are numerous existing systems for identifying fraudulent transactions [18], including supervised and unsupervised learning approaches in common algorithms such as K-nearest neighbor, Naive Bayes, decision trees, support vector machines, and neural networks. Among these methods, this paper aims to give a detailed analysis of all the existing methods and proposes convolutional neural network model in three layers for more accurate prediction of the fraud transaction.

Traditional rule-based approach uses rules that are written by programmers while building the model. This approach works good only for a static set of problems that never gets changed over time. In the case of dynamic problem situations where the datasets are huge and real-time, rules cannot be written over and over again as the dynamic problem can get more complicated over time and circumstances. This drawback of using strict rules for fraud detection may cause high false positives that might lead to loss of genuine customers who make legitimate transactions. So there occurs a need for supervised learning methodologies in which the learning is guided with the available datasets where the model gets trained to make prediction or decision when new data is given to it.

1.2 Supervised Learning Algorithms

K-Nearest Neighbor The K-nearest neighbor (KNN) technique [3] is a straightforward algorithm that stores all available instances before classifying any new ones using a similarity metric. For each transaction, the criterion [12] determining whether it is unauthorized or not is completely based on the categories of its KNN. It is argued in [4] that KNN is an easy, instance-based algorithm that classifies unlabeled instances according to their closest neighbors. This supervised learning technique of KNN can be useful when the data is non-linear, and it is also flexible as it can be useful for classification and regression. But the setback of using KNN is the accuracy which depends on the quality of the data, and KNN is susceptible to the scale of the data and unrelated features.

Naive Bayes The statistical technique of Nave Bayes [1, 19] is based on Bayesian theory. The decision is taken in this case based on the highest occurrence. Bayesian probability calculates unknown probabilities based on known probabilities. The main constraint of Naive Bayes [4] is the presumption of independent predictor features. This algorithm addresses the "zero-frequency problem" where the zero probability is allocated to a categorical variable whose category in the test dataset is absent in the training dataset.

Logistic Regression Dealing with a binary dependent variable, logistic regression [5, 16, 19] is an extensively used technique. The ideal parameters of a non-linear function known as the sigmoid are expressed using this approach. The logistic regression's main limitation is that the dependent feature variable and the independent feature variable must be linear.

Decision Tree The decision tree algorithm [1, 2, 7, 10, 19] can be used for regression and classification. It's a method of encoding discrete-valued target functions. A decision tree, which is commonly used in inductive learning, is used to visualize the learnt function. These decision trees, on the other hand, are more likely to be unstable. This instability occurs when a tiny change in the data causes the optimal decision tree's entire structure to shift.

Support Vector Networks Support vector networks [3, 5, 10, 17, 19] are found to be the most successfully used statistical learning technique. It comparatively works well when there is a distinction between classes where the target class is overlapping. The main drawback of SVM is that it reduces its pace when the count of features for every data point is beyond the count of training datasets.

Neural Networks Neural networks [8, 10] are also often recommended for fraud detection. Neural networks [9] analyze the history and detect real-time credit card frauds. Re-training of algorithm is required for new types of fraud cases [11]. It benefits large-scale real-time operations for its high computational power.

2 Literature Review

Earlier comparative study utilizing logistic regression and support vector machine suggests that creating ingeniously derived attribute transactions would be more accurate for classification that engaged international credit cards (2010). Some limitations like non-availability of specified features for training the algorithm were highlighted in the systems that applied fraud mining using item set which employed UCSD-FICO data mining contest 2009 dataset (2014). Expending network-based addendum focuses on observing the group behavior in the network by exploiting the financial credit card dataset. An algorithm with visual cryptography indicates using historical profile pattern as the feature for detection of fraud (2015).

Comparison of studies involving interpolated networks, imbalanced data, normal network, and trading entropy, genetic algorithm wages ULB datasets, data mining, and financial credit card with forthcoming advancements in considering complex behavior and examining the difference in fraudulent behavior among different types of fraud (2016). In the recent years, there has been a lot of extensive research done on credit card fraudulent transactions using deep learning, unsupervised learning, and tensor flow which demands a huge amount of data. Using AdaBoost, majority voting had less accuracy when compared to other systems, while the system that employed online E-commerce dataset requires models of artificial neural network (2018).

In concurrence with the related works of credit card fraudulent transactions, this paper involves CNN algorithm to expand the range of accessing large-sized real-time datasets with productive tutelage of the model using normalized datasets along with extensive evaluation with the help of standardized parameters.

3 Proposed Work

To determine if a transaction is legal or fraudulent, this suggested model employs a deep learning convolution neural network. Figure 1 depicts a flow diagram of the proposed system for identifying fraud credit card transactions.

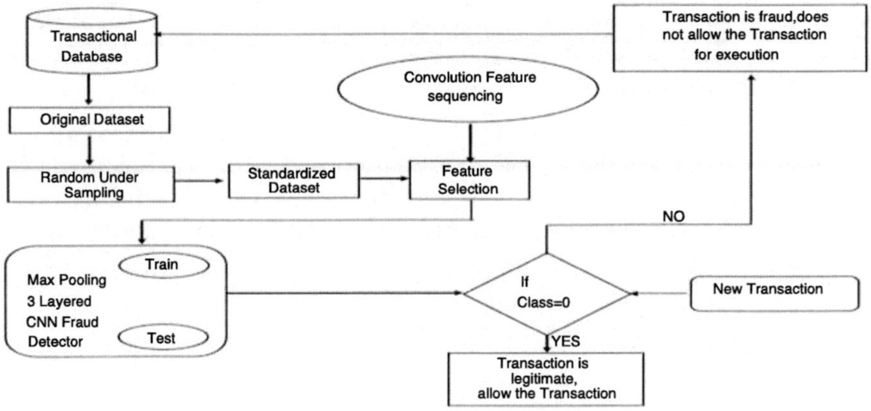

Fig. 1 Overview of the model

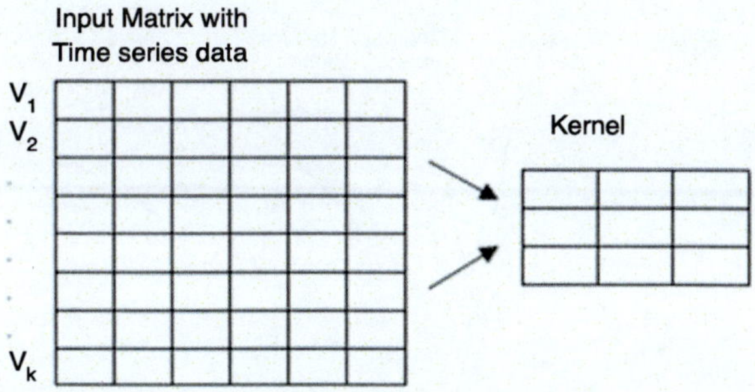

Fig. 2 Smart matrix algorithm

The original dataset from the database is taken to load as input into the proposed CNN system. The original dataset is a highly unbalanced dataset which is obtained from the University Libre De Bruxelles, a European research university. This unbalanced dataset has to be balanced using the random undersampling algorithm which equalizes the number of non-fraud cases equal to that of fraudulent cases. This balanced dataset is normalized for acquiring the standardized dataset using standard scalar. A feature sequencing layer precedes the input layer in the CNN model constructed for sequencing feature for feature selection.

The basic idea of 1D CNN is that the convolution layer does computation with the 1D kernel for efficient feature extraction. It convolutes adjacent features to obtain the derivative features. It is noticed that for multivariate features, order of feature sequence will be different that too for fraud transactions. Hence we introduced feature sequencing concept before the convolution layer to obtain proper sequencing of features. If all the features are arranged based on the parameter's timestamp and location, then it is easy for the convolution layer to extract the optimal features. The feature sequencing layers will randomly transform the arrangement of the features to find an optimal way of arrangement via smart matrix algorithm as shown in Fig. 2.

3.1 Smart Matrix Algorithm Applied for Feature Sequencing

Algorithm: SMA

Input: Time series data

Output: Proper sequenced features

Step 1: Time series data of credit card dataset in the form of seven-tuple information such as source and destination IP address, source and destination port, timestamp, type of service, and location are represented in an input matrix.

Step 2: Row-wise screening happens at the input matrix (Tm) in a random transformation mode with the parameter's timestamp and location the transaction happened.

Step 3: Based on the related data, respective pixel positions of the input matrix (Tm) are multiplied with kernel (k) values to obtain the sequenced feature set.

***Step 4*:** Repeat steps 2 and 4 till all the rows are done to find final proper ordering feature set.

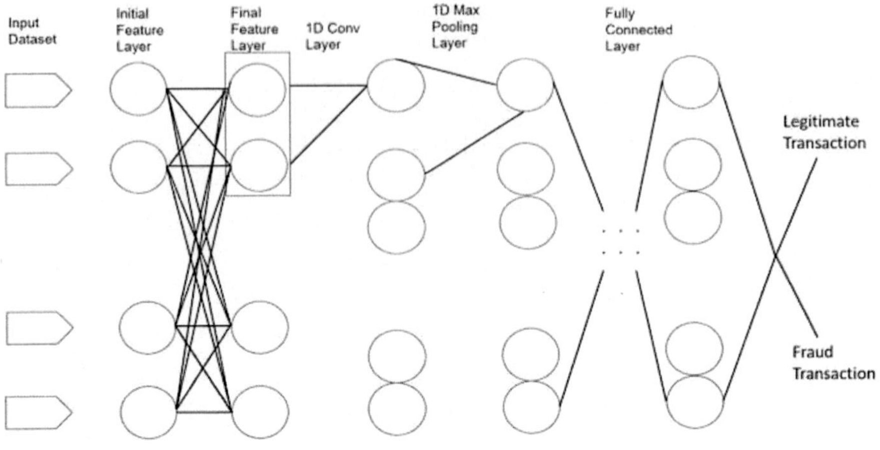

Fig. 3 Three-layered CNN of credit card fraudulent detector

$$\sum_{i=v1}^{vk} Tm * k$$

Figure 3 shows the working flow of our proposed model. Original input dataset represented in time series matrix is fed as the input to the input feature layer which processes the smart matrix algorithm to find the proper arrangement of feature set. Three-layered CNN model is used to extract the features after feature sequencing. Max pooling reduces the size and computational parameter. Then the features are flattened into one column vector as the input to the fully connected layer which performs dropout to consider only the relevant features and ReLu for effective scaling to identify the optimal features. Then softmax classification is applied at the output layer to classify the existence of fraud and non-fraud transactions. If the class variable is equal to zero, the transaction is considered to be legitimate, and if the class variable is equal to one, then the transaction is detected as fraudulent. The result of three-layered convolutional model shows higher accuracy when compared with existing models.

4 Experimental Results

For our experimental results, we have taken the dataset from University Libre De Bruxelles, a European research university (ULB), and commercial European bank. The total number of cases contains 284,807 data in whole, the number of non-fraud cases contains 284,315 data, the no. of fraud cases contains 492 data, and finally it has 31 features in total. Since the dataset is highly imbalanced, random sampling is applied to balance the number of fraud and non-fraud cases.

Data Normalization The process of structuring the database is known as data normalization. The data is organized to look similar along all records and fields. This procedure raises the cohesion of entry types that employs cleansing and segmentation and provides higher quality of data.

Standard Scalar Standard scalar will transform the data into observations with a mean of 0 and a standard deviation. The balanced datasets that are acceptable for training the model are standardized using standard scalar. It scales to unit variance after removing the mean and standardizing the feature. This standardization of data is for making sure that the data is internally consistent. It regulates the input features by removing the mean of the feature values and scaling it to the unit variance.

4.1 Convolutional Feature Sequencing

CNN or conventional neural networks is a type of neural network that convolves features acquired from input data and employs 1D convolutional layers to make it appropriate for processing 1D data in credit card fraudulent detection. Multiple transaction features arranged in different ways will not change the transaction's original definition, but varied feature arrangements will have an impact on the model after convolutional. This model uses a one-dimensional feature vector, and the convolutional layer is tuned using a 1D convolutional kernel feature vector.

4.2 Three-Layered Convolutional Neural Network

One of the most prominent neural networks, CNN (conventional neural networks), combines the specifications from input data and uses 1D convolutional layers to process 1D data in the credit card fraud detection. Max pooling CNN model proposed here has taken three CNN layers with 16, 32, and 64 size for each layer, respectively. The kernel size used in this model is changed to 3. The CNN model of this system used Conv1D for reading the dataset and get trained using them. By increasing the number of epochs which is one complete cycle of the whole dataset, the efficiency of learning rate of the algorithm increases.

Libraries Neural network is constructed by tensor flow. Numpy executes all of the layers imported to build the model from keras simple array operations. Pandas load and process the data, and the results are calculated using a learning curve. Train/test split splits the dataset into two parts: training and testing. StandardScaler raises the data's values to a standard.

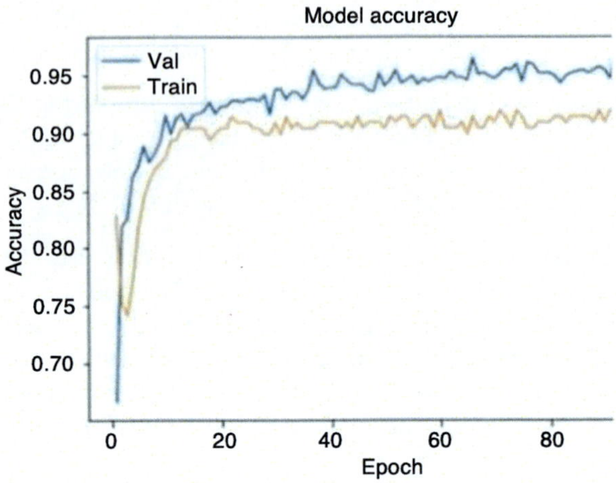

Fig. 4 Model accuracy

For a suitable plain stack of layers with an input and output for each layer, the sequential function is employed. Conv1D() is a 1D convolutional layer that can be used to derive features from a set of fixed features. A total of 31 filters with a convolutional window size of 2 are examined in the first Conv1D() layer. The shape of the input dataset is specified by activation-relu. The initial layer of a neural network must meet this requirement. The rectified linear activation function or ReLU is a linear function that, if the case is positive, returns the input directly, else it returns 0.

Batch normalization () allows for increased independence in self-learning for each layer of the network. Batch normalization increases the stability of a neural network by systematizing the output of the preceding activation layer by eliminating the batch mean and dividing the batch standard deviation. After the training phase, dropout() sets the edges away from concealed units to 0 at random during each update. The probability of the dropped outs is mentioned by the values passed in dropout in the outputs of the layer. Flatten () transforms the data into one-dimensional array and serves as an input for the next layer. The totally linked neural network layer is represented by dense (). Because it is a binary classification task, the output layer is a dense layer with one neuron that predicts a single value. Sigmoid function predicts a binary input and it lies between 0 and 1.

Learning Curve: A learning curve is the plotting of curves for the proposed model whose learning performance is identified over experience or time. Learning curve is a popular diagnostic tool in the machine learning domain that makes the model learn from a gradual increase in training dataset. Learning curves used in this project are plots that show deviations in learning performances with time, indicated by experience. Figure 4 depicts the model performance on the train under fit, over fit, or well-fit mode.

Our proposed algorithm produced 92.31% of accuracy in the training datasets and 96.41% of accuracy in the testing dataset as shown in Fig. 3. We have also performed the comparison of our proposed three-layered CNN model with that of K-nearest neighbor and Naïve Bayes. The performance metrics of these models are evaluated in various parameters like sensitivity, balanced classification rate, Matthew's correlation coefficient, F1 score, false alarm rate, and confusion matrix. In all these parameters, our CNN algorithm performs better when compared to KNN and NB. These performance metrics are represented in the graphical representation against the number of customers performing transactions.

4.3 *Performance Metrics Graphs*

4.3.1 Confusion Matrix

The confusion matrix shows a cross-tab of actual type values and predicted type values and contains the count of observations that fall in each category as shown in Fig. 5. Fraudulent transactions are classified as positive, whereas lawful transactions are classified as negative. So, the terms P+, N–, T+, T–, F+, and F– are defined as follows:

Positives (P+): count of fraudulent transactions
Negatives (N–): count of legitimate transactions
True positives (T+): count of fraud transactions expected to be fraud
True negatives (T–): count of legitimate transactions expected to be legitimate
False positives (F+): count of legitimate transactions expected to be fraud
False negatives (F–): count of fraud transactions expected to be legitimate

Fig. 5 Confusion matrix

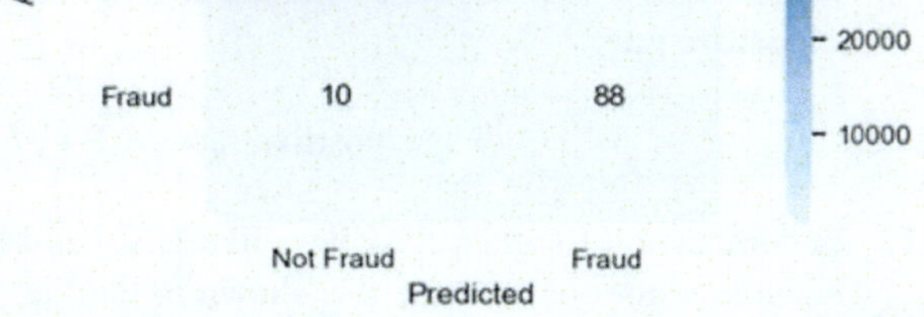

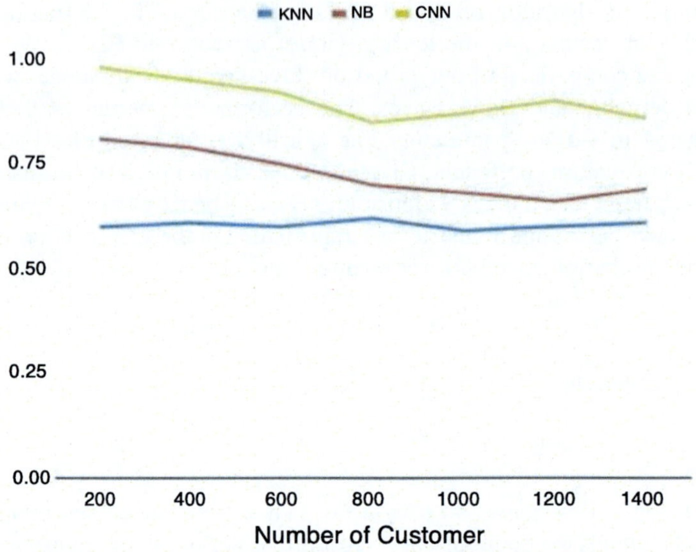

Fig. 6 Sensitivity/true positive rate

4.4 Sensitivity/True Positive Rate

The fraud detection rate is represented by the sensitivity/true positive rate.

Sensitivity is defined as the ratio of the number of fraud transactions projected as fraud to the total number of fraud transactions.

$$\text{Sensitivity} = (\text{T+}) / (\text{P+})$$

Sensitivity graph Fig. 6 shows the efficiency of CNN algorithm on sensitivity correlated with other classifiers such as KNN and NB.

4.5 False Positive Rate/False Alarm Rate (FAR)

The proportion of true negatives that are speculated as positives is referred to as the false positive rate.

$$\text{False positive rate} = (\text{F+}) / (\text{N}-)$$

In comparison to other classifiers like KNN and NB, the efficiency of CNN on FAR with number of customers is shown in the Fig. 7. This statistic should have a low value because an increase in false alarms frustrates customers.

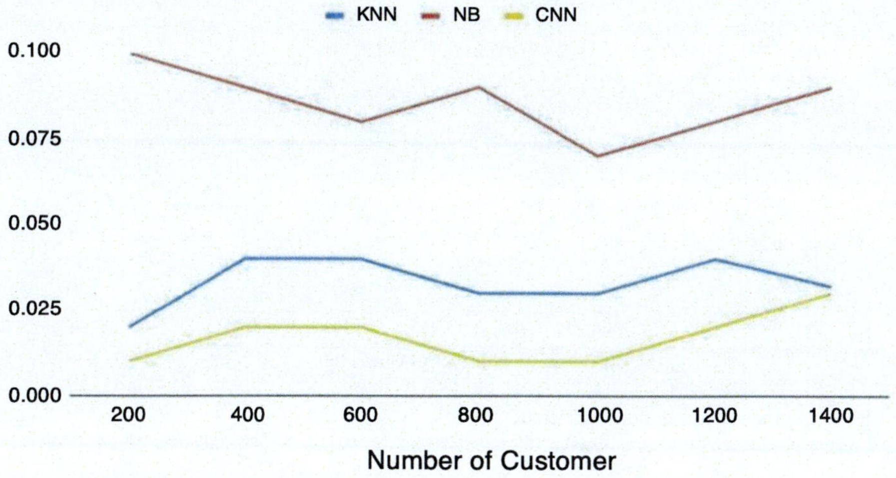

Fig. 7 False positive rate

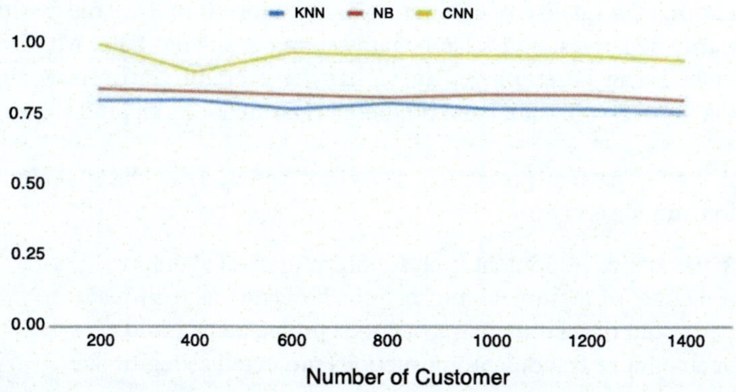

Fig. 8 Balanced classification rate

4.6 Balanced Categorization Rate (BCR)

The average of sensitivity is defined as the balanced categorization rate, which is calculated as follows: BCR = [(T+)/(P+)] + [(T−)/(N−)].

In comparison to classifiers such as KNN and NB, the BCR graph Fig. 8 illustrates the evaluation of CNN algorithm on balanced categorization rate.

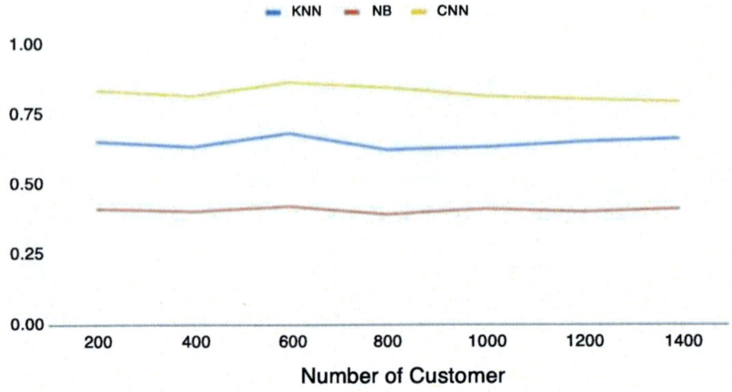

Fig. 9 Matthew's correlation coefficient

4.7 *Matthew's Correlation Coefficient (MCC)*

MCC measures the quality of a binary classification (0 or 1). True positives, false positives, true negatives, and false negatives are considered here. MCC is a steady measure which can be employed in variants like credit card transaction dataset. MCC is enclosed by the actual and predicted classifications and the values lie within −1 and +1.

Cases: +1 → perfect prediction, 0 → not better, 1 → disagreement between assumption and observation

e: Expected count of fraudulent transactions projected as fraud
f: Expected count of legitimate transactions projected as legitimate
g: Expected count of legitimate transactions projected as fraud
d: Expected count of fraudulent transactions projected as legitimate

$$\text{Matthew}'\text{s correlation coefficient} = (e * f) - (g * h) / \sqrt{(e + g)\ (e + h)\ (f + g)\ (f + h)}$$

MCC graph Fig. 9 compares the performance of the CNN algorithm on the MCC to KNN and NB.

4.8 *F1 Score*

The harmonic average of precision (P) and recall (R) is used to determine the F1 score. The value of cross-validation as well as the training score is depicted in Fig. 10. Let R represent recall and P represent precision.

$$\text{F1 Score} = 2 * (R * P) / (R + P)$$

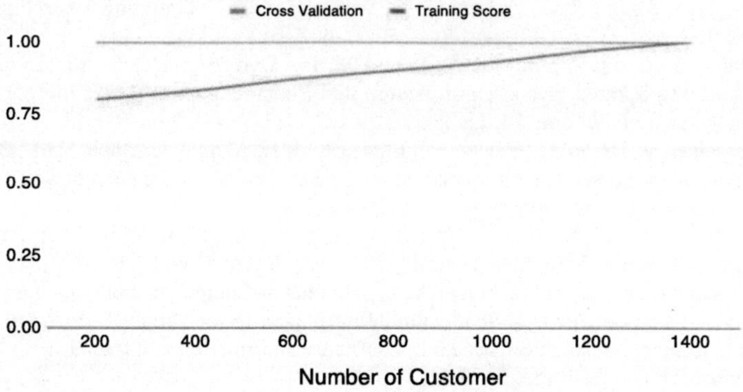

Fig. 10 F1 score

5 Conclusion and Future Work

The result of three-layered convolutional model shows higher accuracy for training and testing data when compared with existing models. The comparative analysis between the CNN model and other supervised learning models using performance metrics shows that CNN model is more efficient in detecting the real-time credit card fraud cases. While our fraud detection model focuses on detecting individual credit card theft, future research will look into group behavior, including the prevalence of unauthorized configurations in the network of credit card holders and retailers. The revealing of transaction sequence characteristics will be the focus of future research. We will do behavior analysis to stop credit card carding in the near future.

References

1. John O., Awoyemi, Adebayo O., Adetunmbi, Samuel A., Oluwadare (2017): Credit card fraud detection using Machine Learning Techniques: A Comparative Analysis on IEEE.
2. Patil, S., Somavanshi, H., Gaikwad, J., Deshmane, A., and Badgujar, R., (2015). Credit Card Fraud Detection Using Decision Tree Induction Algorithm, International Journal of Computer Science and Mobile Computing (IJCSMC), Vol. 4, Issue 4, pp. 92–95, ISSN: 2320-088X.
3. Seeja, K. R., and Zareapoor, M., (2014). FraudMiner: A Novel Credit Card Fraud Detection Model Based on Frequent Itemset Mining, The Scientific World Journal, Hindawi Publishing Corporation, Volume 2014, Article ID 252797, pp. 1–10.
4. Fahmi, M., Hamdy, A. and Nagati, K., (2016). Data Mining Techniques for Credit Card Fraud Detection: Empirical Study, In Sustainable Vital Technologies in Engineering and Informatics BUE ACE1, pp. 1–9, Elsevier Ltd.

5. Bhattacharyya, S., Jha, S., Tharakunnel, K., Westland, J.C.: Data mining for credit card fraud: a comparative study. Decis. Support Syst. 50(3), 602–613 (2011).
6. C.M Nalayini, Jeevaa Katiravan, A New IDS for Detecting DDoS Attacks in Wireless Networks using Spotted Hyena Optimization and Fuzzy Temporal CNN, Journal of Internet Technology, Issue 1, Volume 24, Jan 2023.
7. Van Vlasselaer, V., Bravo, C., Caelen, O., Eliassi-Rad, T., Akoglu, L., Snoeck, M., Baesens, B.: Apate: A novel approach for automated credit card transaction fraud detection using network-based extensions. Decis. Support Syst. 75, 38–48 (2015).
8. CM Nalayini, Jeevaa Katiravan, N Rajesh, Varna K Kumar, R Hamsini, "Pictorial Sequence Authentication using Optimized Disarray System", ICTACC-October 2017, IEEE Explore International Conference on Technical Advancements in Computers and Communications
9. Masoumeh Zareapoor, Pourya Shamsolmoali, Application of Credit Card Fraud Detection: Based on Bagging Ensemble Classifier, International Conference on Intelligent computing, Communication and Convergence (ICCC-2015).
10. Mehak Kamboj, Shankey Gupta, Credit Card Fraud Detection and False Alarms Reduction using Support Vector Machines, International Journal of Advance Research, Ideas and Innovations in Technology, ISSN: 2454-132X (Volume 2, Issue 4), 2016.
11. Kuldeep Randhawa, Chu Kiong Loo, Manjeevan Seera, Chee Peng Lim and Asoke K. Nandi, Credit Card Fraud Detection Using AdaBoost and Majority Voting, IEEE Access, March 28, volume 6, 2018, Digital Object Identifier https://doi.org/10.1109/ACCESS.2018.2806420.
12. Zhenchuan Li; Guanjun Liu; Shuo Wang; Shiyang Xuan; Changjun Jiang Credit Card Fraud Detection via Kernel-Based Supervised Hashing, access no: 18342444, ISBN:978-1-5386-9381-0 date:06 December 2018.
13. https://en.wikipedia.org/wiki/Credit_card_fraud_-_Credit_Card_Fraud-en.Wikipedia.
14. https://public.tableau.com/profile/federal.trade.commission#!/vizhome/Identity TheftReports/TheftTypesOverTime-Federal Trade Commission: Consumer Sentinel Reports.
15. Investopedia-Financial Frauds- Card-Present Fraud by JASON FERNANDO Updated on Dec 4, 2020- https://www.investopedia.com/terms/c/cardpresent-fraud.asp Investopedia-Financial Frauds- Card-Present Fraud by JULIA KAGAN Reviewed by ERIKA RASURE Updated Oct 30, 2020- https://www.investopedia.com/terms/c/cardnotpresent-fraud.asp
16. Shen, R. Tong, Y. Deng, Application of classification models on credit card fraud detection, Service Systems and Service Management, 2007 International Conference on. IEEE 2007, pp. 1–4.
17. C. Cortes and V. Vapnik, "Support-vector networks," Machine Learning, vol. 20, no. 3, pp. 273–297, 1995.
18. NalayiniI, C.M., Gayathri, T. (2022). A Comparative Analysis of Standard Classifiers with CHDTC to Detect Credit Card Fraudulent Transactions. In: Sivasubramanian, A., Shastry, P.N., Hong, P.C. (eds) Futuristic Communication and Network Technologies. Lecture Notes in Electrical Engineering, vol 792. Springer, Singapore. https://doi.org/10.1007/978-981-16-4625-6_99
19. C.M. Nalayini, Dr. Jeevaa Katiravan, "Detection of DDoS Attack using Machine Learning Algorithms", Journal of Emerging Technologies and Innovative Research, Volume 9, Issue 7, July 2022
20. C.M. Nalayini, Dr. Jeevaa Katiravan, Araving Prasad V, "Flooding Attack on MANET – A Survey", International Journal of Trend in Research and Development (IJTRD), ISSN: 2394-9333, Feb 2017
21. Nalayini, C.M., Katiravan, J. (2019). "Block Link Flooding Algorithm for TCP SYN Flooding Attack", International Conference on Computer Networks and Communication Technologies. Lecture Notes on Data Engineering and Communications Technologies, vol 15. Springer, Singapore. https://doi.org/10.1007/978-981-10-8681-6_83, 18 September 2018.

Different Degradation Modes of Field-Deployed Photovoltaic Modules: A Literature Review

Piyali Das, P. Juhi, and Yamem Tamut

1 Introduction

With diminishing conventional energy sources, contributing towards smarter, cleaner, and more sustainable options to produce electricity has become a necessity nowadays. The use of solar energy, i.e., photovoltaic power plants, is a much more reliable option among all the other renewable energy-generating power plants. Though research in the field, to large extent, has revolutionized the world energy market, most countries across the globe are still facing chronic energy poverty. With even more technologies and research coming into existence, the energy crisis problem will increase even further in the future.

For the deployment of photovoltaic (PV) modules in power plants, it is important to understand their performance in the long run [3]. It is observed that behavior of field-deployed PV cells/modules at continuous high voltage leads to deteriorating power in the PV power plants [1–3]. From the data collected from outdoor-installed PV modules at different locations around the world, the module performance mostly depends upon various factors like atmospheric temperature, the temperature of the module, irradiance, atmospheric humidity, system age, type of mounting, size of the installation, etc. [1–3, 5]. Degradation of power in PV modules is majorly caused by some major degradation modes such as potential induced degradation (PID), bypass diode failures in short circuit conditions, high light-induced degradation (LID), hotspots/shaded cells, and cracked cells [1–3, 5, 7]. These factors make the installation of large-scale PV power plants less reliable in comparison to

P. Das (✉)
Department of Electrical Engineering, North Eastern Regional Institute of Science and Technology, Nirjuli, Arunachal Pradesh, India

P. Juhi · Y. Tamut
North Eastern Regional Institute of Science and Technology, Nirjuli, Arunachal Pradesh, India

M. Mishra et al. (eds.), *Applications of Computational Intelligence in Management & Mathematics*, Springer Proceedings in Mathematics & Statistics 417,
https://doi.org/10.1007/978-3-031-25194-8_23

conventional energy-generating power plants. The maximum power, short circuit current, open-circuit voltage, and fill factor (P_{max}, V_{oc}, I_{sc}, FF) of the PV system are performance parameters that influence the I-V characteristics of PV modules when exposed to continuous high voltage and hence affect the performance of PV modules [2, 5].

The National Centre for Photovoltaic Research and Education (NCPRE) established in 2010 at the Indian Institute of Technology, Bombay, along with the National Institute of Solar Energy (NISE), conducted a series of surveys, All India Survey (AIS) of PV Module Reliability in 2013, 2014, 2016, and 2018. According to the study by "All India Survey (AIS) of PV Module Reliability 2018" [1, 5, 7], among the modules spread over 36 different geographical locations in India, many shows usual I-V characteristics. These have been proved through simulations provided in [4, 6, 7]; the ideal I-V curve that we study in the works of the literature differs from the field-deployed PV module characteristics due to assumptions like "no leakage current," "ideal bypass diode at short circuit conditions," and "no damaged or cracked cell in the PV module" [7]. Furthermore, analysis of durability and reliability of around 1250 field-mounted c-Si PV modules from 36 different climate zones were carried out by tracing I-V characteristics, infrared imaging at short circuit conditions, maximum point tracking (MPPT), electroluminescence imaging, and insulated resistance testing [1, 7, 10]. The I-V characteristics of each PV module at different irradiance and temperature levels were modified at standard test conditions (STC) (1000 W/m^2; 25 degrees) [1, 4]. Using these statistics of the survey, calculation of degradation rates of performance parameters (V_{oc}, I_{sc}, P_{max}, FF) of PV module is carried out. Many STC correction procedures and the errors associated with them are demonstrated by researchers Y. R. Golive et al. [4] in detail.

The Florida Solar Energy Center (FSEC), USA, has been studying the performance of PV modules in high-voltage bias systems in hot and humid climates for over a decade now [2]. According to one of the studies conducted on high-voltage biased outdoor PV modules, it is stated that these are prone to power loss due to the leakage current flowing through the PV module packaging material. The information about the reverse leakage current path in the PV module will enable the researchers to draw a clear idea about the cause and solutions for the power loss in the PV module over time [2, 7, 8].

In another study done in tropical rainforest climate "Af" by the Köppen climate classification (in Singapore), the outdoor-placed PV module performance was investigated for 7 years based on the I-V curve obtained at 10-second intervals. From this study, it was found that the PV degradation rate is around −0.3 to 0.47 percent per year. To enhance the PV module lifetime, it becomes crucial to have knowledge about its degradation and failure mechanism, so that it can be easily diagnosed and can be taken care of [9].

The objective of this study is to review the various available state-of-the-art literature that gives insight into the reliability of PV power systems and solutions to various modes of degradation of PV modules over time. There can be several physical, chemical, and environmental factors responsible for various degradation modes (Fig. 1). Some major degradation modes such as PID, bypass diode failures

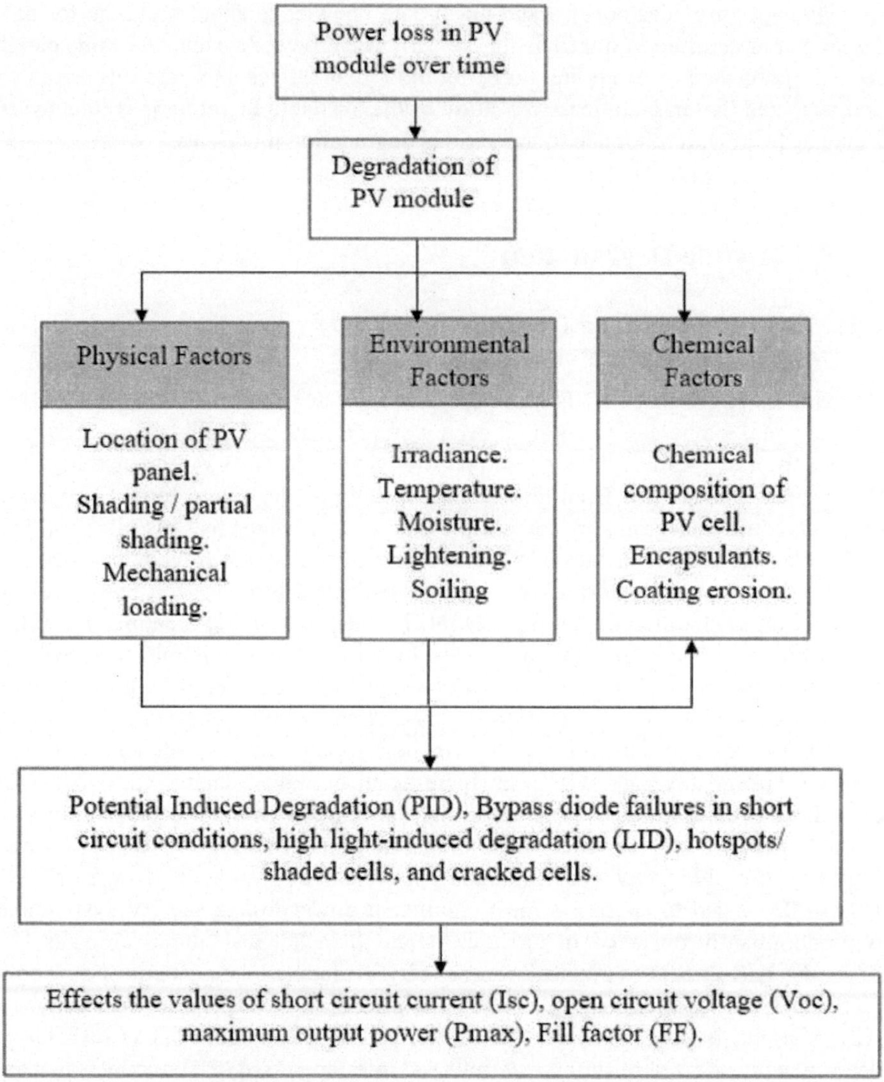

Fig. 1 All the factors responsible for the power loss as seen in PV modules. These consists of three parts: stressors (physical, environmental and chemical), failure modes (PID, LID, bypass diode failure in SSC, hotspots, cracked cells), and its effects

in short circuit conditions, LID, hotspots/shaded cells, and cracked cells are further discussed in this paper. The research intends to make a significant reference to degradation and failure assessment of PV module and henceforth make PV system technologies more dependable. Moreover, different methods used by various researchers to deal with these degradation modes are also discussed.

This paper provides concise and up-to-date knowledge about the topic for new researchers interested in improvising the current PV power systems. For this paper, recently published research literature, not older than 20 years, is carefully explored and analyzed. Later, a comparative study of the available literature is conducted in a tabular form with reference to the various degradation modes used.

2 PV Module Degradation

2.1 Potential Induced Degradation (PID)

Researchers claim that PID is the most dominant degradation mode, with higher humidity and temperature making it even worse. Because of being exposed to high voltage for a long time, a high potential difference up to 1000 V is created between the encapsulants and the front glass frame of the module, due to series connection of PV module and grounding of glass frame due to safety reasons [1, 2, 9, 11]. This potential difference causes a leakage current to flow from the encapsulant to the ground. The magnitude of this leakage current determines the magnitude of degradation of the module due to PID. N.G. Dheere et al. [2] mentioned that the PID occurrence increases with an increase in temperature and humidity as leakage current can be modeled as the function of temperature and humidity. It is observed that the leakage current increases during the hot day or afternoon with the increase in temperature; also during rainy days or dew time of the day (dawn and dusk), leakage current increases due to an increase in humidity. Therefore, it becomes difficult to predict leakage current and find its relation with the decreasing power in the system [2, 11]. However, the total charge transfer during the leakage current flow is responsible for power degradation as concluded by a study done by FSCE, USA. PID is said to cause the most significant power loss in the PV system and is predominantly observed in modules mounted in hot and humid climates [2]. From the AIS 2018, the greatest number of PV modules, i.e., 19 out of 51 "poor performing" or "high degrading" modules, is affected by PID [1]. Electro-luminance (EL) imaging is used to detect the PID effect on the modules. It is difficult to determine the effect of different materials and interfaces used in PV modules that are responsible for leakage current. We have no idea of the path followed by the leakage current inside the module material, which makes it less likely to come up with the improved module packaging ideas. A novel device, custom laminate, is designed at FSCE to diagnose the pathway of leakage current flowing in high-voltage bias PV modules over time. The application of the device to study the distribution of leakage current in the module is discussed [2]. Also, the use of anti-reflective coatings, encapsulated with high volume resistivity, sodium-diffused barriers, or sodium-free glass and alternative inverter configurations, can attenuate the effect of PID on the modules [9]. PID becomes more prevalent as the module ages, whereas it affects only specific cells, without disturbing other cells in the module. It severely damages the cells and repairing is just not possible [11].

2.2 *Bypass Diode Failures (BDF)*

In commercial PV modules, the PV modules are made up of PV cells connected in a series typically around 60–72 cells, with a bypass diode connected to each substring, around 20–24 in a module. When any cell is shaded, it becomes a current-limiting cell. The bypass diodes kick in when any cell in the module becomes a current-limiting cell so that the shaded cell does not become reverse biased and a short circuit condition is developed across the cell. The bypass diode is a protective device in the module intended to prevent power loss in reverse bias conditions by providing an alternative path for the current to flow. During this short circuit condition, many a time the bypass diode fails to act, leading the cell to operate in reverse bias conditions, and hence PV module degradation occurs [6]. Catastrophic failure modes of bypass diode are sudden electrostatic discharge, arching, thermal runaway, etc., which can cause severe damage to the module and hence performance losses. During harsh weather conditions, like lightning, there is high electrostatic discharge causing a high current to flow through the diode over a very small duration of time. This phenomenon leads to thermal runaway in the module; the high heat present in the cell results in current flowing through it causing the cell to heat up even more. This process continues until the temperature is sufficiently high to destroy the bypass diode in the cells. The gradual degradation of the PV module occurs when the module is exposed to continuous high-temperature operations or thermal cycling. This leads to the failure of the bypass diode (destroys the semiconductor junction) in both conditions [9, 14, 15]. A voltage is generated due to a mismatch between PV cells, producing a reverse current to flow into the failed bypass diode. The current flowing into the bypass diode is the sum of short circuit current, I_{sc}, and the current flowing in the PV module. Further, analyzing the short circuit current, open-circuit voltage, and fill factor, the degradation of module performance is mainly attributed to short circuit current (I_{sc}) loss [3].

The failure of bypass diode is the most common mismatch loss in PV system. Authors in [15] have showed that the reverse current due to failure of bypass diode is directly proportional to the number of fault diodes in the circuit and the number of PV cells present in the module. The reverse bias current in the module due to the failed bypass diode can cause electrical-thermal problems in the PV array and hence a significant amount of power loss. Various studies have released methods to minimize the burning out of diodes because of the bypass diode failure through experiments or computer simulations. According to the previous literature, a DC-DC converter can be applied to reduce the mismatch losses due to shading. In yet another research work, the use of current collector optimizer (CCO) topology is suggested in PV modules. The review paper also points out certain methods to recognize the type of faults occurred due to failure of bypass diode in short circuit conditions [15].

2.3 Light-Induced Degradation (LID)

Light-induced degradation (LID) affects the most number of solar cells in the module after PID as seen in the realistic data obtained during AIS 2018 [1]. LID is the power degradation that occurs during the initial stabilization of the PV module when placed under the sunlight, though the effect of LID is relatively small and even the damage done by it is repairable most of the time. The increase in defects caused by LID is responsible for low performance of PV modules. LID results in around 5% of the total power in the PV module. The defect involves meta-stabilities in the semiconductor layer of the module due to high-temperature processing or diffusion between the layers [9].

The LID degradation can be detected by continuously monitoring the system and by periodically comparing the I-V characteristics. Also, the study states that the degradation of P_{max} is not major due to I_{sc} and V_{oc}, but the decrease in the value of the fill factor (FF) leads to LID.

2.4 Cracked/Damaged Cells

The PV cells get damaged over time due to harsh weather conditions or during installations, transportation, or manufacturing. These cells are damaged cells present in the working PV module but create disturbance and hence power loss in the system. Also, the PV module functions in much worst conditions than the laboratory conditions in which they are evaluated. Depending upon the size and number of cracks in the module, power is affected, and more areas of burns or hotspots emerge on the PV module. Mismatch caused due to cell cracks leads to reduced current in the circuit and severely reduced shunt resistance in the circuit leading to unusual-looking I-V characteristic of the module [7].

Due to the appearance of cracks in the circuit, the current is non-uniformly distributed across the module. Consequently, the area around the crack assimilates heat and hence localized heating is developed, commonly known as hotspots [11]. The development of mismatch conditions in the solar cells causes the rise in the temperature of the module. When the module heats up to the extent that the temperature of the solar cell exceeds the critical value, delamination of the solar cell encapsulants may occur. If the reverse bias voltage exceeds the breakdown voltage of the solar cell, thermal breakdown occurs, causing irreversible damage to the cell. This phenomenon leads to the formation of hotspots in the cell. The formation of a hotspot not only decreases the power and efficiency of the cell but also affects performance parameters like open-circuit voltage (Voc), short circuit current (Isc), maximum power (Pmax), and fill factor (FF) [11–13]. The hotspots can occur due to other reasons as well, and the presence of a hotspot in the module can be detected by visual inspection, I-V characteristic analysis, hotspot endurance testing, or individual cell temperature monitoring when the solar cell is in forward bias condition [11]. In AIS 2018, it was observed that higher degradation of Pmax was seen in modules with large number of cell cracks present in the solar module. NCPRE, IIT Bombay, realized the need to conduct a systematic study to acquire knowledge about the impact of cell crack on the performance of PV module over the wide range of mechanical loading conditions [5]. Accelerated testing was developed to test the module in the real field-like conditions including the mechanical parameters that cause the cracks in the solar cell.

However, the use of thin-film Si PV cell is less susceptible to cracking due to the thermomechanical stress experienced by the field-deployed PV modules. The strain level for Si cell is comparatively lower, though damage to the glass substrate may cause the cell cracks and performance loss [9] (Table 1).

3 Conclusion

The world is in the verge of facing extreme energy poverty in the coming years. The best way to face the situation head on is by improving the areas in which the photovoltaic systems lacks. So, the first step in the process is to identify the factors responsible for power depletion due to degradation of PV module. Though there are several physical, chemical, and environmental factors responsible for it, the literature review shows that potential induced degradation (PID), bypass diode failures in short circuit conditions, high light-induced degradation (LID), hotspots/shaded cells, and cracked cells are the most common ones. High reliability can be accomplished by working on these degradation modes. The material of PV cell can be changed in a better alternative to reduce the power loss due to damaged or cracked cells. The material of PV cell should be able to sustain in wide range of temperature, irradiance, moisture, etc. Now, we have clear insight into the areas that require more attention as far as research work is concerned. Building on to this knowledge, the strategies to improve the performance of PV cell can be obtained, and hence in the future, we can make solar PV modules more reliable, efficient, and sustainable.

Table 1 Comparison between research contributions in power loss in PV module under high-voltage bias over time

Author(s) and year	Location of the study area	Problem type (causing degradation of PV cell)	Problem analysis	Observations
Mahmoud Dhimish, Andy M. Tyrrell (2022)	University of York, York, United Kingdom	PID	Electroluminescence (EL) imaging is used to detect the cells affected by PID.	The average power loss is 25%, the surface temperature increases from 25 to 45 °C due to hotspot formation, 60% of the examined PV modules failed the reliability test following IEC61215 standard, and the mean performance ratio is equivalent to 71.16%.
Rajiv Dubey et al. (2021)	36 different sites from India by All India Survey 2018 (using in-house dynamic mechanical loading (DML) tool)	Cell cracks	Effect of mechanical loading cycles on cell cracks	The degradation in power and cell cracks saturates after certain no. of DML cycles. The effect of pressure, no. of cycles, and frequency on power loss and cell cracks is observed.
Chung Geun Lee et al. (2021)	Renewable energy of Korea Institute of Energy Technology Evaluation and Planning (KETEP), Korea	Short circuit failure of bypass diodes	Mismatch factors change the maximum power point (MPP) of PV module.	The most common factors responsible for mismatch loss are bypass diode failure and partial shading.
Y.R.Golive et al. (2020)	36 different sites from India by all India survey (AIS) 2018	PID, LID, over-rating, bypass diode failure in SSC, cell crack, and hotspot	Linear degradation rate of Pmax	The most number of badly degraded cells is seen in the "hot" climatic zones, and most of them are "young."
K. N. Bhavya Jyothi et al. (2020)	–	Determination of applied bias to keep the shaded cells in short circuit conditions (SSC), while bypass diode is inactive	I-V characteristics of shaded and non-shaded cells	Quantum efficiency can provide insight into the losses in PV modules and possible modes of degradation.
Y. R. Golive, Deepanshu Kosta et al. (2020)	36 different sites from India by all India survey (AIS) 2018	PID, bypass diode failure in SSC	Unusual I-V characteristics obtained in field-deployed PV module	Noticeable change in initial, degraded, and reproduced I-V characteristics affected by PID

Wei Luo et al. (2019)	Tropical rain forest "Af" by the Köppen climate classification, Singapore. PV modules installed on rooftop at NUS (national university of Singapore) campus	Corrosion of solder joints, discoloration of encapsulants, adhesion loss between encapsulants and other layers, hotspots, and PID	Performance degradation of various types of c-Si modules Degradation trends of Isc, Voc, and FF for multi-Si modules	Double-glass mono-Si module exhibits a noticeable increase of Voc which indicates the recovery from LID.
Y. R. Golive et al. (2019)	National Centre for photovoltaic research and education (NCPRE), IIT Bombay, Powai, India	Temperature (50–80 °C) and irradiance (700 to 1000 W/m^2)	Contour plot of normalized error in P_{max} for various procedures	Accuracy of various standard test conditions (STC) correction procedure for I-V characteristics of PV modules measured at different irradiance and temperature
Neelkanth G. Dhere et al. (2014)	Denver (hot and humid climate), USA, as observed by Florida solar energy center (FSEC)	Leakage current flowing through the module packaging material under high-voltage bias	Leakage current component	Analysis of correlation between the leakage current flowing through various paths with physical and chemical changes occurring in the packaging material
Ababacar Ndiaye et al. (2013)	Higher polytechnic school (ESP) of the Dakar University in Senegal, extreme western Africa	Short circuit current, open-circuit voltage, maximum power output, and PV module temperature	Comparison and correlation between measured and standardized values of I_{sc}, V_{oc}, and P_{max}	Degradation of I_{sc} is 10%. Degradation of open-circuit voltage is 2% in average, and no degradation of P_{max} is detected after 1 year of observation.
Edson L Meyer et al. (2004)	University of Port Elizabeth, South Africa	Front surface soiling, optical degradation, cell degradation, mismatched cells, light-induced degradation, temperature-induced degradation	Comprehensive visual inspection, light I-V measurement, dark I-V measurement, shunt resistance measurement, hotspot investigation, temperature-dependent measurements, outdoor monitoring	After an initial exposure of 130 sun hours, the performance of the CIS module degraded by more than 20%, a-Si module by 60%, and the a-SiGe module by 13%. The crystalline EFG-Si and mono-Si mod-ules showed no degradation in performance.

References

1. Golive Y. R., Zachariah S., Bhaduri S., Dubey R., Chattopadhyay S., Singh H. K, Kottantharayil A., Shiradkar N., Vasi J., "Analysis and Failure Modes of Highly Degraded PV Modules Inspected during the 2018 All India Survey of PV Module Reliability," 2020 4th IEEE Electron Devices Technology & Manufacturing Conference (EDTM), 2020, pp. 1–4.
2. Dhere N. G., Shiradkar N. S. and Schneller E., "Device for comprehensive analysis of leakage current paths in photovoltaic module packaging materials," 2014 IEEE 40th Photovoltaic Specialist Conference (PVSC), 2014, pp. 2007–2010.
3. Luo W., Khoo Y S, Hacke P., Jordan D., Zhao L., Ramakrishna S., Aberle G., Reindl T., "Analysis of the Long-Term Performance Degradation of Crystalline Silicon Photovoltaic Modules in Tropical Climates," in IEEE Journal of Photovoltaics, vol. 9, no. 1, pp. 266–271, Jan. 2019.
4. Golive Y. R., Singh H. K., Kottantharayil A., Vasi J. and Shiradkar N., "Investigation of Accuracy of various STC Correction Procedures for I-V Characteristics of PV Modules Measured at Different Temperature and Irradiances," 2019 IEEE 46th Photovoltaic Specialists Conference (PVSC), 2019, pp. 2743–2748.
5. Dubey R., Kottantharayil A., Shiradkar N. and Vasi J., "Effect of Mechanical Loading Cycle Parameters on Crack Generation and Power Loss in PV Modules," 2021 IEEE 48th Photovoltaic Specialists Conference (PVSC), 2021, pp. 0799–0802.
6. Jyothi K. N. B., Koshta D., Golive Y. R., Arora B. M., Narasimhan K. L. and Shiradkar N., "Biasing conditions for measurement of Quantum Efficiency of a solar cell in a module," 2020 47th IEEE Photovoltaic Specialists Conference (PVSC), 2020, pp. 0847–0849.
7. Golive Y. R., Koshta D., Rane K. P., Kottantharayil A., Vasi J. and Shiradkar N., "Understanding the Origin of Unusual I-V Curves seen in Field Deployed PV Modules," 2020 47th IEEE Photovoltaic Specialists Conference (PVSC), 2020, pp. 2166–2170.
8. Meyer E. L., Dyk E. E. van, "Assessing the reliability and degradation of photovoltaic module performance parameters," in IEEE Transactions on Reliability, vol. 53, no. 1, pp. 83-92, March 2004.
9. Aghaei, M.; Fairbrother, A.; Gok, A.; Ahmad, S.; Kazim, S.; Lobato, K.; Oreski, G.; Reinders, A.; Schmitz, J.; Theelen, M.;Yilmaz, P.; Kettel, J. "Review of degradation and failure phenomena in photovoltaic modules". Renew. Sustain. Energy Rev. 2022, 159, 112160.
10. Omazic A., Oreski G., Halwachs M., Eder G.C., Hirschl C., Neumaier L., "Relation between degradation of polymeric components in crystalline silicon PV module and climatic conditions: a literature review" Sol Energy Mater Sol Cells, 192 (2019), pp. 123–133.
11. Dhimish, M., Tyrrell, A.M. "Power loss and hotspot analysis for photovoltaic modules affected by potential induced degradation". npj Mater Degrad 6, 11 (2022).
12. Ndiaye A., Kébé C.M.F., Charki A., Sambou V., Ndiaye P.A., "Photovoltaic platform for investigating PV module degradation" Energy Procedia, 74 (2015), pp. 1370–1380.
13. Kaaya I., Ascencio-Vásquez J., Weiss K.A. and Topič M., "Assessment of uncertainties and variations in PV modules degradation rates and lifetime predictions using physical models", Solar Energy 2021, 218, 354–367
14. Ishikura N., Okamoto T., Nanno I., Hamada T., Oke S. and Fujii M., "Simulation Analysis of Really Occurred Accident Caused by Short Circuit Failure of Blocking Diode and Bypass Circuit in the Photovoltaics System," 2018 7th International Conference on Renewable Energy Research and Applications (ICRERA), 2018, pp. 533–536, https://doi.org/10.1109/ICRERA.2018.8566896.
15. Lee C.G., Shin W.G., Lim J.R., Kang G.H., Ju Y.C., Hwang H.M., Chang H.S., Ko S.W. "Analysis of electrical and thermal characteristics of PV array under mismatching conditions caused by partial shading and short circuit failure of bypass diodes" Energy, 218 (2021).

Impact of Security Attacks on Spectrum Allocation in Cognitive Radio Networks

Wangjam Niranjan Singh and Ningrinla Marchang

1 Introduction

As a result of the rise in the number of wireless devices, the bands in which these devices operate on are getting crowded. At the same time, several segments of the licensed spectrum are poorly underutilized. To address this imbalance of overcrowding in some segments of the radio spectrum on one hand and underutilization of some other segments on the other, a new opportunistic network has emerged, which is known as a cognitive radio network (CRN). A CRN is a network of cognitive radios (CRs) [1]. Such radios can sense the environment and adapt accordingly without causing harmful interference to the licensed users. Those users who have the license to operate on certain bands are called primary users (PUs), whereas secondary users (SUs) are dependent on the spectrum band of PUs and use the licensed bands opportunistically when required. The idea of cognitive radio was first coined by Dr. Joseph Mitola. After that, it led to IEEE 802.22 [2], which is a standard aimed for Wireless Regional Area Network (WRAN) using the free resource also called as white spaces in the TV frequency band opportunistically. Resource assignment (a.k.a. spectrum allocation) can be regarded as one of the main functionalities of CRN. It can be defined as the allocation of the free spectrum or spectrum holes to the radio interfaces so as to attain maximum spectrum utilization

W. N. Singh (✉)
Department of Computer Science and Engineering, Assam University, Silchar, Assam, India

Department of Computer Science and Engineering, North Eastern Regional Institute of Science and Technology, Itanagar, Arunachal Pradesh, India

N. Marchang
Department of Computer Science and Engineering, North Eastern Regional Institute of Science and Technology, Itanagar, Arunachal Pradesh, India

M. Mishra et al. (eds.), *Applications of Computational Intelligence in Management & Mathematics*, Springer Proceedings in Mathematics & Statistics 417,
https://doi.org/10.1007/978-3-031-25194-8_24

while maintaining least possible channel interference. Assigning resources in CRN poses other brand-new difficulties that do not appear in conventional wireless radio systems such as WLAN because CRs can dynamically fine-tune the frequency and bandwidth for each transmission according to best available resources, whereas conventional wireless systems use channels of rigid and prearranged bandwidth. Here in this chapter, first we investigate the security attacks that can appear in dynamic resource allocation known as Channel Ecto-Parasite Attack (CEPA) and Network Endo-Parasite Attack (NEPA) in which the compromised nodes switch to the heavily loaded channels. Such activities increase the interference on links that in turn reduces the overall performance of the network. Next, we study the impact of LORA, wherein the channel allocation dependency between the neighboring nodes is exploited. In this attack, compromised nodes transmit the wrong channel assignment information to neighboring nodes resulting in premature and false channel assignments. The main contributions of this chapter are outlined as follows:

1. We propose a distributed spectrum allocation algorithm in CRN based on the channel selection rule of CRTCA [4].
2. We analyze and evaluate the effect of three attacks, viz., NEPA, CEPA, and LORA on the network that uses the proposed algorithm.
3. We present a comparison of the impact of the above three attacks through simulation results.

The rest of the chapter is summarized as follows. In Sect. 2, the related works are mentioned. Section 3 gives the system model. Then the proposed distributed resource allocation algorithm and the attacks based on the algorithm are given in Sect. 4. In Sect. 5, we present the numerical simulation results. Finally, we conclude this chapter in Sect. 6.

2 Related Works

In recent times, we have seen a lot of literature on the resource allocation problem in cognitive radio network and its solutions. The spectrum assignment schemes can be classified based on the following implementation methods: conventional approaches based on centralized control [3, 4], distributed approaches [5, 6] that do not need any central controller, and cluster-based approaches [7, 27] that are amalgams of centralized and distributed approaches that try to avert the limitations of both types. Different resource allocation methods such as genetic algorithms [19, 20], soft computing [21, 28], heuristics [7, 8], game theory [9, 10, 26], linear programming [11, 12, 25], network-graph-based [3, 4], and non-linear programming[13, 14] have been found in the literature. The spectrum allocation algorithms mentioned in [3, 5, 15] are schemes for a single radio interface. Such schemes are uncomplicated, and interference management is easy. However, if the particular resource is redeemed by the PU, in-progress current data communication will be obstructed.

The schemes mentioned in[16, 17] are schemes for users having dual radios. The resource allocation schemes proposed in [8, 11] are for multi-radio users. In such a situation, when a PU redeems a particular channel, network partition does not happen. Naveed and Kanhere [18] analyzed the various security attacks in channel allocation in wireless mesh network. The authors evaluate the impacts of these vulnerabilities, and they come to the conclusion that such vulnerabilities by some malignant node can undoubtedly degrade the overall radio bandwidth and performance of the whole network taken into consideration. Some authors[23, 24] also discuss such kind of vulnerabilities in CRN. In CRN particularly, similar work has been seen in [22], where the authors evaluate the effect of Channel Ecto-Parasite Attack (CEPA) in a centralized spectrum allocation algorithm. Motivated by the preceding literature, we propose to study the effectiveness of Channel Ecto-Parasite Attack (CEPA), Network Endo-Parasite Attack (NEPA), and LOw-cost Ripple effect Attack (LORA) in a distributed spectrum allocation algorithm in CRN.

3 System Model

We consider a CRN scenario of n SUs, where each SU has dual radios that have access to S free resources. The CRN is modeled as an undirected graph $\mathrm{G}_{\mathrm{undirect}}(V, E)$, where each node $x \in V$ corresponds to an SU, and an edge $e = (x, y) \in E$ between node x and node y exists if they are within the transmission range of each other. Note that $\mathrm{G}_{\mathrm{undirect}}(V, E)$ is a connected graph in which any two SUs are connected by either a direct link (i.e., an edge) or a way with multiple SUs. Any two SUs can exchange information among themselves if they are on the same resource (i.e., channel) and are within each other's transmission range. A spectrum allocation $\mathcal{SA}$ creates a new undirected graph $G_{\mathcal{SA}}(V, E_{\mathcal{SA}})$, where $E_{\mathcal{SA}}$ consists of the edges given as below. There exists an edge $e = (x, y; s)$ on channel s if $(x, y) \in E$ and $s \in \mathcal{SA}(x) \bigcap \mathcal{SA}(y)$. The $\mathcal{SA}(x)$ and $\mathcal{SA}(y)$ indicate the group of channels allocated to x and y, respectively. $\mathcal{R}(x)$ represents the number of radios at SU x and $|\mathcal{R}(\mathrm{x})| \leq |S|$. If they share more than one channel, multiple edges may exist between two neighboring SUs. We have assumed that all SUs have the same interference and transmission range. Each of the link between any two SUs can support communication in only one direction at one time, i.e., we have considered half-duplex mode of communication.

4 A Distributed Resource Allocation Method

In this section, we present the design philosophy and the algorithm of a distributed resource allocation method in detail. An SU is assumed to be intelligent enough so that it is able to learn which of its neighboring nodes (SUs) falls under its transmission range. An SU keeps an information record consisting of the following fields: (a) Neighbor_table, (b) Node_id, (c) Neighbor_no, (d) Neighbor Degree_table, and (e) SU_state. The neighbor_table field carries the information about neighbors of the SU. The node_id field indicates the ID of the SU. It is a number that is distinct for each SU. The neighbor_no field indicates the total number of neighbors of the SU. Neighbor Degree_table carries the information about degree, i.e., the number of neighbors of all the SU in the network. SU_state field indicates the current state of the SU. The state transition is as follows (the number inside the brackets denotes the state number): boot(-9)>>passive(-1)>>active(1)>>ready(8)>>allocation(9)>>end(10). The state transition can be understood from the *NODE SU Life Cycle* given below.

NODE SU Life Cycle

1. Boot node with "BOOT" SU_state (value = −9).
2. Complete booting, and change boot state to "PASSIVE" state (value = −1).
3. (a) Start neighbor discovery. Update Neighbor_table.
   During this phase, state will change to "ACTIVE" (value = 1).
   (b) Multicast (Broadcast) node_id and (neigbhor_no + Metric) pair. Every SU will receive the broadcast and update its neighbor degree_table. Metric = id/100.
4. Finish neighbor discovery. Change state to "READY" (value = 8).
5. Wait for some network convergence time T for hearing broadcast from other SUs and updation of node_id and (neigbor_no + Metric) pair in neighbor degree_table. After elapse of time T, change state to "ALLOCATION" (value = 9).
6. When state is in "ALLOCATION" (value = 9), node is ready for channel allocation.
7. Execute channel allocation, and after completion, change state to "END" (value = 10).

Channel Allocation Algorithm

1. Each node with state "ALLOCATION" (value = 9) will check itself if there exists node's (neighbour_no + metric) the highest value in neighbor degree_table.
2. If yes, start channel allocation.
3. After completion, the node will broadcast the channel allocation completion flag.
4. Other nodes will update neigbhor_no with "−1" in the neighbor degree_table for the broadcasted node.

Algorithm 1 (Normal Channel Allocation Rule)

1. **if** $|\mathcal{SA}(x)| < |\mathcal{R}(x)|$ and $|\mathcal{SA}(y)| < |\mathcal{R}(y)|$ then
2. $s \leftarrow$ the least used resource among the resources in S;
3. **else if** $|\mathcal{SA}(x)| = |\mathcal{R}(x)|$ and $|\mathcal{SA}(y)| < |\mathcal{R}(y)|$ then
4. $s \leftarrow$ the least used resource among the resources in $\mathcal{SA}(x)$;
5. **else if** $|\mathcal{SA}(x)| < |\mathcal{R}(x)|$ and $|\mathcal{SA}(y)| = |\mathcal{R}(y)|$ then
6. $s \leftarrow$ the least used resource among the resources in $\mathcal{SA}(y)$;
7. **end if**

In a distributed environment, *"the basic idea is: since the node with a higher degree is likely to interfere with other nodes, it is given higher priority to assign channels.* " Let us discuss how the spectrum allocation is performed by each SU. When the booting process is completed, "BOOT" SU_state (value $= -9$) is changed to "PASSIVE" state (value $= -1$). After that, neighbor discovery is performed by each SU. In the neighbor discovery, each node detects and counts the neighboring nodes within their range and broadcasts that information to all the nodes present in the environment. The SU stores the information about its neighbors in neighbor_table. During this phase, the SU_state is changed to "ACTIVE" (value $=1$). After that, the SU broadcasts its node_id and (neighbor_no + Metric) pair. Every node will receive the broadcast and update its neighbor degree_table. Here, the metric, i.e., Metric = Node_id/100, is used to make distinct nodes for the nodes that have same neighbor number. After finishing neighbor discovery, it changes the SU_state to "READY" (value $=8$). After waiting for some network convergence time T for hearing broadcast from other nodes SUs and updating of node id and (neighbor no + Metric) pair in Degree_table and after elapse of time T, the SU_state changes to "ALLOCATION" (value $=9$). When the state is in "ALLOCATION" (value $=9$), the node is ready for channel allocation.

For channel allocation, each node having the transition state "ALLOCATION" (value $=9$) will check whether its (neighbour_no + metric) value is the highest value in its neighbors degree_table or not. If yes, start channel allocation will perform as per the channel selection rule in Algorithm 1 above. First, if the no. of allocated resources is lesser than the no. of radios at both SUs x and y, the lowest used resource among the resources in S is chosen (line no 2). Second, if the no. of resources equals to the number of radios at x, but the no. of resources is lesser than the no. of radios at SU y, then the lowest used channel from the group of resources allocated to x is chosen (line no 4). Third, if the no. of resources is lesser than the number of radios at x, but the no. of resources is equal to the no. of radios at SU y, then the lowest used resource from the group of resources allocated to y is chosen (line no 6). This has to be performed for every neighboring node within the transmission range for that particular node or SU. For every assignment, the channel usage information has to be broadcasted to all the SUs in the network. After completion, the node will broadcast the channel allocation completion flag. Other nodes will update neigbhor_no with "-1" in the Degree_table for the broadcasted

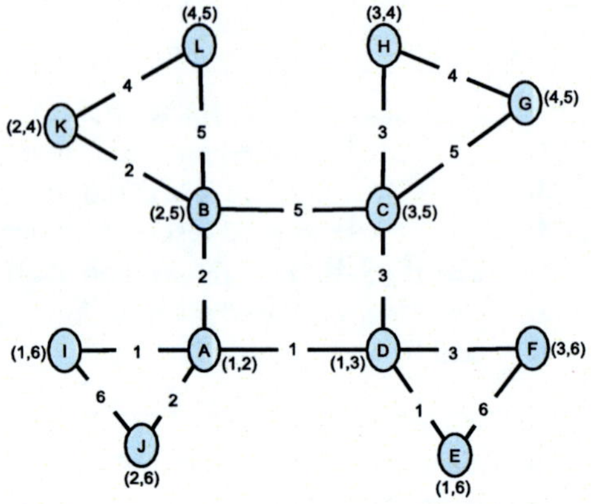

Fig. 1 Normal assignment

node. Then, it will change SU_state to "END," value = 10. This has to be repeated for other remaining nodes. In Fig. 1, we consider 12 SUs having dual radios each, and there are 6 free available resources or channels. The resources are allocated by adopting the resource allocation rule mentioned above. Here, label at links denotes the group of resources shared between two end SUs of that link, and the label along with the SU denotes the group of resources allocated to that SU. Figure 1 shows the resource allocation under the mentioned above distributed resource allocation method.

4.1 Channel Ecto-Parasite Attack (CEPA)

The major intention behind the CEPA is to escalate the interference at the most used resource. We have given how the channel allocation rule is altered by such attack in the following channel allocation rule in CEPA algorithm (as given in Algorithm 2):

Algorithm 2 (CEPA Channel Allocation Rule)

1. **if** $|\mathcal{SA}|(x) < |\mathcal{R}(x)|$ and $|\mathcal{SA}(y)| < |\mathcal{R}(y)|$ then
2. $s \leftarrow$ the most used resource among the resources in S;
3. **else if** $|\mathcal{SA}(x)| = |\mathcal{R}(x)|$ and $|\mathcal{SA}(y)| < |\mathcal{R}(y)|$ then
4. $s \leftarrow$ the most used resource among the resources in $\mathcal{SA}(x)$;
5. **else if** $|\mathcal{SA}(x)| < |\mathcal{R}(x)|$ and $|\mathcal{SA}(y)| = |\mathcal{R}(y)|$ then
6. $s \leftarrow$ the most used resource among the resources in $\mathcal{SA}(y)$;
7. **end if**

With normal resource allocation algorithm, an SU allocates the least used resource to the radio interfaces, but in this type of attack, a malignant SU launches this attack by allocating its interfaces with the highest used resource. Figure 2 shows

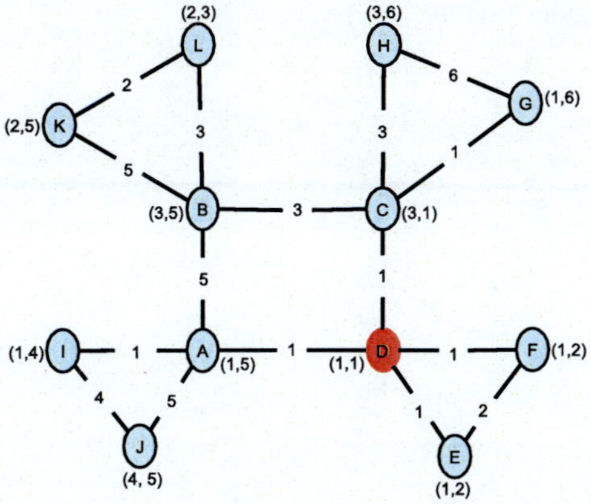

Fig. 2 CEPA attack

the resource allocation under the influence of this attack. Here we have introduced a malignant SU D marked by a node in red color in the network. It can be seen from Fig. 2 that along with the launching of this attack, the resource allocation is altered. We have seen that due to falsified assignment by a malicious SU D, many links have been seen with highest used channels, i.e., channel number 1.

4.2 *Network Endo-Parasite Attack (NEPA)*

Normally, the low-priority (least loaded) resources are allocated by an SU to its radio interfaces. In the NEPA attack, the malicious SU tries to escalate the interference at massively loaded resources by allocating its interfaces with the highly preferred resources (highly used resources); however, the neighbors are not informed about the amendment. We have introduced a malignant SU D denoted by a node in red color in the network. We are able to see from the figure that along with the launching of this attack, the resource allocation is altered (refer Algorithm 3).

Algorithm 3 (NEPA Channel Allocation Rule)

1. **if** $|\mathcal{SA}(x)| < |\mathcal{R}(x)|$ and $|\mathcal{SA}(y)| < |\mathcal{R}(y)|$ then
2. $s \leftarrow$ the high used resource among the resources in S;
3. **else if** $|\mathcal{SA}(x)| = |\mathcal{R}(x)|$ and $|\mathcal{SA}(y)| < |\mathcal{R}(y)|$ then
4. $s \leftarrow$ the high used resource among the resources in $\mathcal{SA}(x)$;
5. **else if** $|\mathcal{SA}(x)| < |\mathcal{R}(x)|$ and $|\mathcal{SA}(y)| = |\mathcal{R}(y)|$ then
6. $s \leftarrow$ the high used resource among the resources in $\mathcal{SA}(y)$;
7. **end if**

With normal resource allocation scheme, an SU allocates the least used resource to the radio interfaces, but in NEPA, a malicious SU launches this attack by

Fig. 3 NEPA attack

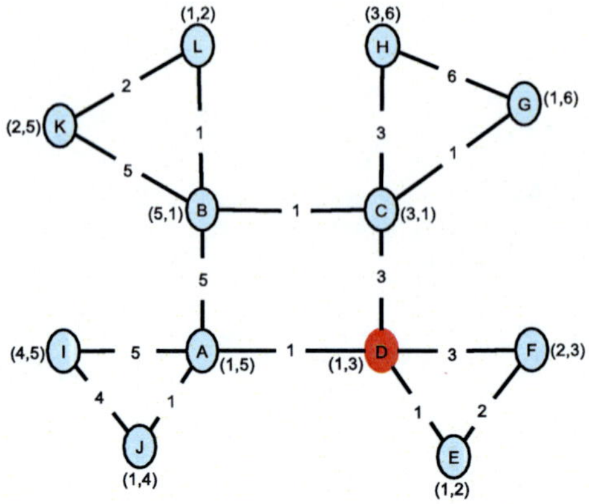

allocating its interfaces with the high-priority resources. Figure 3 shows the resource allocation under the influence of NEPA attack. We have introduced a malignant SU D denoted by a node red in color in the network. We can see from the figure that with the launching of this attack, the resource allocation changes. We have seen that due to falsified assignment by a malicious SU D, many links have been seen with high used channels, i.e., channel number 1. But its impact is less as compared to CEPA attack because all the links of the malicious nodes are not assigned with the highest used channel, i.e., channel no 1.

4.3 LOw-Cost Ripple Effect Attack (LORA)

In CRNs, this sort of attack is launched whenever the malignant SU transmits the wrong misleading resource information through the network and forces other SUs to reallocate their resource. The attack is comparatively more severe than the previous two attacks. The impact is propagated to an oversized portion of the network on the far side of the neighbors of the compromised malignant SU, disrupting the traffic forwarding capability of varied SUs for a considerable time period.

Algorithm 4 (LORA Channel Allocation Rule)

1. **if** $|\mathcal{SA}(x)| < |\mathcal{R}(x)|$ and $|\mathcal{SA}(y)| < |\mathcal{R}(y)|$ then
2. $s \leftarrow$ the least used resource among the resources in S;
3. **else if** $|\mathcal{SA}(x)| = |\mathcal{R}(x)|$ and $|\mathcal{SA}(y)| < |\mathcal{R}(y)|$ then
4. $s \leftarrow$ the least used resource among the resources in $\mathcal{SA}(x)$;
5. **else if** $|\mathcal{SA}(x)| < |\mathcal{R}(x)|$ and $|\mathcal{SA}(y)| = |\mathcal{R}(y)|$ then
6. $s \leftarrow$ the least used resource among the resources in $\mathcal{SA}$(y);
7. **end if**

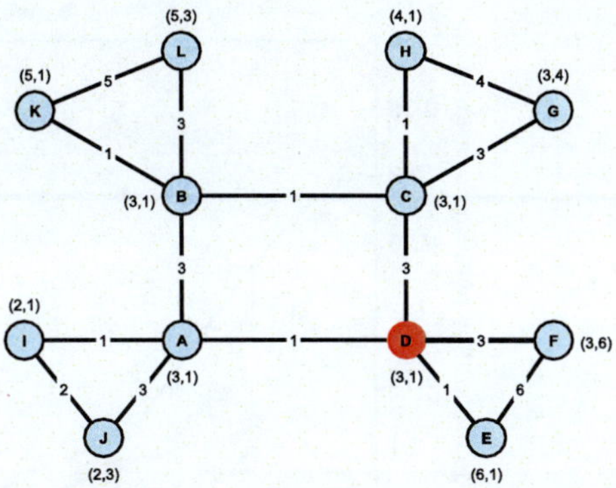

Fig. 4 LORA attack

With the normal resource allocation algorithm, an SU allocates the least used resource to the radio interfaces. In LORA also, the SU will allocate as normal using the normal assignment. But, after the completion of assignment of a particular SU, a compromised SU launches this attack by sending wrong channel usage information of *S*, thereby forcing the other SUs to reallocate their resource. Figure 4 shows the resource allocation under LORA attack. We have introduced a malignant compromised SU D denoted by a node in red color in the network. We are able to see from the figure that along with the launching of this attack, the resource allocation is altered. We have seen that due to falsified assignment by a malicious SU D, many links have been seen with high used channels, i.e., channel number 1. Due to sending wrong channel usage information by the malicious node SU D, not only link of the malicious node is assigned with channel number 1 but its impact is propagated to other links also, so many links of the other nodes get assigned with the channel number 1.

5 Simulation and Performance Evaluation

We have analyzed the CEPA, NEPA, and LORA attacks through simulation-based experiments. We simulate the network topology shown in Fig. 1. We inject 18 constant bit rate (CBR) flows into the network. The origin and destination for every flow are picked up randomly, and Dynamic Routing Protocol (DSR) is employed for routing. First, we employ the probability of network partition as a performance metric to analyze the effectiveness of the above attacks. The probability of network partition can be defined as the probability that the network is partitioned when a channel is reclaimed, i.e., redeemed by the PU. And the average probability of network partition can be defined as the probability that the network is partitioned when different available channels are reclaimed by the PU. Figure 5 presents the average probability of network partition vs. the no. of channels under normal resource assignment and different attacks. Here, the number of nodes (n) is 12, and

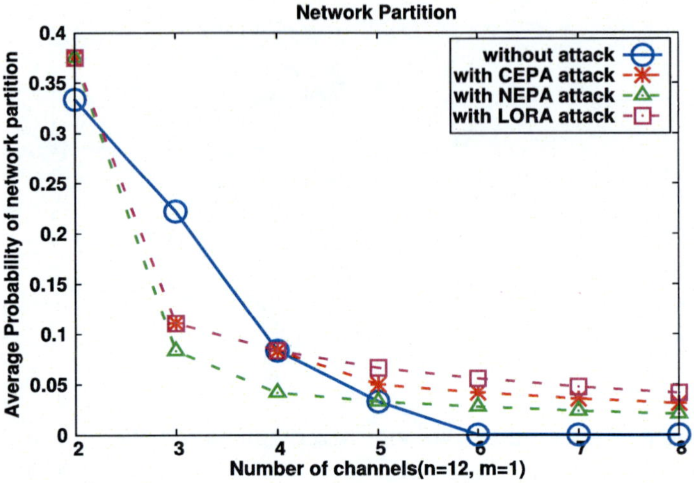

Fig. 5 Average probability of network partition vs. the number of channels

the number of malicious nodes (*m*) is 1. As expected, we observe that as the no. of channels, i.e., resources, increases, the probability of network partition decreases. However, there is a sharp difference in the two situations: when there is an attack and when there is no attack. Once there is no security attack, the average probability reduces to 0 from the point when the number of channels is 6. In contrast, when there is a NEPA attack, the average probability of partition remains at around 0.04 from the point when the number of channels is 6. When there is a CEPA attack, we find a better average probability of partition that remains at around 0.06 from the point when the number of channels is 6. But when there is LORA attack, we find the highest average probability of partition that remains at around 0.09 from the point when the number of channels is 6. We have seen sudden changes in the average probability of network partition for under security attacks. This is because many links have been assigned with the wrong channels due to the malicious behavior. So, when a PU reclaims this channel, network partition occurs.

Figure 6 shows the probability of network partition vs. the number of malicious nodes (*m*) under different attacks. Here, we have considered when only 1 channel, i.e., channel no. 1, is reclaimed by the PU. We have seen that as the no. of malicious nodes increases, the probability of network partition increases as expected. When there is a NEPA attack, the probability of partition increases from 0.18 if the no. of malicious nodes is from 1 to 0.21. When there is a LORA attack, we find a more severe impact as compared to the NEPA attack. The probability of partition increases from 0.18 if the no. of malicious nodes is 1 to 0.5. But when there is a CEPA attack, we find the highest probability of partition, i.e., the probability of partition increases from 0.25 if the no. of malicious nodes is 1 to 1.0 and the no. of malicious nodes is 12. We observe the highest severe impact in the CEPA attack. This is due to the

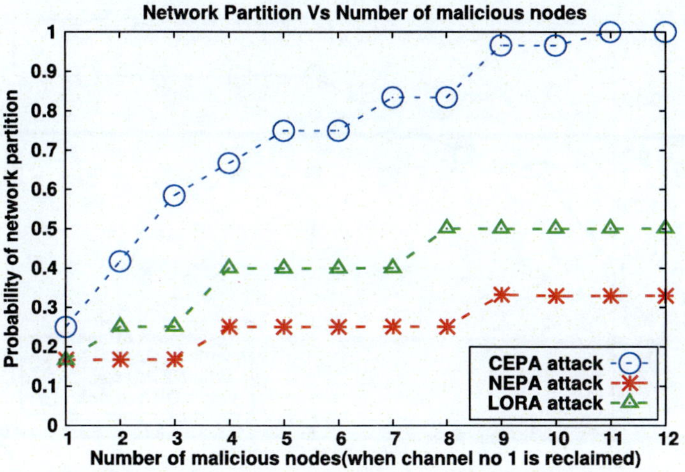

Fig. 6 Probability of network partition vs. the number of malicious nodes

nature of the attack: many edges or links have been assigned with the highest used channel, i.e., many links have been using this channel for communication. So, when a PU redeems this channel, network partition is more likely to occur.

Next, to analyze the network performance evaluation, we measure throughput and average delay. Throughput tells us how much data are transferred from a source at any given time, and it can also be defined as the rate of successful transmission of data packets. Average delay is defined as the average time taken by the data packets to propagate from origin point to destination point across the network. Figure 7 shows the throughput (in Mbps) vs. the number of channels under normal resource allocation scheme and different attacks. We employ the situation when the PU reclaims channel no. 1. We observe that throughput underneath the influence of attack is much less than when there is no attack present. This is because reclamation of channel no. 1 by PU has resulted in a network partition. Hence, some packets are not able to reach the aimed destination point. Therefore, throughput is reduced. We also observe under CEPA, the lowest throughput is found as compared to the other two attacks. This is because, in CEPA attacks, maximum links have been assigned to the highest used channel, i.e., channel number 1. So, when a PU redeems this channel, network partition occurs and, hence, throughput decreases. We also found that for a higher no. of channels, say from 5, the throughput is almost same for LORA and NEPA attacks. This is because for a higher number of channels the probability of network partition is the same in both NEPA and LORA attacks.

Figure 8 shows the average delay (in microsec) vs. the no. of channels under normal resource allocation scheme and different attacks. We consider the situation when the PU redeems channel no 1. We have seen that average delay increases under the effect of attack. This is because when PU redeems channel no. 1, network

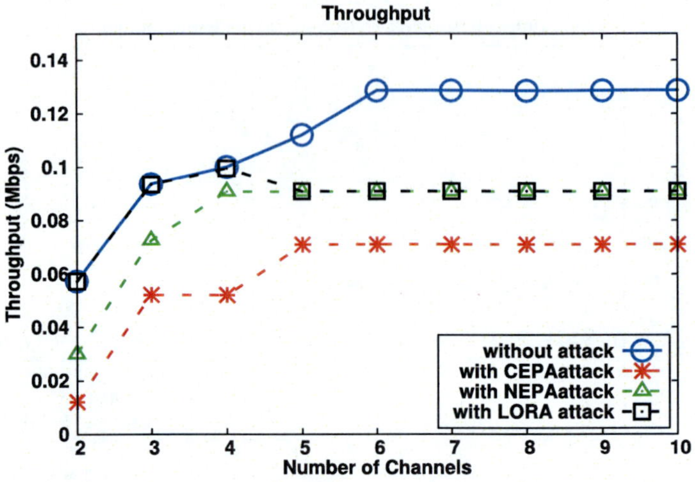

Fig. 7 Throughput vs. the number of channels

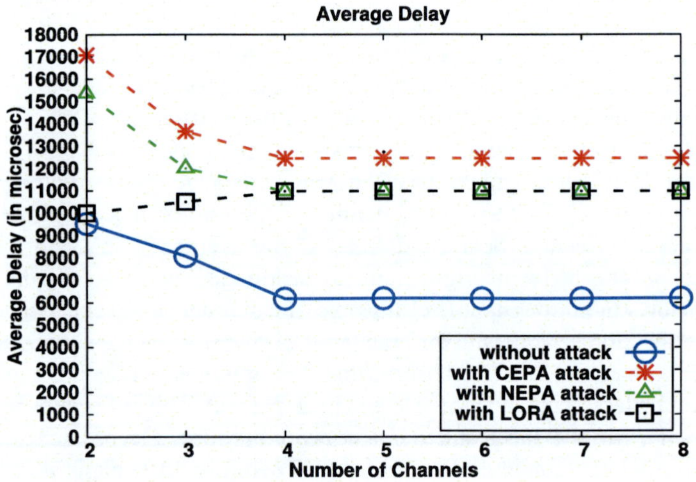

Fig. 8 Average delay vs. the number of channels

partition occurs. Hence, some packets will search for another route to reach the aimed destination. Consequently, the average delay increases. We also observe in CEPA that the highest average delay is found as compared to the other two attacks. We also found that for a higher number of channels, say from 5, the average delay is almost the same for LORA and NEPA attacks. This is because, for a higher number of channels in the case of both NEPA and LORA attacks, the probability of network partition is the same.

6 Conclusions

In this chapter, we propose a distributed resource allocation algorithm in CRN and also exposed security vulnerabilities in resource allocation in CRN, viz., CEPA, NEPA, and LORA. It was seen through numerical simulation results that having such vulnerability in resource assignment deteriorates the overall performance of the network, and also it results in network partition. It was found through experimental simulations that LORA has the highest average probability of network partition as compared to CEPA and LORA attack. CEPA has a more severe impact in terms of network throughput and delay as compared to the NEPA and LORA attacks in some cases. In the majority of the cases, NEPA shows the least severe impact as compared to the other two attacks. In future, we wish to investigate ways of detecting these attacks.

References

1. Mitola J.: Cognitive radio: An integrated agent architecture for software defined radio. PhD thesis in Royal Institute of Technology (KTH) (2000).
2. Carlos, C., Challapali, K., Brie D., Shankar N.S.: IEEE 802.22: the first worldwide wireless standard based on cognitive radios. In Proc. First IEEE International Symposium on New Frontiers in Dynamic Spectrum Access Networks, (DySPAN), pp. 328–337 (2005)
3. Xin, C., Ma L., Shen C.C.: A Path-Centric Channel Assignment Framework for Cognitive Radio Wireless Networks. Mobile Networks and Applications. 13(5):463–476 (2008)
4. Zhao, J., Cao, G.: Robust topology control in multi-hop cognitive radio networks. In: Proceedings of the 2012 IEEE INFOCOM, Orlando, FL, pp. 2032–2040 (2012)
5. T, T.L., Le, L.B.: Channel Assignment With Access Contention Resolution for Cognitive Radio Networks. IEEE Transactions on Vehicular Technology 61(6):2808–2823 (2012)
6. Hashem, M., Barakat, S.I., AttaAlla, M.A.: Distributed channel selection based on channel weight for cognitive radio network. In: Proceedings of the 10th International Computer Engineering Conference (ICENCO), Giza, pp. 115–120 (2014)
7. Alsarahn, A., Agarwal, A.: Channel assignment in cognitive wireless mesh networks. In: IEEE 3rd International Symposium on Advanced Networks and Telecommunication Systems (ANTS), New Delhi, pp. 1–3 (2009)
8. Kim, W., Kassler, A.J., Felice, M.D., Gerla, M.: Urban-X: Towards distributed channel assignment in cognitive multi-radio mesh networks. In: Proceedings of the IFIP Wireless Days, Venice, pp. 1–5 (2010)
9. Zhang, H., Yan, X.: Advanced Dynamic Spectrum Allocation Algorithm Based on Potential Game for Cognitive Radio. In: Proceedings of the 2nd International Symposium on Information Engineering and Electronic Commerce, Ternopil, pp. 1–3 (2010)
10. Wu, Z., Cheng, P., Wang, X., Gan, X., Yu, H., Wang, H.: Cooperative Spectrum Allocation for Cognitive Radio Network: An Evolutionary Approach. In: IEEE International Conference on Communications (ICC), Kyoto, pp. 1–5 (2011)
11. Irwin, R.E, MacKenzie, A.B, DaSilva, L.A.: Resource-minimized channel assignment for multi-transceiver cognitive radio networks. IEEE J. Sel. Areas Commun 31(3):442–450 (2013)
12. Yu, L., Liu, C., Hu, W.: Spectrum allocation algorithm in cognitive ad-hoc networks with high energy efficiency. In: The International Conference on Green Circuits and Systems, Shanghai, pp. 349–354 (2010)

13. Chen-li, D., Guo-an, Z., Jin-yuan, G., Zhi-hua, B.: A route tree-based channel assignment algorithm in cognitive wireless mesh networks. In: Proceedings of the International Conference on Wireless Communications and Signal Processing, Nanjing, pp. 1–5 (2009)
14. Pareek, U., Lee, D.C.: Resource allocation in bidirectional cooperative cognitive radio networks using swarm intelligence. In: IEEE Symposium on Swarm Intelligence, Paris, pp. 1–7 (2011)
15. Lee, D.H., Jeon, W.S.: Channel assignment and routing with overhead reduction for cognitive radio-based wireless mesh networks. In: International Conference on Wireless Communications and Signal Processing (WCSP), Nanjing, pp. 1–5 (2011)
16. Wang, J., Yuqing, H.: A cross-layer design of channel assignment and routing in Cognitive Radio Networks. In: Proceedings of the 3rd International Conference on Computer Science and Information Technology, Chengdu, pp. 542–547 (2010)
17. Anifantis, E., Karyotis, V., Papavassiliou, S.: A Markov Random Field framework for channel assignment in Cognitive Radio networks. In: IEEE International Conference on Pervasive Computing and Communications Workshops, Lugano, pp. 770–775 (2012)
18. Naveed, A., Kanhere, S.S.: NIS07-5: Security Vulnerabilities in Channel Assignment of Multi-Radio Multi-Channel Wireless Mesh Networks. In: IEEE Globecom, San Francisco, CA, pp. 1–5 (2006)
19. Elhachml, J., and Guennon, Z.: Cognitive radio spectrum allocation using genetic algorithm. EURASIP Journal on Wireless Communication and Networking 133(1): 1–11 (2016)
20. Morabit, Y.E., Mrabti, F., and Abarkan, E.H.: Spectrum allocation using genetic algorithm in cognitive radio networks. Third International Workshop on RFID and Adaptive Wireless Sensor Networks (RAWSN), Agadir, pp. 90–93 (2015)
21. Chowdhury, S.A., Benslimane, A., and Akhter, F.: Throughput maximization of cognitive radio network by conflict-free link allocation using neural network. IEEE International Conference on Communications (ICC), Paris, pp. 1–6 (2017)
22. Singh, W.N. and Marchang, N.: Security Vulnerability in Spectrum Allocation in Cognitive Radio Network. In: Kamal R., Henshaw M., Nair P. (eds) International Conference on Advanced Computing Networking and Informatics. Advances in Intelligent Systems and Computing, Springer, Singapore 870:215–224 (2019)
23. Parvin, Sazia., Hussain, F.K., Hussain, O.K., Han, S., Tian, B., Chang, E.: Cognitive radio network security: A survey. Journal of Network and Computer Applications 35(6):1691–1708 (2012)
24. Singh, W.N. and Marchang, N.: A review on spectrum allocation in cognitive radio network. Int. J. Communication Networks and Distributed Systems 23(2):172–193 (2019)
25. Martinovic, J., Jorswieck, E., Scheithauer, G., and Fischer, A.: Integer Linear Programming Formulations for Cognitive Radio Resource Allocation. IEEE Wireless Communications Letters 6(4):494–497 (2017)
26. Xiao, Z. et al.: Spectrum Resource Sharing in Heterogeneous Vehicular Networks: A Non-cooperative Game-Theoretic Approach with Correlated Equilibrium. IEEE Transactions on Vehicular Technology 67(10):9449–9458 (2018)
27. Alam, S., Malik, A.N., Qureshi, I.M., Ghauri, S.A., and Sarfraz, M.: Clustering-Based Channel Allocation Scheme for Neighborhood Area Network in a Cognitive Radio Based Smart Grid Communication. IEEE Access 6:25773–25784 (2018)
28. Ali, A. et al.: Hybrid Fuzzy Logic Scheme for Efficient Channel Utilization in Cognitive Radio Networks. IEEE Access 7:24463–24476 (2019)

Human Activity Detection Using Attention-Based Deep Network

Manoj Kumar and Mantosh Biswas

1 Introduction

In this era, video-based activity recognition is a challenging field, especially for finding and recognising activities in a surveillance stream's video sequence. Human activity identification in video has a wide range of applications, including content-based video retrieval [1], security surveillance systems [2], activity recognition, and human–computer interaction [3]. Due to the exponential growth of digital content on a daily basis, effective AI-based intelligent systems are required for monitoring and recognising human activities. Activity recognition aims to recognise and detect people and suspicious actions in videos, as well as give relevant information to assist interactive programmes. Despite significant breakthroughs in illumination variations, camera movements, complicated backgrounds, and occlusions, action recognition still faces various challenges when it comes to safeguarding the safety and security of people, including violence detection, industrial surveillance, and person recognition. Recognising various human behaviours in films requires both geographical and temporal information. Most strategies used feature processing to signal the spatial aspects of motion for describing the related human activity in video sequences in the last decade. Due to the motion style and complex background, the handcrafted feature approach in action detection is largely database-oriented and cannot suit the universal condition. To get reliable information, representative motion features and traditional methods are gradually increased from 2D to 3D. In

M. Kumar (✉)
JSS Academy of Technical Education, Noida, U.P., India
e-mail: manojkumar@jssaten.ac.in

M. Biswas
NIT Kurukshetra, Thanesar, Haryana, India
e-mail: mantoshbiswas@nitkkr.ac.in

M. Mishra et al. (eds.), *Applications of Computational Intelligence in Management & Mathematics*, Springer Proceedings in Mathematics & Statistics 417,
https://doi.org/10.1007/978-3-031-25194-8_25

order to capture dynamic information in a sequence of frames at the same time, such procedures translated spatial qualities into 3D spatiotemporal features [11]. Deep learning is a popular method for learning high-level distinguishing key features and creating video-based action identification. To learn characteristics from video frames, currently available deep learning systems for human activity recognition (HAR) use straightforward CNNs processes with pre-trained models in convolution operation.

To train a classification model, these layers of convolutional neural networks extract and learn spatial features. Traditional CNN models, on the other hand, perform worse in sequential data than handcrafted features [4]. DenseNet, VGG, and Inception are examples of standard CNN models which extract spatial information from a input frame. All these models are good at obtaining geographical data, but not so well at capturing temporal data, which is crucial for capturing movement information for the human activity recognition in a sequence of video stream. For example, Dai et al. [5] employed LSTM to recognise actions using characteristics learnt from a CNN with information of spatiotemporal features. The construction of different modules for learning spatiotemporal aspects in video stream by fusing processes to capture motion features in sequential data is required for video-based high-level HAR algorithms [6]. The LSTM, which was established for long-duration video clips to learn and interpret temporal properties for HAR in surveillance systems, has lately been used to address the spatiotemporal issues [7]. To address the HAR's existing constraints and limitations, most researchers have devised a two-stream technique for activity recognition that combines spatiotemporal features for joint feature training.

Due to a lack of information on style, motion, and clutter backdrop for human action identification, precise action detection in real-world recordings remains challenging based on these facts. Conventional techniques failed to handle these challenges due to worries about handling continuous activities, dense surroundings due to noise, and occlusion [8]. Similarly, contemporary HAR approaches used RNNs, LSTMs, and gated recurrent units to handle the sequence learning problem, without concentrating on specific features in video, which is critical for maintaining a relationship between consecutive frames. We presented a HAR system with attention that learns spatiotemporal characteristics and uniquely focuses on distinctive features in long-term video clips for recognising activities in video stream, which is ideal for a framework for surveillance. We employ DenseNet, a pre-train CNN, to upgrade the learnt features in this system, as well as the attention-weighted LSTM to focus on the important information in the input frame and recognise motion in a movie. We found multiple pre-eminences in the suggested system for activity recognition. The CNN uses convolution to extract spatial features, which is then learned by the LSTM to realise the contents in order to better recommend normal human activities. We gather high-level discriminative information from every fifth frame of a movie and send it to the attention mechanism, which adjusts the attention weights for selected characteristics in the sequence once again. Because of these properties, experiments show that the recommended technique is suitable for human activity in surveillance.

The remainder of the article is organised as follows: Sect. 2 briefly describes the relevant literature, and Sect. 3 shows the suggested human activity recognition architecture and its components. The experiments and results are presented in Sect. 4, while Sect. 5 draws conclusions as well as prospective future research directions.

2 Related Works

Action identification is an important area of research in the computer vision domain, with numerous strategies utilising standard artificial intelligence, machine learning (ML), and neural networks (NN) to recognise human activities in video sequences. Traditional machine learning techniques were largely applied in the past decade to construct effective systems for the human activity from extracting the features. Deep learning algorithms are now being used by academics to calculate both temporal and spatial features from a stack of frames to classify the action. The next sections provide a review of existing methods in the literature that are connected to this topic.

The three primary phases of machine learning-based human activity recognition systems are (i) handcrafted feature extraction, (ii) feature representation, and (iii) feature classification using an appropriate machine learning algorithm [3]. There are two types of feature extraction strategies used in computer vision: local feature-based approaches and global feature-based framework. The features as interest spots, independent patches, and gesture features are known as local features as well as area of interest as a global feature. The majority of handmade features extractors are domain specific and tailored for certain datasets. They can't be used for learning all-purpose features [9]. To reduce the processing time of their systems, some writers adopted key frame-based approaches [10]. They are, however, limited by human cognition and continue to have limitations, such as being time-consuming, labour-intensive, and the selection of features being a lengthy process [11]. The authors turned to deep learning to offer creative and efficient methodology for improving vision-based HAR method after recognising the limitations and drawbacks of handcraft-based HAR.

In contrast to basic machine learning architecture, deep learning proposes an architecture that learns high-level discriminative visual characteristics and concurrently classifies them. Standard CNN architectures use convolutional operations to determine the best features and change parameters based on the data [12]. In most cases, the CNN-based deep feature learning algorithms used to extract visual information from 2D data are not adequate for 3D data. Instead of utilising 2D filters, some researchers have proposed employing 3D filters to learn features from frames [13]. When compared to 2D CNNs and handmade feature-based approaches, deep models outperformed the findings in video analysis (action detection, object tracking, and video retrieval). Author [14] presented a two-stream solution for video analysis that relied on a CNN to overcome the challenge of extracting motion signals from several video frames. Using a multi-stream CNN model, Tu et al. [15] trained areas surrounding humans that detected multiple activities in a movie. To learn the spatial and temporal features of video sequences, the authors [16] employed a two-stream RNN-based fused network. These deep learning algorithms, on the other hand, are designed to learn and recognise short-term temporal information and are therefore inadequate for long-term sequences. Following RNN's success, an improved variation known as LSTM was created to encode long-term dependencies [7]. In a variety of applications, such as action identification, speech processing, and weather prediction, LSTM networks are now commonly employed [7] to classify long sequential data. As a result, big data redundancy is a problem; therefore Xu et al. [6] devised a deep learning-based method to avoid big data redundancy, employing reinforcement learning for resource allocation based on content-centric IoT [17].

The attention mechanism was recently established and has shown promise in a variety of temporal tasks such as action identification and video captioning. A stack of frames plays a significant function at different phases in activity video contents such as running, walking, jogging, and so on to display diversified attention to the viewers' first glance [18]. Attention mechanisms in video and picture captioning applications achieved better result on common datasets, which focus to learn and characterise the specific features. Using attention mechanisms, in [19], the authors proposed a method for extracting crucial frames from a film in order to distinguish activity. The two-stream attention model [20], which pays attention to sequential input in an indirect manner, is currently widely used in a number of applications.

3 Proposed Methodology

The suggested architecture of human activity recognition and its essential components, such as pre-train CNN and LSTM, are explained in this section. We use a pre-train convolution network, i.e. DenseNet, to extract features and an LSTM layer with an attention layer to distinguish a human activity in video frames. We use a DenseNet pre-train network to extract the deep features from the input data. The input data size for DenseNet is 224×224; the network establishes an relationship among them and gives an output of size $8 \times 8 \times 1920$. Here one frame is represented by one row, i.e. 1×1920. We have taken 40 frames in a video. By pre-train network one video is represented by 40×1920 feature set. The feature is passed into a LSTM with 512 cells, which learns sequential information before using an attention mechanism. We again adjust the attention weights based on the learnt high-level features to make it easier to detect and focus on important cues so that we can recognise human activities in a video. Finally, a full convolution network with SoftMax activation is applied to uniquely identify the human activity. The proposed system's overall design is depicted in Fig. 1, and the essential components are described in detail in the following sections.

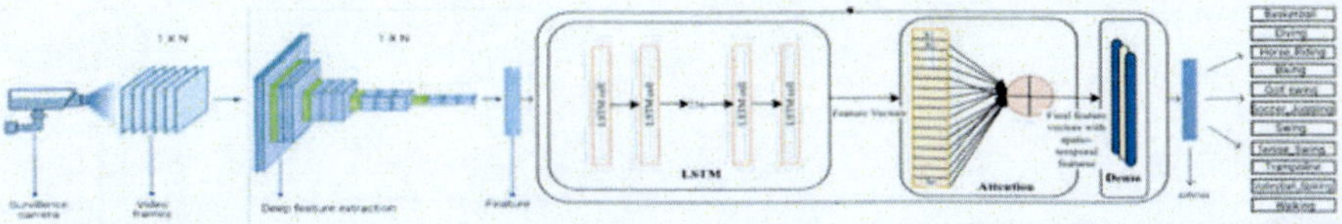

Fig. 1 Proposed framework to recognise human activity

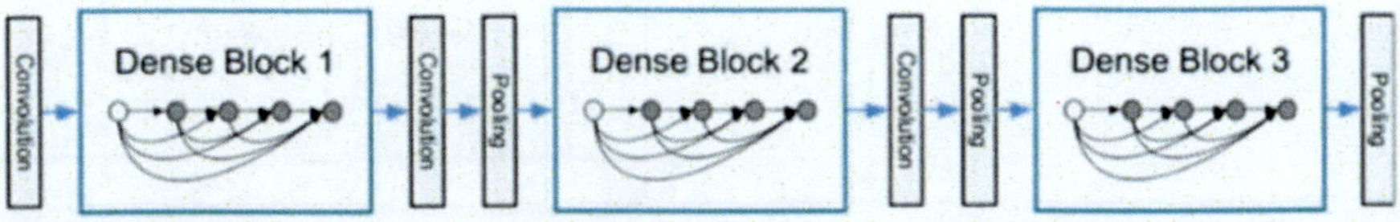

Fig. 2 A five-layer dense block with growth rate of k = 5.

3.1 DenseNet

DenseNet is a convolutional neural network that uses dense blocks to connect all levels directly, allowing for dense connections between layers. To retain the feed-forward nature, each layer gets extra inputs from all preceding levels and passes on its own feature maps to all subsequent layers (Fig. 2).

3.2 LSTM

Long short-term memory (LSTM) networks are a type of RNN that can learn long-term dependencies. This is currently frequently utilised and functions exceptionally effectively on a number of applications (Fig. 3).

The key to LSTMs is the cell state, which is represented by the horizontal line running through the top of the picture. It's quite easy for info to flow along it. The LSTM may erase or add information to the cell state, which is carefully controlled by gates. Gates are a type of device that selectively allows information to pass through. A sigmoid neural net layer and a pointwise multiplication mechanism make them up. An LSTM contains three of these gates to protect and regulate the cell state. The forget gate layer is used to determine whether or not specific data should be ignored. It examines h ((t-1)) and x t, and for each number in the cell state C ((t-1)), it returns a number between 0 and 1.

$$f_t = \sigma\left(W_t\left(h_{(t-1)} + x_t\right) + b_f\right) \tag{1}$$

where W_t and b_f are weight and bias at state t and f_t is output of forget gate layer which varies from 0 to 1.

Fig. 3 Architecture of LSTM

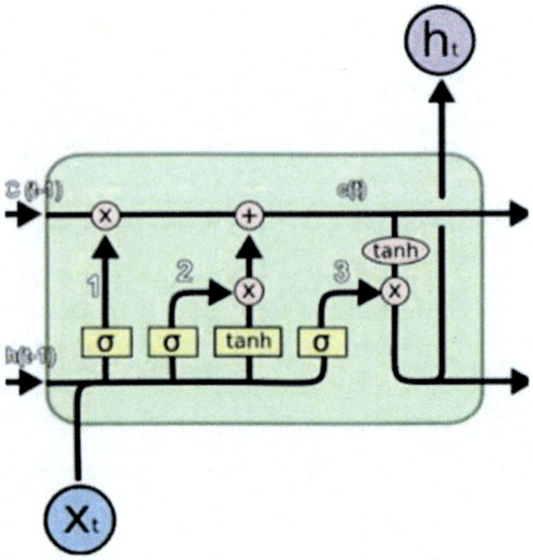

$$i_t = \sigma\left(W_i * \left[h_{(t-1)}, x_t\right] + b_i\right) \tag{2}$$

$$\acute{Ç}_t = \tan\mathrm{h}\left(W_c * \left[h_{(t-1)}, x_t\right] + b_c\right) \tag{3}$$

$$C_t = f_t * C_{t-1} + i_t * \acute{Ç}_t \tag{4}$$

$$O_t = \sigma\left(W_o * \left[h_{(t-1)}, x_t\right] + b_o\right) \tag{5}$$

$$h_t = O_t * \tanh\left(C_t\right) \tag{6}$$

The new cell state C_t depends on cell state, C_{t-1}, calculated by multiplying the forget gate layer with previous cell state C_{t-1} and adding $i_t * \acute{Ç}_t$. This is the new candidate value, scaled by how much we decided to update each state value. Finally, the output will be filtered and will be based on the cell state. First, a sigmoid layer is used to determine which bits of the cell state will be output. After passing through activation tanh, the cell state is multiplied by the output of the sigmoid gate.

3.3 Attention Mechanism

Although LSTM networks are good at capturing long-range dependency, they lack a mechanism for determining which parts of the input series are crucial for producing a more precise classification. The attention mechanism can be used to solve this issue. As seen in Fig. 1, the attention layer receives the h_1, h_2, $h_3 \ldots h_n$ vectors produced by the LSTM network as input, which are then encoded by the attention encoders into the information vectors x_1, x_2, $x_3 \ldots \ldots$, and x_n.The weighted sum of the encoder RNN output is used in this process to calculate the context vectors.

In order to calculate the context vectors c_1, c_2, c_3 . . . . ., and c_n, we use Eq. (7).

$$c_t = \sum_{t=0}^{n} a_t.x_t \tag{7}$$

Here, c_t stands for the encoded information vector and a_t for the attention score. Calculations for the attention scores are made using (8) and (9). In (9), the feed-forward network is denoted by the function F_{att}, the encoded information vector is denoted by x_t, and the prior cell state vector is represented by d_{t-1}.

$$softmax\,(a_t) = \frac{\exp\,(o_t)}{\sum_{t=0}^{n} \exp\,(o_t)} \tag{8}$$

$$o_t = F_{att}\,(x_t.d_{t-1}) \tag{9}$$

The context vector, c_t; the output of the prior time step, y_{t-1}; and the prior cell state vector, d_{t-1}, are all used to determine the output of this attention layer at every time step t.

4 Implementation

In this section, we evaluated the presented model's utility on datasets UCF11 [21] which outperformed state-of-the-art approaches. The testing findings are run on a laptop with an Intel Core i7-8550U eighth Generation Processor 4.0 GHz and 8 GB memory to confirm our approach. Overall accuracy, class-wise accuracy, confusion matrix, and training and testing loss are the evaluation criteria. Finally, we showed the comparison of the suggested model to the SOTA approaches.

Because of the variations in illumination, crowded backgrounds, and camera motions, UCF11 is a challenging dataset for recognising human actions. The UCF11 dataset contains 1600 movies in 11 different activity categories, such as riding, shooting, leaping, swimming, and so on, all of which are shot at 25 frames per second (fps). Figure 3 shows a bar chart representation of human activities from the UCF11 dataset (Fig. 4).

Table 1 shows that the proposed approach outperformed effective event models [22], motion trajectories [7], improved trajectory [23], and hierarchical clustering multi-task [5] for this dataset, with accuracy of 89.43%, 89.70%, 89.50%, 96.90%, and 97.9%, respectively.

Accuracy of the model is shown by Fig. 5a and training loss vs testing loss is shown by Fig. 5b. Figure 6 represents the class-wise accuracy of all activities in dataset UCF11, where all activity recognition accuracy is more than 96%.

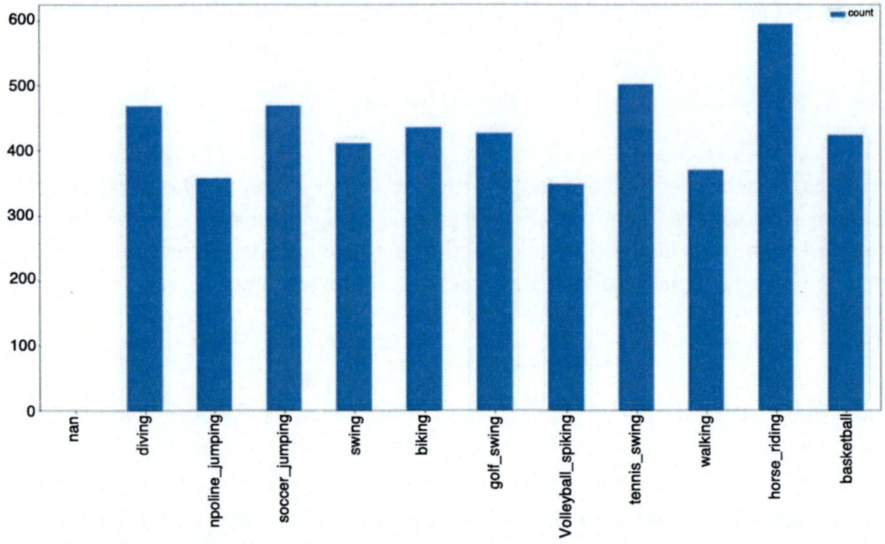

Fig. 4 Activity of UCF11

Table 1 Comparison with state-of-the-art approaches

Method	Accuracy (%)
Patel et al. [22]	89.43
Meng et al. [7]	89.70
Gharaee et al. [23]	89.50
Dai et al. [5]	96.90
Our proposed	**97.90**

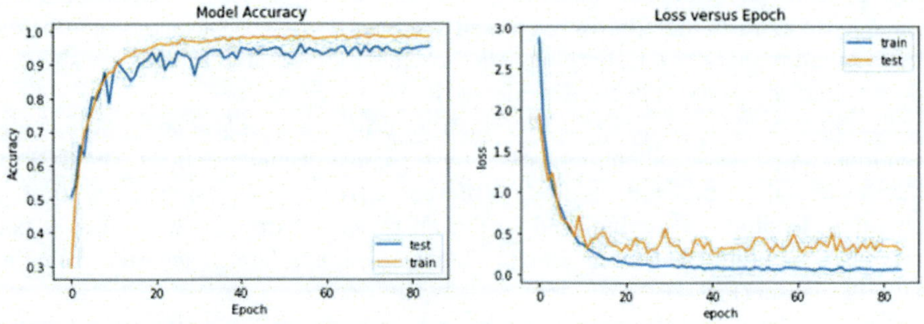

Fig. 5 (**a**) Accuracy of proposed model. (**b**): Train and Test loss plot of model

5 Conclusion

Spatiotemporal properties are critical for distinguishing different activities in surveillance video data, such as human activity detection. In this paper, we created an attention-based framework for recognising human activity that uses

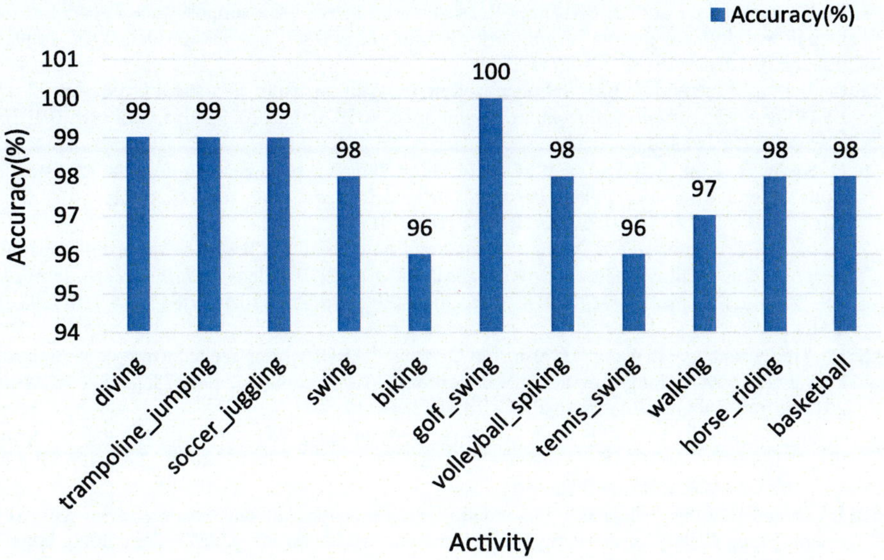

Fig. 6 Class-wise accuracy of UCF11

both temporal and spatial data from a sequence of frames. We used a pre-train DenseNet network to represent the high-level deep properties from the video frames. Furthermore, using these deep properties, the LSTM network was used to learn the temporal information. To help determine the temporal information in better depth, an attention layer was added to the LSTM, which enhanced performance at every level. To increase the categorisation performance of human activities in videos, SoftMax functions are used. On the benchmark datasets UCF11, we did comprehensive tests. The suggested system gains a recognition accuracy of 97.90%, representing a nearly 1% improvement over existing techniques.

References

1. N. Spolaôr, H. D. Lee, W. S. R. Takaki, L. A. Ensina, C. S. R. Coy, and F. C. Wu, "A systematic review on content-based video retrieval," *Eng. Appl. Artif. Intell.*, vol. 90, p. 103557, Apr. 2020, doi: https://doi.org/10.1016/J.ENGAPPAI.2020.103557.
2. A. Keshavarzian, S. Sharifian, and S. Seyedin, "Modified deep residual network architecture deployed on serverless framework of IoT platform based on human activity recognition application," *Futur. Gener. Comput. Syst.*, vol. 101, pp. 14–28, Dec. 2019, doi: https://doi.org/10.1016/J.FUTURE.2019.06.009.
3. A. Das Antar, M. Ahmed, and M. A. R. Ahad, "Challenges in sensor-based human activity recognition and a comparative analysis of benchmark datasets: A review," *2019 Jt. 8th Int. Conf. Informatics, Electron. Vision, ICIEV 2019 3rd Int. Conf. Imaging, Vis. Pattern Recognition, icIVPR 2019 with Int. Conf. Act. Behav. Comput. ABC 2019*, pp. 134–139, May 2019, https://doi.org/10.1109/ICIEV.2019.8858508.

4. R. Khemchandani and S. Sharma, "Robust least squares twin support vector machine for human activity recognition," *Appl. Soft Comput. J.*, vol. 47, pp. 33–46, Oct. 2016, https://doi.org/10.1016/J.ASOC.2016.05.025.
5. C. Dai, X. Liu, and J. Lai, "Human action recognition using two-stream attention based LSTM networks," *Appl. Soft Comput. J.*, vol. 86, p. 105820, Jan. 2020, https://doi.org/10.1016/J.ASOC.2019.105820.
6. H. Kwon, Y. Kim, J. S. Lee, and M. Cho, "First Person Action Recognition via Two-stream ConvNet with Long-term Fusion Pooling," *Pattern Recognit. Lett.*, vol. 112, pp. 161–167, Sep. 2018, https://doi.org/10.1016/J.PATREC.2018.07.011.
7. B. Meng, X. J. Liu, and X. Wang, "Human action recognition based on quaternion spatial-temporal convolutional neural network and LSTM in RGB videos," *Multimed. Tools Appl.*, vol. 77, no. 20, pp. 26901–26918, Oct. 2018, https://doi.org/10.1007/S11042-018-5893-9/TABLES/4.
8. M. Xin, H. Zhang, H. Wang, M. Sun, and D. Yuan, "ARCH: Adaptive recurrent-convolutional hybrid networks for long-term action recognition," *Neurocomputing*, vol. 178, pp. 87–102, Feb. 2016, https://doi.org/10.1016/J.NEUCOM.2015.09.112.
9. B. Saghafi and D. Rajan, "Human action recognition using Pose-based discriminant embedding," *Signal Process. Image Commun.*, vol. 27, no. 1, pp. 96–111, Jan. 2012, https://doi.org/10.1016/J.IMAGE.2011.05.002.
10. J. Lee and H. Jung, "TUHAD: Taekwondo Unit Technique Human Action Dataset with Key Frame-Based CNN Action Recognition," *Sensors*, vol. 20, no. 17, p. 4871, Aug. 2020, https://doi.org/10.3390/S20174871.
11. X. S. Wei, P. Wang, L. Liu, C. Shen, and J. Wu, "Piecewise Classifier Mappings: Learning Fine-Grained Learners for Novel Categories with Few Examples," *IEEE Trans. Image Process.*, vol. 28, no. 12, pp. 6116–6125, Dec. 2019, https://doi.org/10.1109/TIP.2019.2924811.
12. J. Schmidhuber, "Deep learning in neural networks: An overview," *Neural Networks*, vol. 61, pp. 85–117, Jan. 2015, https://doi.org/10.1016/J.NEUNET.2014.09.003.
13. T. M. Lee, J. C. Yoon, and I. K. Lee, "Motion Sickness Prediction in Stereoscopic Videos using 3D Convolutional Neural Networks," *IEEE Trans. Vis. Comput. Graph.*, vol. 25, no. 5, pp. 1919–1927, May 2019, https://doi.org/10.1109/TVCG.2019.2899186.
14. K. Simonyan and A. Zisserman, "Very Deep Convolutional Networks for Large-Scale Image Recognition," *3rd Int. Conf. Learn. Represent. ICLR 2015 - Conf. Track Proc.*, Sep. 2014, https://doi.org/10.48550/arxiv.1409.1556.
15. Z. Tu *et al.*, "Multi-stream CNN: Learning representations based on human-related regions for action recognition," *Pattern Recognit.*, vol. 79, pp. 32–43, Jul. 2018, https://doi.org/10.1016/J.PATCOG.2018.01.020.
16. H. Gammulle, S. Denman, ... S. S.-2017 I. W., and undefined 2017, "Two stream lstm: A deep fusion framework for human action recognition," ieeexplore.ieee.org, Accessed: May 13, 2022. [Online]. Available: https://ieeexplore.ieee.org/abstract/document/7926610/
17. X. He, K. Wang, H. Huang, T. Miyazaki, Y. Wang, and S. Guo, "Green Resource Allocation Based on Deep Reinforcement Learning in Content-Centric IoT," *IEEE Trans. Emerg. Top. Comput.*, vol. 8, no. 3, pp. 781–796, Jul. 2020, https://doi.org/10.1109/TETC.2018.2805718.
18. S. Kulkarni, S. Jadhav, and D. Adhikari, "A Survey on Human Group Activity Recognition by Analysing Person Action from Video Sequences Using Machine Learning Techniques," pp. 141–153, 2020, https://doi.org/10.1007/978-981-15-0994-0_9.
19. J. Wen, J. Yang, B. Jiang, H. Song, and H. Wang, "Big Data Driven Marine Environment Information Forecasting: A Time Series Prediction Network," *IEEE Trans. Fuzzy Syst.*, vol. 29, no. 1, pp. 4–18, Jan. 2021, https://doi.org/10.1109/TFUZZ.2020.3012393.
20. M. Ma, N. Marturi, Y. Li, A. Leonardis, and R. Stolkin, "Region-sequence based six-stream CNN features for general and fine-grained human action recognition in videos," *Pattern Recognit.*, vol. 76, pp. 506–521, Apr. 2018, https://doi.org/10.1016/J.PATCOG.2017.11.026.
21. J. Liu, J. Luo, ... M. S.-I. conference on computer vision and, and undefined 2009, "Recognizing realistic actions from videos 'in the wild,'" ieeexplore.ieee.org, Accessed: May 13, 2022. [Online]. Available: https://ieeexplore.ieee.org/abstract/document/5206744/

22. C. I. Patel, S. Garg, T. Zaveri, A. Banerjee, and R. Patel, "Human action recognition using fusion of features for unconstrained video sequences," *Comput. Electr. Eng.*, vol. 70, pp. 284–301, Aug. 2018, https://doi.org/10.1016/J.COMPELECENG.2016.06.004.
23. Z. Gharaee, P. Gärdenfors, and M. Johnsson, "First and second order dynamics in a hierarchical SOM system for action recognition," *Appl. Soft Comput.*, vol. 59, pp. 574–585, Oct. 2017, https://doi.org/10.1016/J.ASOC.2017.06.007.

Software Vulnerability Classification Using Learning Techniques

Birendra Kumar Verma , Ajay Kumar Yadav, and Vineeta Khemchandani

1 Introduction

Today, software's are diverse, complex, and an integral part of our lives. The best minds around the world are involved in software development, but even the best can make mistakes which could generate software vulnerabilities. These flaws or defects in the software construction can be exploited by attackers to obtain some privileges in the system. Thus, software vulnerabilities offer a possible entry point for attackers into the system. Despite growing research and increasing knowledge about vulnerabilities, we witness a growing tendency in the number of reported vulnerabilities. In 2021 alone, the overall number of new vulnerabilities increased to 20,161 compared to 18,375 in 2020 and 17,308 in 2019. The common vulnerability and exposure (CVE) observed a rise of around 58% more vulnerability registration from 2019 to 2020. The year 2021 witnessed an increase of 65% in vulnerabilities in third-party components as compared to the previous year. Software vulnerabilities have been scored by different organizations having their own methodologies. Several software vulnerability scoring systems have been implemented by several non-profitable organizations and vendors involved in system security over the past decades. They have limited in scope and have no interoperability between the systems. In this paper, we proposed a framework for the statistical and natural language processing-based classification of software vulnerabilities and their severity. In the statistical method, we applied standard preprocessing methods such as dealing with null

B. K. Verma (✉) · A. K. Yadav
Banasthali Vidyapith, Jaipur, Rajasthan, India
e-mail: birendraverma@jssaten.ac.in

V. Khemchandani
JSS Academy of Technical Education, Noida, Uttar Pradesh, India
e-mail: vkhemchandani@jssaten.ac.in

M. Mishra et al. (eds.), *Applications of Computational Intelligence in Management & Mathematics*, Springer Proceedings in Mathematics & Statistics 417,
https://doi.org/10.1007/978-3-031-25194-8_26

values, encoding of categorical data, etc., and in the second phase, we applied PCA for feature extraction. In the next phase, we split the data set into train and test data set, and in the last phase, we evaluated the performance of different classifiers and compare them. In the NLP-based technique, we applied the NLTK toolkit for preprocessing the textual data present in the description attribute which describes the particular vulnerability. For feature extraction we used term frequency metrics and inverse document frequency metrics. For feature selection we again applied PCA in NLP section. In the next phase, we applied the same methods as in statistical-based method and compared the performance of different classifiers. In the preprocessing phase of description textual data, we did tokenization, case changing, and lemmatization and removed URLs, punctuation symbols, stop words, numbers, and special symbols.

2 Previous Related Work

Karen Scarfone and Peter Mell [1] proposed and analyzed the effectiveness of CVSS version 2 and its deficiencies. Experiment was performed on both CVSS version 1 and CVSS version 2 using large datasets, and various characteristics of both the systems were also compared. Goals were met based on changes in the system, but some changes had a minor effect on scoring while making the system more complex [2]. Ju An Wang et al. [3] created a method to capture and utilize the fundamental characteristics of threat analysis, security mechanisms, threat analysis, etc. to measure the effects of vulnerabilities on systems. All vulnerabilities have been populated in OVM with various rules and knowledge discovery programs to provide an efficient and reliable security mechanism [4]. Qixue Xiao et al. in [5] studied vulnerabilities in deep learning frameworks such as torch and tensor flow. These frameworks contain heavy dependencies and are quite complex as compared to small code size of deep learning models [6]. Attackers can exploit these frameworks and cause DOS attacks that crash or hang the application or cause system compromise. Security of deep learning frameworks is improved in this paper. Haipeng Chen et al. in [7] proposed a system known as VEST (vulnerability exploit scoring and timing) which is used for beforehand or early prediction of when a vulnerability will be exploited. It predicts its attributes and score [8]. The framework used to implement this method is flask, and other deep learning could be used for better performance. Georgios Spanos and Lefteris Angelis [9] presented an advanced version of vulnerability characteristic assignment based on manual procedure. Text analysis and multi-target classification techniques were used to create a model and achieve this task. This model gives an estimated vulnerability score and gives characteristics of the vulnerabilities also [10]. The data set contains 99,091 records, and the results were able to achieve scores and prediction of vulnerability by a significant improvement. Random forest, boosting, and decision tree were also used to perform classification in this model [11]. Christian Frühwirth and Tomi Männistö [12] created a method that can be particularly used by practitioners who don't have access to large amounts of data. This allows them to improve the current systems with the help of using public datasets. They found that adding context information in a 720-vulnerability announcement greatly improved the priority analysis of vulnerability. Their research helped organizations in terms of security management, security investment, and various other processes [13]. Laurent Gallon [14] conducted a study to determine the effects of environmental metrics to CVSS scores from NVD. It focuses on variation in CVSS systems [15]. Various authors simulated possible combinations of vulnerabilities with environmental factors using predefined formulas. Shuguang Huang et al. in [16] presented vulnerability classification to solve the problem of taxonomy overlap in software vulnerability on NVD based on text clustering. COI (cluster overlap index) is used to calculate sample mean, bisecting means, and bathos clustering algorithms [17]. From 40,000 records only 45 vulnerabilities are selected according to the descriptor dominance index. These vulnerability taxonomies are used to transform their works from individual to vulnerability taxonomy research. Su Zhang et al. in [18] studied the process of applying data mining methods on NVD datasets to predict the time to next vulnerability for a particular software system. Experimentation on various features was done, and different algorithms were applied to examine the power of prediction of the NVD data. NVD data has a poor prediction capability. Reasons were given for the poor performance of the NVD data, and risk estimation is the method in which NVD dataset can be used as it gives better performance. Kai Shuang et al. [19] proposed a CDWE method also known as convolution-deconvolution word embedding, a method used to do end-to-end multi-fusion embedding that combines task-specific information and context-specific information. The efficiency and generalization ability of CDWE was demonstrated by applying it to NLP representative tasks [20]. CDWE models outperform and achieve state-of-the-art results on both text classification and machine translation [21]. Polysemous unaware problem was solved using text deconvolution saliency to verify the efficiency of CDWE. Tadas Baltrusaitis et al. [22] studied the recent improvements in the field of multimodal machine learning, and he presented it in a common taxonomy. This method identifies broader challenges that are faced by multimodal machine learning such as alignment, translation, fusion, representation, and co-learning apart from typical early and late fusion categorization. Jinfu Chen et al. [23] presented a framework for classification of vulnerability severity using TF-IGM also known as term frequency-inverse gravity moment and information gain feature selection based on five machine learning algorithms or applications containing 27,248 security vulnerabilities. The TF-IGM model is a great metric for vulnerability classification, the feature selection process is highly significant, and it improves classification of vulnerabilities [24]. Jia Liu et al. [25] proposed methodologies of urban big data fusion on the basis of deep learning into mainly three categories: DL-output-based fusion, DL-input-based fusion, and DL-double-stage-based fusion. The methods mentioned above use deep learning to learn presentation of features from multi-source big data [26]. Further, a detailed explanation of the three categories was done and various examples were explained. Misbah Anjum et al. [27] proposed a method to determine the relative vulnerability of sensitive network protection services to attacks. Fuzzy best worst method is used to prioritize the identified vulnerabilities. SQL injection (SQLI), information gain (IG), and gain of privileges are the highly severe vulnerabilities that need to be resolved at the earliest possibility. Jukka Ruohonen [28] studied the time delays occurring in CVEs (common vulnerability and exposures) in the NVD datasets using CVSS scores [29]. Based on results of 80,000 archived vulnerabilities, it was concluded that (i) the time delays are not statistically influenced by CVSS content time delays CVSS content [30], (ii) but on the other hand, they are strongly affected by a decreasing annual trend. Linear regression is used to answer three questions: (1) time delay is correlated with CVSS content, (2) but the correlation is spurious, the decreasing annual trend affecting the time delays (RQ1), and (3) it also makes the effects of CVSS content negligible.

3 Methodology

In this section we describe the proposed framework for classifying the software vulnerabilities and their severity using two techniques; one is statistical, and another one is NLP based. Figure 1 represents the overall architecture of the proposed model. In the statistical method, we applied standard preprocessing methods such as dealing with null values, encoding of categorical data, etc., and in the second phase, we applied PCA for feature extraction. In the next phase, we split the data set into train and test data set, and in the last phase, we evaluated the performance of different classifiers and compare them. In the NLP-based technique, we applied the NLTK toolkit for preprocessing the textual data present in the description attribute which describes the particular vulnerability. For feature extraction we used term frequency metrics and inverse document frequency metrics. For feature selection we again applied PCA in NLP section. In the next phase, we applied the same methods as in statistical-based method and compare the performance of different classifiers. In the preprocessing phase of description textual data, we did tokenization, case changing, lemmatization and removed URLs, punctuation symbols, stop words, numbers, and special symbols.

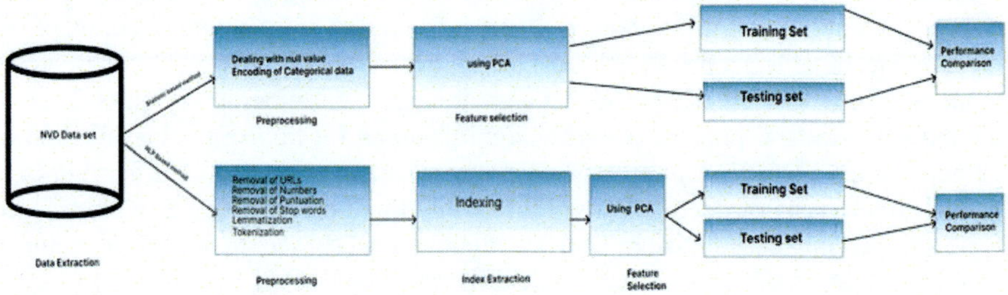

Fig. 1 Model architecture

3.1 *Dataset Description*

Dataset has been taken from the website (https://nvd.nist.gov/vuln/data-feeds) in the form of json file (Table 1). Python script has been used to convert it into csv file. Csv file has been uploaded into pandas data frame. Initially it has 159,979 rows and 32 columns. Every row has unique common vulnerability exposures such as CVE-2021-0202, and all 32 rows have unique categorical features such as attack complexity, attack vector, availability impact, confidentiality impact, and so on. In the preprocessing step, every categorical feature has been changed into separate columns such as attack complexity changed into attack complexity high and attack complexity low because it has two categories, high and low [31]. Similarly, every mandatory feature has been changed into their category. Finally, 56 features have been used before applying PCA (Fig. 1).

3.2 *Tokenization*

In the tokenization process we divide the textual data into small words or phrases such as "Attacker," "Vulnerability," "Allows," "Via," "remote," and "arbitrary."

3.3 *Case Changing*

In this process textual data changes either in lower or upper case we changed it in lower case as ["attacker," "vulnerability," "allows," "via," "remote," "arbitrary"]. This process is important because we are giving more importance to the meaning of the words so that words having the same meaning should have the same case.

3.4 *Removal of Punctuation Symbols, Stop Words, Numbers, and Special Symbols*

In this step to get the refined information from the description data, we have to remove the URLs, numbers, punctuation symbols, hyperlinks, numbers, special characters, and stop words. In addition to this, WordNetLemmatizer is used to get the refined information on description data set. We used NLTK toolkit that contains several functions for NLP.

Table 1 Dataset description (https://nvd.nist.gov/vuln/data-feeds)

CVE	Vector String	Exploitability Score	Impact Score	Base Score	Exploitability Score.1	Impact Score	Base Score	Severity	Description
CVE-2021-0109	CVSS:3.1/AV:L/AC:L/PR:L/UI:N/S:U/C:H/I:H/A:H	1.8	5.9	7.8	3.9	6.4	4.6	MEDIUM	Insecure inherited permissions for the Intel(R)...
CVE-2021-0202	CVSS:3.1/AV:N/AC:L/PR:N/UI:N/S:U/C:N/I:N/A:H	3.9	3.6	7.5	10	2.9	5	MEDIUM	On Juniper Networks MX Series and EX9200 Serie...
CVE-2021-0203	CVSS:3.1/AV:N/AC:L/PR:N/UI:N/S:C/C:N/I:N/A:H	3.9	4	8.6	8.6	2.9	4.3	MEDIUM	On Juniper Networks EX and QFX5K Series platfo...
CVE-2021-0204	CVSS:3.1/AV:L/AC:L/PR:L/UI:N/S:U/C:H/I:H/A:H	1.8	5.9	7.8	3.9	10	7.2	HIGH	A sensitive information disclosure vulnerabili...
CVE-2021-0205	CVSS:3.1/AV:N/AC:L/PR:N/UI:N/S:C/C:N/I:N/A:L	3.9	1.4	5.8	8.6	2.9	4.3	MEDIUM	When the "Intrusion Detection Service" (IDS) f...
Total	**159,979.56**								

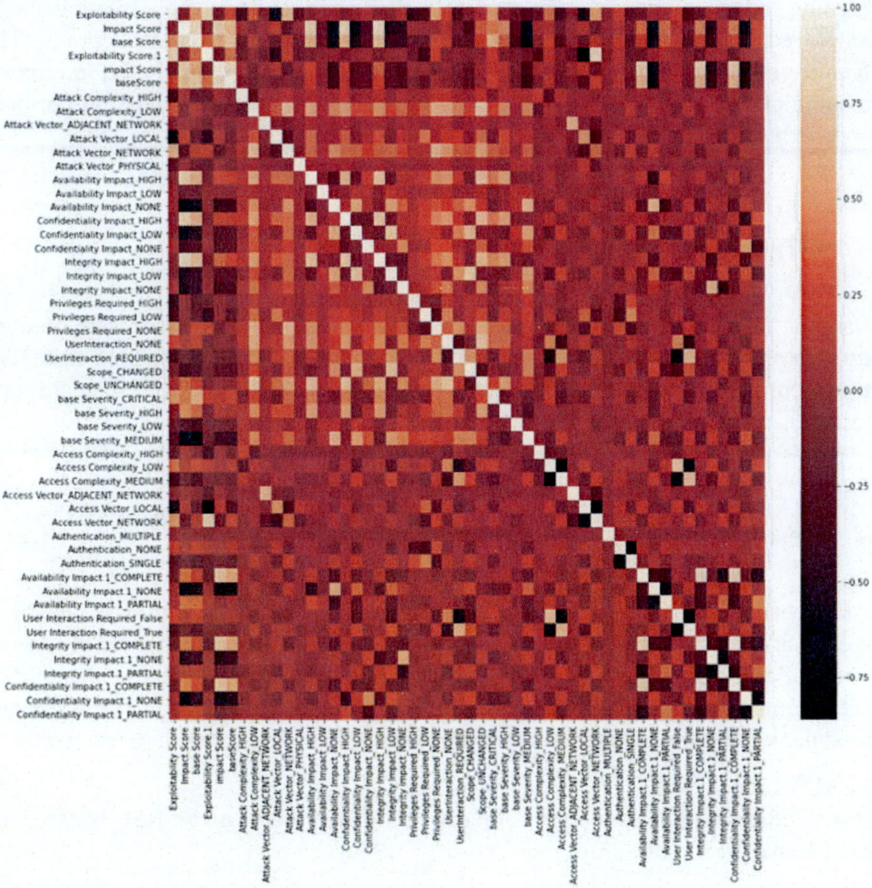

Fig. 2 Correlation matrix (heatmap)

3.5 Word Lemmatizer

This is the process in which the context of the word is preserved, for example, "allows," "allowing," etc. are converted to "allow"; this could be archived by NLTK WordNetLemmatizer function, for example, ["attack," "vulnerability," "allow," "via," "remote," "arbitrary"]. It improves the quality of feeds to the network.

3.6 Feature Selection and Extraction Computation and Target Label Fixation

In statistical method we applied PCA for feature extraction; out of 51, 16 features have been selected as independent features. We applied these 16 features into training and testing phase of our model. In NLP-based technique our training model employed 90,647 features computed by TfidfVectorizer across the corpus. The feature vector is computed using vector count from the text present in the description column of vulnerability data set in the form of token. The target label of our both approaches is same as base score a continuous value.

3.7 *Train and Test Methods*

Random forest regressor, decision tree regressor, and linear regressor were used for training and testing on features derived from preprocessing phase on both cases. For the performance evaluation of our model, we used ten cross-fold and standard train test split methods.

4 Results and Discussion

All experiments have been carried out on *Jupyter notebook*, a web-based environment that consists of codes and data. It provides a flexible environment to the user in terms of configuration and supports the wide range of data flow in the area of data science, machine learning, and scientific computing. JupyterLab is extensible and modular write plugins that add new components and integrate with existing ones.

Scikit Learn
It is a Python module used for machine learning, distributed under BSD license, and built above the top of Scipy.

Pandas
It is an open-source library and distributed under BSD license; it provides high-performance computation. For programing language like Python, data analysis and data structure uses are easy to use and learn.

Natural Language Processing Kit: NLTK (Natural Language Toolkit)
It is a suite that has programs and libraries for language processing. It consists of packages that are capable of human language translation to machines and answers with appropriate response.

Table 2 presents the classification results using linear regression, decision tree regressor, and random forest regressor on 80% of training and 20% of testing (Figs. 3 and 4). The performance of the classifiers is assessed based on explained variance, r2, MAE, MSE, and RMSE values (Table 2).

Table 2 The results based on 20% of testing and 80% of training

Model	Training					Testing				
Linear regression	Explained variance: 0.7756	r2: 0.7756	MAE: 0.2445	MSE: 0.0796	RMSE: 0.2821	Explained variance: 0.7775	r2: 0.7775	MAE: 0.2448	MSE: 0.0796	RMSE: 0.2821
Decision tree regressor	Explained variance: 0.9688	r2: 0.9688	MAE: 0.133	MSE: 0.0418	RMSE: 0.2044	Explained variance: 0.9588	r2: 0.978	MAE: 0.134	MSE: 0.0408	RMSE: 0.2103
Random forest regressor	Explained variance: 0.9995	r2: 0.9995	MAE: 0.002	MSE: 0.0006	RMSE: 0.0246	Explained variance: 0.9973	r2: 0.9973	MAE: 0.0053	MSE: 0.0036	RMSE: 0.0598

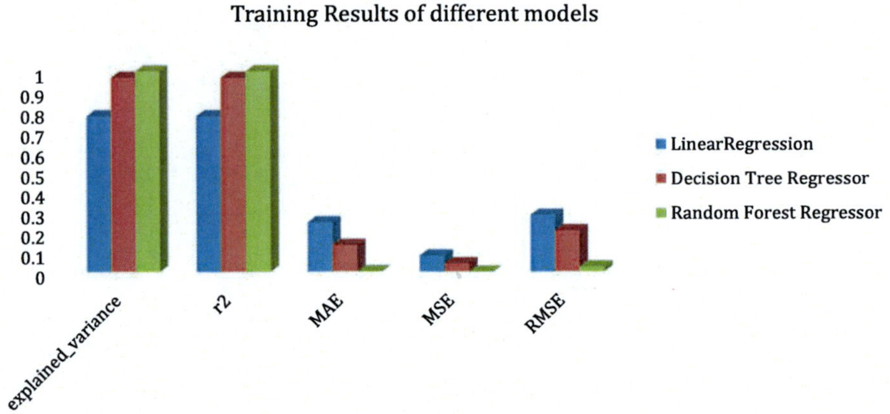

Fig. 3 Training results of linear, decision tree, and random forest regressor

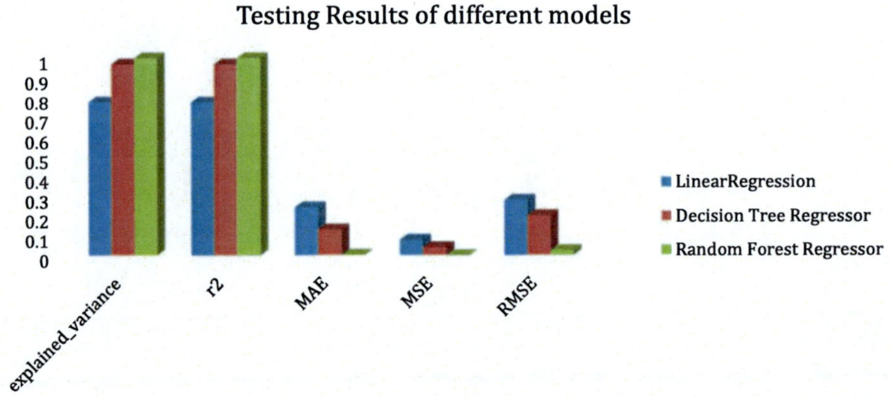

Fig. 4 Testing results of linear, decision tree, and random forest regressor

5 Conclusion

In the proposed work, we have implemented an automatic software vulnerability classification and scoring based on two methods; one is statistical, and another is NLP. In NLP inverse document frequency and term frequency have been used. Different machine learning algorithms have been used for classification and testing phases. The performance of the classifiers has been assessed based on explained variance (0.9973), *r*-squared (0.9973), mean absolute error (0.0052), mean squared error (0.0035), and root mean squared error values (0.0594). Best learning algorithms have been identified by using GridSearchCV on 159,979 different CVEs. Random forest showed the best score as shown by the experiment as 99.99% as depicted in Table 3.

Table 3 Best model selection using GridSearchCV

Model	Best score	Best parameters
Random forest	0.999905	{'max_depth': 13, 'min_samples_leaf': 1, 'min_samples_split': 2}
Decision tree	0.999848	{'max_depth': 49, 'min_samples_leaf': 1, 'min_samples_split': 7}
Linear regression	0.999514	{}

References

1. K. Scarfone and P. Mell, "An analysis of CVSS version 2 vulnerability scoring," *2009 3rd Int.* Symp. *Empir. Softw. Eng. Meas. ESEM 2009*, pp. 516–525, 2009, https://doi.org/10.1109/ESEM.2009.5314220.
2. S. H. Houmb, V. N. L. Franqueira, and E. A. Engum, "Quantifying security risk level from CVSS estimates of frequency and impact," in *Journal of Systems and Software*, Sep. 2010, vol. 83, no. 9, pp. 1622–1634, https://doi.org/10.1016/j.jss.2009.08.023.
3. J. A. Wang, M. M. Guo, and J. Camargo, "An ontological approach to computer system security," *Inf. Secur. J.*, vol. 19, no. 2, pp. 61–73, 2010, https://doi.org/10.1080/19393550903404902.
4. R. Syed, "Cybersecurity vulnerability management: A conceptual ontology and cyber intelligence alert system," *Inf. Manag.*, vol. 57, no. 6, Sep. 2020, https://doi.org/10.1016/j.im.2020.103334.
5. Q. Xiao, K. Li, D. Zhang, and W. Xu, "Security risks in deep learning implementations," *Proc. - 2018 IEEE Symp. Secur. Priv. Work. SPW 2018*, pp. 123–128, 2018, https://doi.org/10.1109/SPW.2018.00027.
6. S. Chatterjee and S. Thekdi, "An iterative learning and inference approach to managing dynamic cyber vulnerabilities of complex systems," *Reliab. Eng. Syst. Saf.*, vol. 193, Jan. 2020, https://doi.org/10.1016/j.ress.2019.106664.
7. H. Chen, J. Liu, R. Liu, N. Park, and V. S. Subrahmanian, "VEST: A system for vulnerability exploit scoring & timing," *IJCAI Int. Jt. Conf. Artif. Intell.*, vol. 2019-Augus, pp. 6503–6505, 2019, https://doi.org/10.24963/ijcai.2019/937.
8. L. Allodi and F. Massacci, "A preliminary analysis of vulnerability scores for attacks in wild: The EKITS and SYM datasets," *Proc. ACM Conf. Comput. Commun. Secur.*, pp. 17–24, 2012, doi:https://doi.org/10.1145/2382416.2382427.
9. G. Spanos and L. Angelis, "Impact Metrics of Security Vulnerabilities: Analysis and Weighing," *Inf. Secur. J.*, vol. 24, no. 1–3, pp. 57–71, Jul. 2015, https://doi.org/10.1080/19393555.2015.1051675.
10. L. Allodi and F. Massacci, "Comparing vulnerability severity and exploits using case-control studies," *ACM Trans. Inf. Syst. Secur.*, vol. 17, no. 1, 2014, https://doi.org/10.1145/2630069.
11. W. Zheng *et al.*, "The impact factors on the performance of machine learning-based vulnerability detection: A comparative study," *J. Syst. Softw.*, vol. 168, Oct. 2020, https://doi.org/10.1016/j.jss.2020.110659.
12. C. Frühwirth and T. Männistö, "Improving CVSS-based vulnerability prioritization and response with context information," *2009 3rd Int. Symp. Empir. Softw. Eng. Meas. ESEM 2009*, pp. 535–544, 2009, https://doi.org/10.1109/ESEM.2009.5314230.
13. M. Abedin, S. Nessa, E. Al-Shaer, and L. Khan, "Vulnerability analysis For evaluating quality of protection of security policies," *Proc. 2nd ACM Work. Qual. Prot. QoP'06. Co-located with 13th ACM Conf. Comput. Commun. Secur. CCS'06*, pp. 49–52, 2006, https://doi.org/10.1145/1179494.1179505.
14. L. Gallon, "On the impact of environmental metrics on CVSS scores," *Proc. - Soc. 2010 2nd IEEE Int. Conf. Soc. Comput. PASSAT 2010 2nd IEEE Int. Conf. Privacy, Secur. Risk Trust*, pp. 987–992, 2010, https://doi.org/10.1109/SocialCom.2010.146.

15. P. Mell and K. Scarfone, "Improving the common vulnerability scoring system," *IET Inf. Secur.*, vol. 1, no. 3, pp. 119–127, 2007, https://doi.org/10.1049/iet-ifs:20060055.
16. S. Huang, H. Tang, M. Zhang, and J. Tian, "Text clustering on national vulnerability database," *2010 2nd Int. Conf. Comput. Eng. Appl. ICCEA 2010*, vol. 2, pp. 295–299, 2010, https://doi.org/10.1109/ICCEA.2010.209.
17. H. Holm, M. Ekstedt, and D. Andersson, "Empirical analysis of system-level vulnerability metrics through actual attacks," *IEEE Trans. Dependable Secur. Comput.*, vol. 9, no. 6, pp. 825–837, 2012, https://doi.org/10.1109/TDSC.2012.66.
18. S. Zhang, X. Ou, and D. Caragea, "Predicting Cyber Risks through National Vulnerability Database," *Inf. Secur. J.*, vol. 24, no. 4–6, pp. 194–206, Dec. 2015, https://doi.org/10.1080/19393555.2015.1111961.
19. K. Shuang, Z. Zhang, J. Loo, and S. Su, "Convolution–deconvolution word embedding: An end-to-end multi-prototype fusion embedding method for natural language processing," *Inf. Fusion*, vol. 53, no. June 2019, pp. 112–122, 2020, https://doi.org/10.1016/j.inffus.2019.06.009.
20. J. A. Morente-Molinera, X. Wu, A. Morfeq, R. Al-Hmouz, and E. Herrera-Viedma, "A novel multi-criteria group decision-making method for heterogeneous and dynamic contexts using multi-granular fuzzy linguistic modelling and consensus measures," *Inf. Fusion*, vol. 53, no. June 2019, pp. 240–250, 2020, https://doi.org/10.1016/j.inffus.2019.06.028.
21. D. Wijayasekara, M. Manic, and M. Mcqueen, "Vulnerability Identification and Classification Via Text Mining Bug Databases."
22. T. Baltrusaitis, C. Ahuja, and L. P. Morency, "Multimodal Machine Learning: A Survey and Taxonomy," *IEEE Trans. Pattern Anal. Mach. Intell.*, vol. 41, no. 2, pp. 423–443, 2019, https://doi.org/10.1109/TPAMI.2018.2798607.
23. J. Chen, P. K. Kudjo, S. Mensah, S. A. Brown, and G. Akorfu, "An automatic software vulnerability classification framework using term frequency-inverse gravity moment and feature selection," *J. Syst. Softw.*, vol. 167, p. 110616, 2020, https://doi.org/10.1016/j.jss.2020.110616.
24. V. E. Balas, *Advances in Intelligent Systems and Computing 634 Soft Computing Applications*, vol. 2, no. Sofa. 2016.
25. J. Liu, T. Li, P. Xie, S. Du, F. Teng, and X. Yang, "Urban big data fusion based on deep learning: An overview," *Inf. Fusion*, vol. 53, no. June 2019, pp. 123–133, 2020, https://doi.org/10.1016/j.inffus.2019.06.016.
26. G. Spanos and L. Angelis, "A multi-target approach to estimate software vulnerability characteristics and severity scores," *J. Syst. Softw.*, vol. 146, pp. 152–166, Dec. 2018, https://doi.org/10.1016/j.jss.2018.09.039.
27. M. Anjum, P. K. Kapur, V. Agarwal, and S. K. Khatri, "A Framework for Prioritizing Software Vulnerabilities Using Fuzzy Best-Worst Method," *ICRITO 2020 - IEEE 8th Int. Conf. Reliab. Infocom Technol. Optim. (Trends Futur. Dir.*, pp. 311–316, 2020, https://doi.org/10.1109/ICRITO48877.2020.9197854.
28. J. Ruohonen, "A look at the time delays in CVSS vulnerability scoring," *Appl. Comput. Informatics*, vol. 15, no. 2, pp. 129–135, Jul. 2019, https://doi.org/10.1016/j.aci.2017.12.002.
29. H. Holm and K. K. Afridi, "An expert-based investigation of the Common Vulnerability Scoring System," *Comput. Secur.*, vol. 53, pp. 18–30, Jun. 2015, https://doi.org/10.1016/j.cose.2015.04.012.
30. P. Johnson, R. Lagerstrom, M. Ekstedt, and U. Franke, "Can the common vulnerability scoring system be trusted? A Bayesian analysis," *IEEE Trans. Dependable Secur. Comput.*, vol. 15, no. 6, pp. 1002–1015, Nov. 2018, https://doi.org/10.1109/TDSC.2016.2644614.
31. L. Castrejon, Y. Aytar, C. Vondrick, H. Pirsiavash, and A. Torralba, "Learning Aligned Cross-Modal Representations from Weakly Aligned Data," *Proc. IEEE Comput. Soc. Conf. Comput. Vis. Pattern Recognit.*, vol. 2016-Decem, pp. 2940–2949, 2016, https://doi.org/10.1109/CVPR.2016.321.

Service Recovery: Past, Present, and Future

Akuthota Sankar Rao and Damodar Suar

1 Introduction

A provider's actions convert an aggrieved customer into a satisfied customer after failure [4]. Service providers need to take proper steps when customers experience service failure [6]. Finally, it is a thought-out, planned process of returning aggrieved or dissatisfied customers to a state of being satisfied with service provider actions [1]. From the 1990s onwards, researchers and market practitioners started to find out the impact of service recovery on customers using different variables and methodologies in various types of sectors.

The perceived justice framework has been widely used to address customers' complaints in the service sector [12, 15]. Foremost in this framework, distributive justice concentrates on the compensation of the complainers [13]. It refers to the tangible outcomes in terms of replacement, refund, and apology [8]. Procedural justice refers to the system procedures, policies, and tools used to address customers' complaints [9], e.g. refund policies and time required for refund. Interactional justice refers to the extent of the fair treatment of complaints. It means the employee's manner in which the recovery is operationalised [7, 16]. It includes employee courtesy, politeness, and adequacy of language level.

To the authors' best knowledge, the article is the most comprehensive study on service recovery, covering all possible data sources from the last 30 years of literature. The study looked at the scientific publications available between 1990 and 2020 in the top leading journals. The article improves the body of literature, particularly in terms of theories. The study begins by extensively analysing the

A. S. Rao (✉) · D. Suar
Department of Humanities and Social Sciences, Indian Institute of Technology Kharagpur, Kharagpur, India
e-mail: ds@hss.iitkgp.ernet.in

M. Mishra et al. (eds.), *Applications of Computational Intelligence in Management & Mathematics*, Springer Proceedings in Mathematics & Statistics 417,
https://doi.org/10.1007/978-3-031-25194-8_27

fictitious methodologies and processes that were empirically examined in the earlier studies on service recovery. This analysis makes the study distinct from earlier studies. Next, the study tries to fill the gap by investigating and classifying service recovery articles to feature the significant contributions that have been made as previous articles in journals. Last, the study highlights the possible shortcomings and provides the future directions in the service recovery stream.

2 Research Domain

Service recovery is a prime and key function for managers to keep their customers satisfied. So as to have a fruitful service recovery, customers' expectations should be recognised and met. Customers will be loyal and hold a high emotional association with service providers when they can provide their services or address customer complaints properly after failures. Loyal customers mean highly visiting the store and positively speaking about the provider to their friends.

Even though recent notable meta-analytical studies [3, 10, 11] provide new directions to research in the service recovery stream, there are a few streams that are not explored where required to give high consideration. Even though most of the studies have been distributed on service recovery stream, a few studies are limited in scope by way of scholars repeatedly examining this conceptual framework or their measures in various country contexts, sector contexts, etc. Considering the significance of the service recovery and existing articles, the current study has embraced an entirely different way to deal with the service recovery stream. The present comprehensive review gives a large group of service recovery data including the theories and variables that have been used in previous articles, which give and get other than the fundamental key variables of interest and methodologies considered in the study analysis and the articles identified with administration recuperation distributed between the years 1990 and 2020. Despite the fact that the study found a review article in the field of service recovery [5], it still requires more information in this area. With the anticipated results of the current study, we are trying to overcome the limitations of the available literature.

3 Method

Data for this study was collected from the Scopus online databases to find the scientific articles published on service recovery from 1994 to 2018. By using this time frame, the study intends to investigate the comprehensive and systematic review on service recovery. Nonetheless, the current study has investigated the citation analysis with the counts as of May 1, 2021. The study pursued a "systematic review process" [14] of empirical articles, which were taken at the time of literature search through online by using the keywords "service recovery" and "complaint handling" so that the study could concentrate on pure empirical service recovery articles with theoretical and methodological applications. It will help us highlight the service recovery theories and dimensions that are frequently used.

We restricted ourselves to finding publications from the journals included either in Scopus or in the Australian Business Deans Council (ABDC) official list of journals in Australia. Next, the study cross-checked these selected articles through the journal website to see which ones fall under the category of business, management, and accounting. Our final collection had around 1490 articles. Notwithstanding, we chose to confine the articles with keywords like "service recovery" and "complaint handling" incorporated into the title/abstract/keywords to reduce the sample bias. Finally, we selected 480 publications.

From this point on, we constructed a database of publications with the data gathered, with the plan to record the rise in service recovery publications over a time period of 25 years. The study categorised this time frame into five sections. The study found 57 publications were published during 1994–1998, 142 publications during 1999–2003, 242 publications in 2004–08, 451 publications in 2009–2013, and 598 articles during 2014–2018. This data shows that interest in service recovery research has significantly grown over time. As a result, the quantity of service recovery publications also increased substantially during the time period 2009–2018 compared to the remaining four time periods. More interestingly, the quantity of service recovery articles published during the time period of 2009–2018 is almost two-thirds of the total, compared to the past 15-year time period. Figure 1 shows the number of articles published per year.

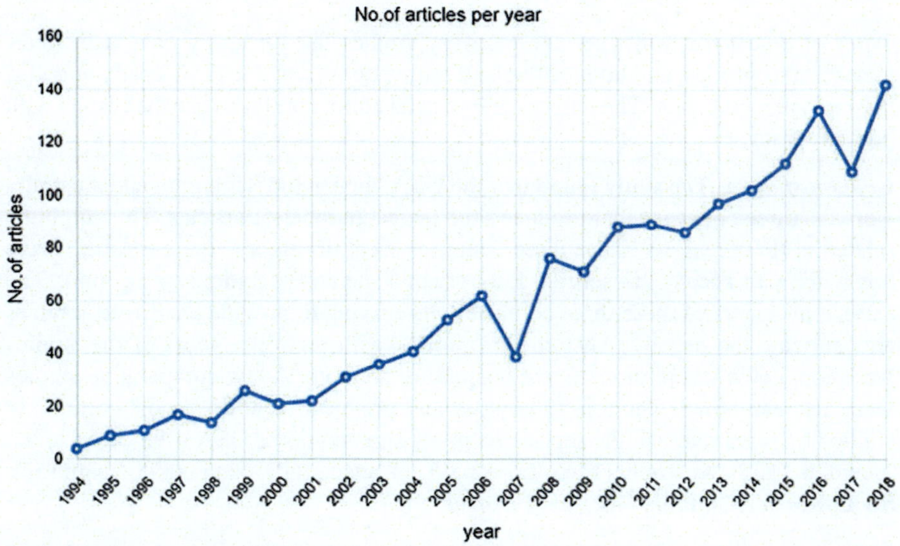

Fig. 1 Number of publications, year-wise on SR field

3.1 Journals

The study systematically examines service recovery research, analysing 1490 articles published during 1994–2018. We found that the maximum number of articles (67) was published by the *Journal of Services Marketing*. It indicates that this journal is the oldest operating business journal compared to other journals. The *Journal of Service Research* published the second highest number (54). These were followed by the *Journal of Business Research* (38), *Managing Service Quality* (35), *Service Industries Journal* (32), *Journal of Cleaner Production* (31), *International Journal of Contemporary Hospitality Management* (29), *Journal of the Academy of Marketing Science* (27), *European Journal of Marketing* (23), *International Journal of Hospitality Management* (22), and *Journal of Travel and Tourism Marketing* (22).

Other journals that have published the most articles on service recovery were the *Journal of Hospitality and Tourism Research* (21), *International Journal of Service Industry Management* (20), *Service Business* (19), *Journal of Retailing and Consumer Services* (17), *International Journal of Bank Marketing* (15), and *Journal of Air Transport Management* (14). Based on this, we observed that all the top journals were key outlets for service recovery research publications.

4 Results

4.1 Review of Methods

Types of data used Most service recovery studies used primary data with interviews, focus groups, and both offline and online questionnaires. It is costly and also hard to collect the data. The authors and journal editors prefer to use the results with high accuracy.

Country studied The study found that the USA, the United Kingdom, and Australia were the three countries that produced the largest number of articles. This might be related to three reasons: these three countries were developed; the customers were not willing to accept the service failures; and their expectations were high. The service providers of these three countries were trying to provide error-free services and mitigate the negative customer responses with recovery. Most of the studies were conducted where tourism and hospitality are high. A few comparative studies were also examined. The top 20 popular and produced countries most commonly studied by researchers in the service recovery stream are the USA (498), the United Kingdom (155), Australia (126), Taiwan (70), Canada (60), China (57), Spain (53), South Korea (52), India (46), and Germany (45).

Statistical methods The commonly used method was structural equation modelling (SEM) with the support of SPSS software. Most of the studies used descriptive statistics, regression analysis, ANOVA, F-test, etc. Although a few of those studies did not adopt the regression models, the study revealed that the majority of the publications used F-tests and ANOVA tests. So, the study identified the dependent variables (DVs), which are most commonly used in service recovery research, and then independent variables (IVs).

Variables studied Most of the studies used a regression framework. So, we identified the antecedent and consequence variables of service recovery that are mostly used in the service recovery stream.

Antecedent variable The most commonly used antecedent variables for service recovery were service failures, complaints, failure severity, employee proficiency, failure type, and speed of recovery.

Consequence variables Customer emotions, affection, commitment, recovery satisfaction, overall satisfaction, word of mouth, repurchase intention, revisit intention, customer loyalty, corporate image, and relationship quality.

Citation analysis Citation analysis, also provided along with the analysis of methods, theories, and journals, is a method to find the most popular and influential articles in the service recovery stream [2, 14]. During the last three decades, several popular and influential service recovery articles have been published. The study used the data available from Scopus to assign the ranking of the articles, which mainly depends on the number of total citations received. Because the total citations are proof of the article's degree of influence and popularity in the scientific field, the study calculated the average weighted citation score (AWCS) of the article because it shows the citations of the article per year. The ten most popular and influential articles are shown in Table 1 in descending order based on the article's total citations and AWCS.

5 Directions and Future Research

The current study explains different types of theoretical and methodological approaches to research on service recovery. Despite the fact that extensive research is taking place on service recovery, noteworthy shortcomings can be found in the past articles. The results show only 45% of the articles inspected build, review, develop, or contribute to theory; the lion's share tends to depict the findings from the quantitative articles. In addition, the focal point of most of the articles was the outcome variables of service recovery in a single field, not by any comparative studies with existing variables, in spite of the fact that there are some eminent and prominent articles dependent on the direct impact of service recovery on customer responses such as loyalty, satisfaction, trust, and positive word of mouth. This is on the grounds that most analysts have depended upon primary data.

In total, customer satisfaction and loyalty are found to be the most widely used dependent constructs in service recovery research. In spite of the fact that researchers have presented new theories and variables, it seems various other

Table 1 Most popular and influential articles

Rank	Authors	Title	Year	Source title	Cited by	AWCS
1	Smith A.K., Bolton R.N., Wagner J.	A model of customer satisfaction with service encounters involving failure and recovery	1999	*Journal of Marketing Research*	1236	58.86
2	Tax S.S., Brown S.W., Chandrashekaran M.	Customer evaluations of service complaint experiences: Implications for relationship marketing	1998	*Journal of Marketing*	1222	55.55
3	Hart C.W., Heskett J.L., Sasser Jr. W.E.	The profitable art of service recovery	1990	*Harvard Business Review*	765	25.50
4	Anderson E.W., Fornell C., Rust R.T.	Customer satisfaction, productivity, and profitability: Differences between goods and services	1997	*Marketing Science*	629	27.35
5	Maxham III J.G., Netemeyer R.G.	A longitudinal study of complaining customers' evaluations of multiple service failures and recovery efforts	2002	*Journal of Marketing*	498	27.67
6	Maxham III J.G., Netemeyer R.G.	Modelling customer perceptions of complaint handling over time: The effects of perceived justice on satisfaction and intent	2002	*Journal of Retailing*	432	24.00
7	Maxham III J.G.	Service recovery's influence on consumer satisfaction, positive word-of-mouth, and purchase intentions	2001	*Journal of Business Research*	406	21.37
8	Goodwin C., Ross I.	Consumer responses to service failures: Influence of procedural and interactional fairness perceptions	1992	*Journal of Business Research*	402	14.36
9	Smith A.K., Bolton R.N.	The effect of customers' emotional responses to service failures on their recovery effort evaluations and satisfaction judgements	2002	*Journal of the Academy of Marketing Science*	401	22.28
10	Hess Jr. R.L., Ganesan S., Klein N.M.	Service failure and recovery: The impact of relationship factors on customer satisfaction	2003	*Journal of the Academy of Marketing Science*	398	23.41

theories have not been considered at a similar level of consideration as justice theory and social exchange theory and related research. The study found that few authors have discussed customer overall satisfaction (recovery satisfaction and initial satisfaction). Very few meta-analytical studies have been conducted in the service recovery stream. There is a scope for taking both satisfactions as dependent variables and conducting meta-analytical studies. The study also found most articles concentrated on a single stream in a particular area. It would make sense to conduct a comparative study either for a group of nations or two nations with similar or dissimilar features and compare the findings. There is a loop to fulfil the theory-based rationale for the selection of countries and to justify the methodological approach. The future study can examine the antecedents and outcomes of service recovery simultaneously and develop a service recovery scale for individual sectors. The new research can expand the previous research to explore the cost of expenses for service recovery.

In total, customer satisfaction and loyalty are found to be the most widely used dependent constructs in service recovery research. In spite of the fact that researchers have presented new theories and variables, it seems various other theories have not been considered at a similar level of consideration as justice theory and social exchange theory and related research. The study found that few authors have discussed customer overall satisfaction (recovery satisfaction and initial satisfaction). Few meta-analytical studies have been conducted in the service recovery stream. There is a scope for taking both satisfactions as dependent variables and conducting meta-analytical studies. The study also found most articles concentrated on a single stream in a particular area. It would make sense to conduct a comparative study either for a group of nations or two nations with similar or dissimilar features and compare the findings. There is a loop to fulfil the theory-based rationale for the selection of countries and to justify the methodological approach. The future study can examine the antecedents and outcomes of service recovery simultaneously and develop a service recovery scale for individual sectors. Future research can expand on the previous research to explore the cost of expenses for service recovery.

6 Conclusion

This study has identified the most popular and commonly used variables, statistical tools, country analysis, and journals on service recovery. The systematic review of service recovery shows a complete picture of the field that can help future scholars. The study also acknowledged the high quality of scholarly work that was found based on the citations. Since 2008, the scholarly interest and popularity of the stream have significantly grown during service recovery. As a result, it is a high integration of various theoretical viewpoints. It corroborates the fact that the importance of the service recovery process has increased to achieve a competitive advantage in the market.

References

1. Bell, C. R., & Zemke, R. (1990). The performing art of service management. *Management Review, 79*(7), 42–46.
2. Canabal, A., & White, G. O. (2008). Entry mode research, past and future. *International Business Review, 17*(3), 267–284.
3. Gelbrich, K., & Roschk, H. (2011). A meta-analysis of organizational complaint handling and customer responses. *Journal of Service Research, 14*(1), 24–43.
4. James A. Fitzsimmons and Mona J. Fitzsimmons: Service management: operations, strategy, information technology, 2011, 7th edition, p 136.
5. Krishna, A., Dangayach, G. S., & Jain, R. (2011). Service recovery: Literature review and research issues. *Journal of Service Science Research, 3*(1), 71.
6. Kau, A. and Loh, E.W. (2006), "*The effects of service recovery on consumer satisfaction: a comparison between complainants and non-complainants*", *Journal of Services Marketing*, Vol. 20 No. 2, pp. 101–11.
7. Kuo, Y. F., & Wu, C. M. (2012). Satisfaction and post-purchase intentions with service recovery of online shopping websites: Perspectives on perceived justice and emotions. *International Journal of Information Management, 32*(2), 127–138.
8. Maxham III, J. G., & Netemeyer, R. G. (2002). A longitudinal study of complaining customers' evaluations of multiple service failures and recovery efforts. *Journal of marketing, 66*(4), 57–71.
9. Nikbin, D., Ismail, I., Marimuthu, M., & Armesh, H. (2012). Perceived justice in service recovery and switching intention: Evidence from Malaysian mobile telecommunication industry. *Management Research Review, 35*(3/4), 309–325.
10. Orsingher, C., Valentini, S., & de Angelis, M. (2010). A meta-analysis of satisfaction with complaint handling in services. *Journal of the Academy of Marketing Science, 38*(2), 169–186.
11. Roschk, H., & Gelbrich, K. (2014). Identifying appropriate compensation types for service failures: A meta-analytic and experimental analysis. *Journal of Service Research, 17*(2), 195–211.
12. Smith, A. K., Bolton, R.N., & Wagner, J. (1999). A model of customer satisfaction with service encounters involving failure and recovery. *Journal of Marketing Research, 36*(3), 356–372.
13. Sparks, B. A., & McColl-Kennedy, J. R. (2001). Justice strategy options for increased customer satisfaction in a services recovery setting. *Journal of Business Research, 54*(3), 209–218.
14. Terjesen, S., Hessels, J., & Li, D. (2016). Comparative international entrepreneurship: A review and research agenda. *Journal of Management, 42*(1), 299–344.
15. Tax. S. S., Brown, S.W., & Chandrashekaran, M. (1998). Customer evaluations of service complaint experiences: Implications for relationship marketing. *Journal of Marketing, 62*(2), 60–76.
16. Voorhees, C. M., & Brady, M. K. (2005). A service perspective on the drivers of complaint intentions. *Journal of Service Research, 8*(2), 192–204

An In-depth Accessibility Analysis of Top Online Shopping Websites

Nishtha Kesswani and Sanjay Kumar

1 Introduction

According to the World Health Organization, one billion people, or 15% of the total population in the World, include persons with disabilities [1]. Significant challenges that a person with special needs faces less or no accessibility include the physical environment, transportation, lack of assistive technologies, and prejudice in society. While on the one hand, there is a proliferation of Information and Communication enabled technologies, on the other hand, the accessibility of these technologies to persons with special needs is overlooked. Out of the total population of persons with disabilities, the ones having a visual impairment are a large number. We can imagine the problems that a person with visual impairment would face while accessing a computer without a screen reader. Even if a screen reader is there and the documents are not designed to be accessible, the problems would escalate.

It is worthwhile to note that 52% of the adults with disabilities are literate [2]. Thus, we need just a step forward to enhance the capabilities of these people who are accessing the World through the Internet by creating more and more accessible websites. Web accessibility allows persons with disabilities to access websites, tools, and technologies [3]. And thus, persons with auditory, cognitive, neurological, speech, and visual or physical disabilities are facilitated for accessing the Web.

The World Wide Web Consortium (W3C) has proposed Web Content Accessibility Guidelines (WCAG), Authoring Tool Accessibility Guidelines (ATAG), User-Agent Accessibility Guidelines (UAAG), Accessible Rich Internet Applications (WAI-ARIA) standards to control the accessibility of websites, web authoring tools, web browsers, and dynamic web applications. WCAG is the universally

N. Kesswani (✉) · S. Kumar
Department of Computer Science, Central University of Rajasthan, Ajmer, Rajasthan, India
e-mail: nishtha@curaj.ac.in

M. Mishra et al. (eds.), *Applications of Computational Intelligence in Management & Mathematics*, Springer Proceedings in Mathematics & Statistics 417,
https://doi.org/10.1007/978-3-031-25194-8_28

adopted standard for making websites accessible for all. The accessibility principles POUR specified under the WCAG check if the websites are Perceivable, Operable, Understandable, and Robust [4]. As in the current evolving retail scenarios, more and more consumers across all sections of society are using online shopping websites for their daily buying needs, it is important that these guidelines must be followed while designing the websites.

Notably, several researchers have tried to investigate accessibility in diverse domains. However, none of these studies has focused on the accessibility of online shopping websites. Physical access to stores and shopping malls is in a poor state. It would be good if persons with disabilities can access the stores online. In order to facilitate online access to the websites, it is essential that the websites are designed to be accessible. The accessibility analysis of the online shopping websites is required so that online shopping websites can also target persons with disabilities, and an essential section of society is not overlooked. Thus, this can emerge as a win–win situation for both the customers who can access the store online and the businesses that can increase sales by targeting all segments of society. The present research tries to fill this gap and contends that the shopping decisions of persons with disabilities are primarily affected by how accessible online shopping websites are. Addressing these open issues would not only advance scholarly knowledge on online shopping website accessibility but also facilitate the persons with special needs.

The major contributions of this chapter are as follows:

1. The accessibility of the top 40 online shopping websites of the World has been analyzed based on WCAG. Major accessibility problems with these websites have been identified.
2. The websites have been classified based on accessibility into high, medium, and low accessibility.
3. Karl Pearson's coefficient of correlation among different types of problems faced by users was calculated to check if there is a correlation between different kinds of accessibility problems.
4. The correlation between the number of visits on a website and the number of problems faced by users was also calculated using Spearman's rank coefficient of correlation.
5. Kendall's W Test was performed to check if there is an agreement between the ranking of the websites and the number of accessibility problems.

2 Preliminaries

The Web Content Accessibility Guidelines [4] define four principles and three conformance levels, as shown in Fig. 1. The checkpoints in the WCAG guidelines are divided into three conformance levels. Priority 1 includes those checkpoints that a web content developer must satisfy as these are essential requirements for some persons to access the Web. Priority 2 consists of those checkpoints that should be

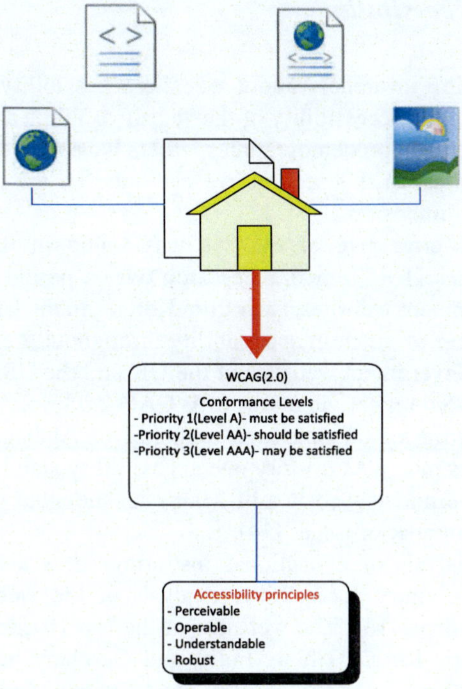

Fig. 1 WCAG principles and conformance levels

satisfied as these may impose significant barriers on access. Priority 3 includes those checkpoints that a developer may address to improve accessibility.

3 Related Work

3.1 Online Shopping

In many countries, more than 80% of the Internet users shop online [5]. Due to the customer-centered approach, there is existing research on the consumer experience with digital services [6], customer satisfaction, and its impact on the sales volume [7]. Apart from this, inhibitors for mobile purchasing have also been investigated in the literature [8]. Attributes such as accessibility, security, flexibility, fault recovery, and interaction affect customer satisfaction in online shopping [9]. Along with these factors, online shopping websites must be designed to be accessible for all so as to overcome existing barriers. This research tries to explore the unaddressed area of online shopping website accessibility.

3.2 Website Accessibility

Due to the increasing awareness about website accessibility, several researchers have tried to analyze the accessibility of the websites. While some researchers have focused their study based on demographics, others have studied the accessibility of different types of sites. In this section, we elaborate on the research done by the researchers in both dimensions.

An accessibility error rate of 69.38% was found in the Kyrgyz Republic Government websites [10]. Temporal research over a period of time from 2004–2007 of four Government websites, each from Korea and the United States, revealed that there was an improvement in accessibility throughout the study [11]. Analysis of the Traffic and Government websites of the UK and the USA over a period from 1999 to 2012 indicated higher violations of WCAG 2.0 level A [12].

Accessibility testing of 60 state-level websites of Alabama [13] revealed that 19% of the websites met WAI priority one accessibility standard. When compared to the Federal Government websites of Malaysia, the state government websites indicated a higher number of issues [14].

Accessibility analysis of educational institutions has also been done in the literature. A cross-country accessibility analysis of the websites of educational institutes has been done [15]. The websites of the top 10 educational institutes in the United Kingdom, Russia, China, India, and Germany have been explored in the research. And it was found that most of the websites of educational institutions lagged far behind in accessibility.

Gonçalves et al. [16] had conducted an accessibility test of Portuguese enterprise websites. It was found that the accessibility of the websites that were evaluated was significantly worse. They have also prepared an improvement proposal that suggests the recommendations for the improvement of these websites.

Some researchers have also done accessibility analysis across other domains. Accessibility analysis of library websites has been done in the literature [17]. The researchers have done accessibility testing with the users and checked it against the technical accessibility audit. The results indicated that the two did not match. And it was found that library websites were not accessible to screen reader users. Another study of library website accessibility has been done by Billingham [18]. The researcher has shown how the website of the Edith Cowan University library was improved after testing its accessibility against the WCAG guidelines.

Accessibility of official tourism websites of 210 countries included in the World Tourism Organization Report has also been investigated [19]. As per the WCAG 2.0 guidelines, the primary focus is whether the websites were navigable, adaptable, compatible, and provided text alternatives. Rau et al. [20] have done a temporal study of web accessibility in China from 2009 to 2013. They evaluated 38 popular websites in 2009 and fifty websites in 2013 based on WCAG 1.0 standard using the automated tool HERA. And the results indicated that none of the websites was able to fulfill even Priority 1 for the minimum requirement of website accessibility. The study also suggests that e-government websites have shown significant improvement during these four years.

An analysis of 30 Australian Business to Consumer websites indicated that there is less focus on the accessibility of the websites [21]. Other factors such as the speed, navigability, and content of corporate websites have also been analyzed in the literature. The research indicates that the navigability of the websites makes users more comfortable[22]. The online shopping experience of fast food mobile commerce has also been explored [23]. The website quality and website equity have been investigated to demonstrate the buying habits of Chinese customers. While several studies are focusing on mobile and online shopping, there is a dire need to address the issue of website accessibility.

The existing designs can be improved upon only when the designs are evaluated, and the shortfalls in the existing designs are identified. The current piece of work tries to investigate the extent to which online shopping websites comply with the WCAG guidelines. This would help in building an inclusive society. Though the accessibility of the websites has been analyzed across diverse domains such as education, tourism, e-government, and library websites, online shopping website accessibility is unexplored. The current research tries to fill this research gap.

4 Research Methods

4.1 Data Collection and Research Tools

In the current research, an accessibility analysis of the home pages of the top online shopping websites of the World was done. The home pages of the websites have been analyzed since the inaccessibility of the home page will stop users from accessing other parts of the website [24].

The top 40 online shopping websites ranked on the basis of the number of visitors were taken from Alexa website ranking [25]. Alexa ranks websites based on daily time spent on the site, daily page views per visit, and total sites linking to the website.

Popular accessibility testing tools TAW [3] and AChecker [26] have been used for the accessibility analysis. This was followed by manual analysis of the accessibility. The websites have been analyzed on the basis of WCAG 2.0 guidelines. Though the WCAG 2.1 guidelines are there, these have not been taken into consideration as these are currently not supported by automated tools.

Another important component of accessibility is checking the color contrast so that persons having color deficits are able to access the websites. The color contrast of the background and foreground colors of the websites was checked using CheckMyColors [27].

4.2 Data Analysis and Procedure

After the accessibility analysis of the websites on the basis of WCAG 2.0, the websites have been classified into three classes: High, medium, and low accessibility. For the purpose of classification, the technique used by Ismail and Kuppusamy [28] has been used. The average accessibility of all the websites has been calculated using Eq. 1.

$$\sigma = \sum_{i=1}^{N} P_i / N \tag{1}$$

where N indicates the total number of websites and P_i indicates the total number of problems associated with a website i.

The median range of accessibility is calculated using Eq. 2.

$$\rho = \sigma \pm \lambda \tag{2}$$

where $\lambda = 0.25 * \sigma$ indicates the offset.

5 Results

The results of the study have been divided into the following sub-sections:

1. The top 40 online shopping websites have been analyzed against the WCAG 2.0 accessibility guidelines.
2. The websites have been classified into 3 classes: high, medium, and low accessibility.
3. A correlation analysis has been performed to investigate if there is any correlation between one problem and another problem related to accessibility. Another correlation analysis was performed to check if there is a correlation between the number of visits of a website and its accessibility. Kendall's W Test was performed to check if there is any correlation between the rank of the websites and their accessibility.

5.1 Accessibility Analysis

The results of accessibility analysis on level AA on TAW are shown in Table 1. The results indicate that out of the four principles, Robust contributes maximum toward the total number of problems followed by Perceivable, Operable, and Understandable. A robust website should support different kinds of user agents

Table 1 Principle-wise total number of problems reported by TAW

Number of websites	Perceivable	Operable	Understandable	Robust	Total problems
40	1415	716	254	3102	5471

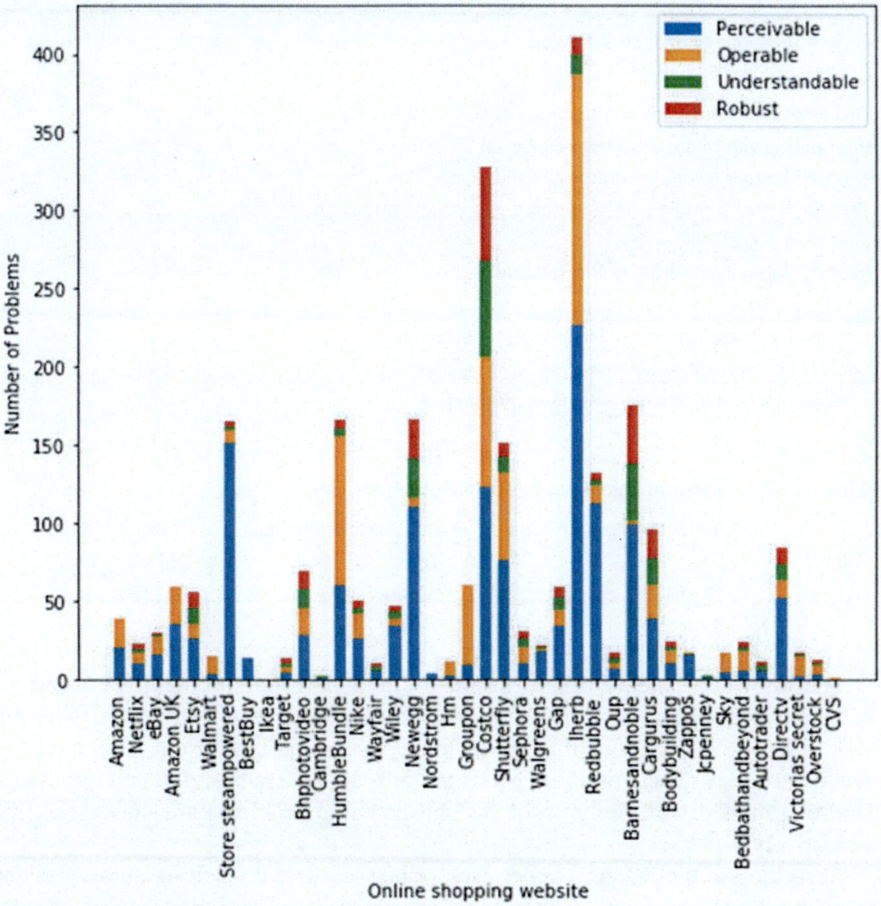

Fig. 2 Accessibility problems in different websites

including assistive technologies. The results indicate that the least emphasis is on making the websites accessible to different kinds of user agents including assistive technologies. And in this era when users are accessing the Web through different agents, it is important that the websites should be accessible with all kinds of user agents.

The number of accessibility problems pertaining to POUR principles of WCAG for all the websites that have been analyzed is shown in Fig. 2. It is evident from the figure that the website of iHerb has the highest number of problems, while the website of Ikea is the best in terms of accessibility.

Table 2 Total number of problems reported by AChecker

Number of websites	Known problems	Likely problems	Potential problems
40	698	46	12,158

Table 3 Percentage of sites with violations

Success criteria	Percentage
Text alternative exists for non-text content	25
Information and relationships are conveyed	25
Text and images have a contrast ratio of at least 4.5:1	12.5
Text can be resized up to 200% without loss of content or functionality	20
Keyboard accessibility	5
Identify link purpose from link text alone	15
Headings and labels define purpose	12.5
Language of page can be programmatically determined	5
Labels and instructions are provided for user input	40
Parsing web page elements is as per requirements	25

Table 4 Total number of color contrast-related problems

	Luminosity contrast ratio	Brightness difference	Color difference
Total	10,533	10,481	10,899
Average	319.181	317.606	330.272

The known problems, likely problems, and potential problems reported by AChecker are tabulated in Table 2. The results indicate that these 40 websites have at least 698 problems that are definitely accessibility barriers. This indicates that on average, each of these websites has 17 accessibility-related problems. The total numbers of likely problems and potential problems are 46 and 12,158, respectively.

According to the WCAG, under each principle, there are certain guidelines and success criteria. The detailed analysis of the percentage of violations under different success criteria is shown in Table 3. The results indicate that for 40% of the websites, labels and instructions are missing for user inputs. Due to this, users may have difficulty in accessing the websites. Another major problem is that there is a lack of text alternatives for non-text content that makes it difficult to read the pages with the help of a screen reader.

Color contrast plays an important role when it comes to persons having color deficits. The results indicate that the websites have on average more than 300 color-related problems. The total and an average number of failures on different parameters are tabulated in Table 4.

5.2 Website Classification

The classification of the websites was done on the basis of accessibility data collected from the automated tools and manual analysis. The total number of problems was $\sum_{i=1}^{N} P_i = 16280$, and for N = 40 websites, the average accessibility problems come out to be 407. With the offset value of 101.75(0.25 × 407), the median range is [508.75..305.25].

Based on these values, the websites were classified into high, medium, and low accessibility. The classification results are shown in Fig. 3. The color codes green, yellow, and red correspond to the high, medium, and low accessibility of the websites. The classification results indicate that 50% of the websites have high accessibility, 20% have medium, and 30% have low accessibility.

Online_shopping_website	Domain Name	Total Problems	Accessibility
Amazon	(http://www.amazon.com)	720	Low
Netflix	(http://www.netflix.com)	163	High
eBay	(http://www.ebay.com)	1136	Low
Amazon UK	(http://www.amazon.co.uk)	202	High
Etsy	(http://www.etsy.com)	849	Low
Walmart	(http://www.walmart.com)	360.5	Medium
Store steampowered	(http://www.store.steampowered.com)	125	High
BestBuy	(http://www.bestbuy.com)	39.5	High
Ikea	(http://www.ikea.com)	62.5	High
Target	(http://www.target.com)	353.5	Medium
Bhphotovideo	(http://www.bhphotovideo.com)	336	Medium
Cambridge	(http://www. Cambridge.org)	3	High
HumbleBundle	(http://www.humblebundle.com)	238.5	High
Nike	(http://www.nike.com)	246.5	High
Wayfair	(http://www.wayfair.com)	125	High
Wiley	(http://www.wiley.com)	100	High
Newegg	(http://www.newegg.com)	1368.5	Low
Nordstrom	(http://www. Nordstrom.com)	98.5	High
Hm	(http://www.hm.com)	7.5	High
Groupon	(http://www.groupon.com)	224.5	High
Costco	(http://www.costco.com)	793.5	Low
Shutterfly	(www.shutterfly.com)	305	Medium
Sephora	(www.sephora.com)	225	High
Walgreens	(www.wallgreens.com)	309	Medium
Gap	(www.gap.com)	335.5	Medium
Iherb	(www.iherb.com)	845.5	Low
Redbubble	(www.redbubble.com)	527	Low
Oup	(www.oup.com)	62.5	High
Barnesandnoble	(www.barnesandnoble.com)	635.5	Low
Cargurus	(www.cargurus.com)	910	Low
Bodybuilding	(www.bodybuilding.com)	94	High
Zappos	(www.zappos.com)	773.5	Low
Jcpenney	(www.jcpenney.com)	355.5	Medium
Sky	(www.sky.com)	430.5	Medium
Bedbathandbeyond	(www.bedbathandbeyond.com)	473.5	Low
Autotrader	(www.autotrader.com)	12.5	High
Directv	(www.directv.com)	148.5	High
Victorias secret	(www.victoriassecret.com)	498.5	Low
Overstock	(www.overstock.com)	70.5	High
CVS	(www.cvs.com)	26.5	High

Fig. 3 Classification of websites based on accessibility

5.3 Correlation Analysis

In this section, correlation analysis has been done to study correlation between different types of accessibility problems. Hypothesis has been formulated, and the correlation analysis has been done.

H1: Particular type of problems related to a website has association with other types of problems related to that website.

Karl Pearson's coefficient [29] of correlation was calculated to find out the strength of association among different types of problems faced by the users while accessing the websites. The correlation between different types of problems is shown in Table 5.

It was found that the number of problems related to the website being perceivable and the number of problems related to websites being operable were highly correlated and strongly associated with each other. It means that when websites are perceivable, they are also operable. On the other hand, when websites are not perceivable, they are also not operable.

It is also observed that there is very little association between the number of problems related to the website being perceivable and the number of problems related to the robustness of the websites. So the problems related to the robustness of the websites are not related to problems related to the website being perceivable. It shows not all types of accessibility problems arise together while visiting a website.

H2: The number of visits made by users on a website has associated with the number of problems faced by users while accessing that website.

Spearman's rank correlation [30] between the number of visits made by users and the number of problems faced by the user was calculated. The number of visits to each site was taken from Alexa. It was found that there is not enough association between the number of visits made by the users on the particular website and the number of problems faced by him related to a particular parameter. So, visits to a website do not have any association with a number of problems faced by users while accessing the website. The Spearman's rank correlation is shown in Table 6.

H3: There is agreement between ranking of websites on the basis of the number of problems faced by users while accessing them.

Kendall's W test [31] is a non-parametric test that is used for determining the degree of association among several (K) sets of the ranking of N cases. In the case

Table 5 Karl Pearson's coefficient of correlation among different types of problems faced by users

Problems related to the website being	Perceivable	Operable	Understandable	Robust
Perceivable	1.000			
Operable	0.662	1.000		
Understandable	0.545	0.345	1.000	
Robust	0.112	0.207	0.466	1.000

Table 6 Spearman's rank coefficient of correlation between the number of visits on a website and the number of problems faced by users related to a particular parameter

	Rank based on the number of problems faced by users on a particular parameter			
	Perceivable	Operable	Understandable	Robust
Rank based on the number of visits on a particular website	0.158	0.092	−0.103	0.188

Table 7 Kendall's W test

K	10
M	4
T	150
W	0.437008
R	0.249344
Chi-square	15.73228
Df	9
P value	0.072687

of two sets of ranking, Spearman's coefficient of correlation is used, but Kendall's coefficient of concordance (W) is considered appropriate when association among three or more sets of ranking has to be sought. The coefficient of concordance (W) is an index of divergence of the actual agreement shown in data from the perfect agreement.

From test statistics of Table 7, it is found that there is no concordance among the ranking of websites on the basis of different types of problems that are faced by the users (P value is .0072687 at 5 percent level of significance). The value of Kendall's W is 0.4370008, which shows that there is not enough agreement between the rankings of websites on the basis of problems faced by users while accessing them. The value of W can be between 0 and 1, and moving toward one reflects that there is enough agreement among rank assigned on any object. In this case, W is 0.4370008, which is not good enough to say an agreement between the ranking of different websites on the basis of problems faced by users. The value of the coefficient of correlation, which is 0.249344, is also very low and insignificant. Hence, it can be said users face different types of problems while accessing different websites even if the websites have a high ranking.

6 Recommendations for Improvement

In this section, recommendations for improvement in the accessibility of online shopping websites are discussed.

As evident from Table 3, 25% of the websites in the study have missing text alternatives. This problem can be solved by adding text alternatives to non-text content such as images. For instance, using the alt attribute of img tag would make the images more accessible as the screen reader can read the alternative text mentioned in the alt attribute. Another major issue is to identify the purpose of the link from the text that can be seen in 15% of the websites that have been analyzed. This issue can be resolved by adding text to the anchor tag or adding a title attribute to it.

If these recommendations are followed, this would facilitate in attracting consumers with disabilities to online shopping websites. This would help in building a win–win situation for both the online shopping websites and persons with disabilities.

7 Discussion

Both human and automated evaluation indicated that over 50% of the websites had medium or low accessibility. The outcomes of the current research have three-fold contributions:

1. *Academic contribution*
   Although diverse aspects of online shopping such as the ease of use of mobile applications [32], privacy [33], and the role of social media marketing [34] have been explored in the literature, the area of online shopping website accessibility has not been explored much—a potential research gap addressed in this chapter. The outcomes of the current research can aid in future advancements in the area of online shopping website accessibility. This can further be extended by developing tools and techniques for better accessibility. The study provides advancements in the field of online shopping and retail business.
2. *Managerial decisions for online shopping industry*
   Consumers' reactions to price discounts in online shopping [35] and gender-based comparison of online and offline shopping [36] have been mentioned in the literature. Another important perspective for increasing the customer base for online shopping websites can be increasing the accessibility of the websites. The results of this chapter provide deep insight into online shopping website accessibility. The correlation results of the current study indicate that there is a positive correlation between the accessibility problems related to Perceivable and Operable guidelines. Thus, similar kinds of problems can be addressed together, and this can help businesses in addressing a larger and more diverse range of customers.
3. *Contributions to the society*
   Persons with disabilities face problems in shopping malls due to physical barriers [37]. Due to physical barriers, there is a tendency for online shopping among persons with disabilities [38]. The outcomes of the current research would help in building an inclusive society providing access to all. With a large population using online shopping websites, the outcomes of this study would help in facilitating the less privileged ones. Accessibility of a website would be

a major deciding factor in whether persons with disabilities would opt for online shopping or not.

8 Conclusion

Website accessibility is an important aspect that should be taken into consideration while designing online shopping websites. In this chapter, the classification of the websites has been done based on the accessibility of the websites. The results indicate that 50% of the websites have low or medium accessibility. Correlation analysis indicates that the problems related to the accessibility of the websites are not correlated to the visits made by the users on these websites. Also, there is a positive correlation between the accessibility problems related to Perceivable and Operable guidelines. And not much correlation exists between problems related to Perceivable and Robust principles.

Conflict of Interest Statement
On behalf of all authors, the corresponding author states that there is no conflict of interest.

References

1. WHO, "World report on disability." http://www.who.int/disabilities/world_report/2011/report/en/, 2011.
2. UNESCO, "Education and disability: Analysis of data from 49 countries." http://uis.unesco.org/en/news/education-and-disability-analysis-data-49-countries, 2018.
3. "Test de accesibilidad web." https://www.tawdis.net/#, 2020.
4. WCAG, "Web Content Accessibility Guidelines (WCAG) 2.0." https://www.w3.org/TR/WCAG20/, 2020.
5. "The UNCTAD B2C E commerce index 2019." https://unctad.org/en/PublicationsLibrary/tn_unctad_ict4d14_en.pdf, 2019.
6. A. A. Shaikh, M. D. Alharthi, and H. O. Alamoudi, "Examining key drivers of consumer experience with (non-financial) digital services—an exploratory study," *Journal of Retailing and Consumer Services*, vol. 55, p. 102073, 2020.
7. L. Nicod, S. Llosa, and D. Bowen, "Customer proactive training vs customer reactive training in retail store settings: Effects on script proficiency, customer satisfaction, and sales volume," *Journal of Retailing and Consumer Services*, vol. 55, p. 102069, 2020.
8. S. Sohn and M. Groß, "Understanding the inhibitors to consumer mobile purchasing intentions," *Journal of Retailing and Consumer Services*, vol. 55, p. 102129, 2020.
9. G. Tontini, "Identifying opportunities for improvement in online shopping sites," *Journal of Retailing and Consumer Services*, vol. 31, pp. 228–238, 2016.
10. R. Ismailova, "Web site accessibility, usability and security: a survey of government web sites in Kyrgyz Republic," *Universal Access in the Information Society*, vol. 16, no. 1, pp. 257–264, 2017.
11. S. Hong, P. Katerattanakul, and S. J. Joo, "Evaluating government website accessibility: A comparative study," *International Journal of Information Technology & Decision Making*, vol. 7, no. 03, pp. 491–515, 2008.
12. V. L. Hanson and J. T. Richards, "Progress on website accessibility?," *ACM Transactions on the Web (TWEB)*, vol. 7, no. 1, pp. 1–30, 2013.

13. N. E. Youngblood, "Revisiting Alabama state website accessibility," *Government Information Quarterly*, vol. 31, no. 3, pp. 476–487, 2014.
14. W. Isa, M. R. Suhami, N. I. Safie, S. S. Semsudin, *et al.*, "Assessing the usability and accessibility of Malaysia e-government website," *American Journal of Economics and Business Administration*, vol. 3, no. 1, pp. 40–46, 2011.
15. N. Kesswani and S. Kumar, "Accessibility analysis of websites of educational institutions," *Perspectives in Science*, vol. 8, pp. 210–212, 2016.
16. R. Gonçalves, J. Martins, and F. Branco, "A review on the Portuguese enterprises web accessibility levels–a website accessibility high level improvement proposal," *Procedia Computer Science*, vol. 27, no. 0, pp. 176–185, 2014.
17. K. Yoon, R. Dols, L. Hulscher, and T. Newberry, "An exploratory study of library website accessibility for visually impaired users," *Library & Information Science Research*, vol. 38, no. 3, pp. 250–258, 2016.
18. L. Billingham, "Improving academic library website accessibility for people with disabilities," *Library Management*, 2014.
19. T. Domínguez Vila, E. Alén González, and S. Darcy, "Website accessibility in the tourism industry: an analysis of official national tourism organization websites around the world," *Disability and Rehabilitation*, vol. 40, no. 24, pp. 2895–2906, 2018.
20. P.-L. P. Rau, L. Zhou, N. Sun, and R. Zhong, "Evaluation of web accessibility in China: changes from 2009 to 2013," *Universal Access in the Information Society*, vol. 15, no. 2, pp. 297–303, 2016.
21. O. Sohaib and K. Kang, "Assessing web content accessibility of e-commerce websites for people with disabilities," 2016.
22. B. Hernández, J. Jiménez, and M. J. Martín, "Key website factors in e-business strategy," *International Journal of Information Management*, vol. 29, no. 5, pp. 362–371, 2009.
23. U. Akram, A. R. Ansari, G. Fu, and M. Junaid, "Feeling hungry? let's order through mobile! examining the fast food mobile commerce in China," *Journal of Retailing and Consumer Services*, vol. 56, p. 102142, 2020.
24. J. Lazar and K.-D. Greenidge, "One year older, but not necessarily wiser: an evaluation of homepage accessibility problems over time," *Universal Access in the Information Society*, vol. 4, no. 4, pp. 285–291, 2006.
25. Alexa, "Alexa website ranking." https://www.alexa.com/siteinfo, 2019.
26. AChecker, "AChecker." https://achecker.ca/checker/index.php, 2020.
27. CheckMyColors, "CheckMyColors." https://www.checkmycolours.com/, 2020.
28. A. Ismail and K. Kuppusamy, "Accessibility of Indian universities' homepages: An exploratory study," *Journal of King Saud University-Computer and Information Sciences*, vol. 30, no. 2, pp. 268–278, 2018.
29. J. Benesty, J. Chen, Y. Huang, and I. Cohen, "Pearson correlation coefficient," in *Noise reduction in speech processing*, pp. 1–4, Springer, 2009.
30. J. Hauke and T. Kossowski, "Comparison of values of Pearson's and Spearman's correlation coefficients on the same sets of data," *Quaestiones geographicae*, vol. 30, no. 2, pp. 87–93, 2011.
31. M. G. Kendall and B. B. Smith, "The problem of m rankings," *The Annals of Mathematical Statistics*, vol. 10, no. 3, pp. 275–287, 1939.
32. X. Li, X. Zhao, W. Pu, *et al.*, "Measuring ease of use of mobile applications in e-commerce retailing from the perspective of consumer online shopping behaviour patterns," *Journal of Retailing and Consumer Services*, vol. 55, p. 102093, 2020.
33. R. Bandara, M. Fernando, and S. Akter, "Explicating the privacy paradox: A qualitative inquiry of online shopping consumers," *Journal of Retailing and Consumer Services*, vol. 52, p. 101947, 2020.
34. F. Kawaf and D. Istanbulluoglu, "Online fashion shopping paradox: The role of customer reviews and Facebook marketing," *Journal of Retailing and Consumer Services*, vol. 48, pp. 144–153, 2019.

35. D. Sheehan, D. M. Hardesty, A. H. Ziegler, and H. A. Chen, "Consumer reactions to price discounts across online shopping experiences," *Journal of Retailing and Consumer Services*, vol. 51, pp. 129–138, 2019.
36. R. Davis, S. D. Smith, and B. U. Lang, "A comparison of online and offline gender and goal directed shopping online," *Journal of Retailing and Consumer Services*, vol. 38, pp. 118–125, 2017.
37. A. Bashiti and A. A. Rahim, "Physical barriers faced by people with disabilities (PWDS) in shopping malls," *Procedia-Social and Behavioral Sciences*, vol. 222, pp. 414–422, 2016.
38. T. L. Childers and C. Kaufman-Scarborough, "Expanding opportunities for online shoppers with disabilities," *Journal of Business Research*, vol. 62, no. 5, pp. 572–578, 2009.

Index

M. Mishra et al. (eds.), *Applications of Computational Intelligence in Management & Mathematics*, Springer Proceedings in Mathematics & Statistics 417,
https://doi.org/10.1007/978-3-031-25194-8